BIODEMOGRAPHY

BIODEMOGRAPHY

An Introduction to Concepts and Methods

James R. Carey
Deborah A. Roach

PRINCETON UNIVERSITY PRESS

PRINCETON AND OXFORD

Copyright © 2020 by Princeton University Press

Published by Princeton University Press
41 William Street, Princeton, New Jersey 08540
6 Oxford Street, Woodstock, Oxfordshire OX20 1TR

press.princeton.edu

All Rights Reserved

Library of Congress Control Number: 2019937611
ISBN 978-0-691-12900-6

British Library Cataloging-in-Publication Data is available

Editorial: Alison Kalett, Kristin Zodrow, and Abigail Johnson
Production Editorial: Mark Bellis
Text and Jacket Design: Lorraine Doneker
Production: Jacqueline Poirier
Publicity: Matthew Taylor and Katie Lewis
Copyeditor: Jennifer McClain

Jacket Credit: English plantain, *Plantago lanceolata* / iStock

This book has been composed in Sabon

Printed on acid-free paper. ∞

Printed in the United States of America

1 3 5 7 9 10 8 6 4 2

Pure science is like a beautiful cloud of gold and scarlet that diffuses wondrous hues and beams of light in the west. It is not an illusion, but the splendor and beauty of truth. However, now the cloud rises, the winds blow it over the fields, and it takes on darker, more somber colors. It is performing a task and changing its party clothes—think of it as putting on its work shirt. It generates rain that irrigates the fields, soaking the land and preparing it for future harvests. In the end it provides humanity with its daily bread. What began as beauty for the soul and intellect ends by providing nourishment for the humble life of the body.

José Echegaray, Literature Nobel Laureate (1904), cited in "Advice for a Young Investigator" by Santiago Ramón y Cajal (1916, p. 26), Physiology and Medicine Nobel Laureate (1906)

SUMMARY OF CONTENTS

CONTENTS

FOREWORD

JAMES W. VAUPEL

Demography is an interdiscipline. Indeed, demography is the quintessential interdiscipline, making contributions to and drawing from research in mathematics and statistics, the social sciences, the biological sciences, the health sciences, engineering, cultural studies, and policy analysis.

The title of this book is *Biodemography: An Introduction to Concepts and Methods*—and it is a major contribution to the burgeoning field of biodemography, clear and accessible enough to be a stimulating introduction, focused on concepts and methods and with many concrete examples, often intriguing and thought provoking. It covers the broad range of biodemography, structuring and fostering understanding of material previously covered, if at all, in more specialized or advanced publications.

In addition, this book introduces readers to the fundamental concepts and methods of human demography, currently by far the largest component of the interdiscipline of demography. Other excellent introductions exist, as authors James R. Carey and Deborah A. Roach note, but this volume provides an alternative that addresses new advances and that provides novel illustrations.

Furthermore—and this is impressive and unprecedented—this book provides an overview of the full range of demography as an interdiscipline. It covers a panoply of mathematical and statistical concepts and methods, drawing from the social, natural, and health sciences. It forges links with disciplines as diverse as policy analysis, reliability engineering, actuarial prognosis, and cultural studies. It provides cogent examples of how demographic concepts and methods can be used to study *any* kind of population, not just living populations of individuals (humans or in nonhuman species) but also populations structured in hierarchies and families as well as populations of organizations and other nonliving entities.

Demography rests on a bedrock of mathematics and statistics. The key mathematical and statistical concepts and methods are expertly elucidated in this book; authors Carey and Roach have decades of experience in marshaling these concepts and methods in their biodemographic research. Carey has contributed innovative methods, for example, to estimate age from data on time to death and to summarize how individuals allocate their life spans to life history stages. Both Carey and Roach have pioneered major additions to age-specific demographic data on fertility, mortality, and morbidity—Carey in massive studies of Medflies, Mexflies, mayflies, parasitoid wasps, and other insects; Roach in unprecedented experiments on thousands of *Plantago lanceolata* plants over more than two decades.

Secure on this bedrock, demographers are able to contribute influential perspectives and factual findings to the airy, windy heights of public policy discussions. This important role is touched on several times in this book, especially in the discussion in chapter 10 of biological control, population harvesting, and conservation. The treatment of pressing public policy issues is balanced and nuanced, and the value of demographic concepts and findings in shedding light on the issues is cogently laid out—in contrast to the heated arguments and fiery rhetoric that often characterize public discussions of, say, how Medflies should be controlled in California or how elephant populations should be culled in South Africa.

The main thrust of this book is to advance the field of biodemography, a field with deep historical roots in demography that is currently thriving and that seems likely to become a component of demography as important as its social science components. Dobzhansky famously observed that nothing in biology makes sense except in the light of evolution. Because evolution is driven by—and drives—birth and death rates, it is equally valid that nothing in evolution makes sense except in the light of demography. And to a considerable extent vice versa—much in demography, especially age patterns of fertility and mortality, makes sense only in the light of evolution. Two related kinds of research at the intersection of demography and biology are important. First, populations of a nonhuman species can be studied using concepts and methods also used to study human populations. Second, broad analyses can be conducted across many species to uncover basic regularities, as well as key differences, among species that govern life, including human life. Such knowledge of particular species and of overarching patterns can be used to shed light on fundamental evolutionary processes. This is basic science. On the other hand, the twin sources of knowledge can also be used to advance conservation biology and the protection of endangered species. This is applied science. Carey and Roach cover the range of studies of particular species, comparative studies across species, studies of evolutionary processes, and studies of practical importance in managing species and averting species extinction.

In the biological sciences, molecular biology has over recent decades become more and more prominent. It is possible for demographers to make contributions to the study of populations of molecules and cells, for example, the study of the origin and growth of a cancer, and some research on this has started. Historically, a closer link ties population biology (including ecology and life history biology) with demography. The pioneers of these fields—Aristotle and Darwin—are still heroes of biology. Interest in studying populations of individuals is certainly growing; this book provides the tools—the concepts and methods—for analyzing populations. One sign of the renaissance of population studies in biology was the founding, a few years ago, of the Evolutionary Demography Society, of which Roach is past president.

The term *biodemography* is sometimes used to describe two different fields of study—biological demography, as discussed above, and biomedical demography, which is focused on human health. This book gives greater attention to biological demography but does not neglect biomedical demography. In particular, two sections of chapter 8 cover basic aspects of health demography—namely, active life expectancy and multiple-decrement life tables. Attention to health demography continues in chapter 10, but with an emphasis on the health of nonhuman species. Other aspects of biomedical demography are also covered, including some epidemiological examples, in chapter 11.

A major strength of this book is its wealth of more than 200 illustrations. The authors have chosen these illustrations with care and thought to enhance conceptual and methodological clarity. They use a variety of graphs and schematics that are useful not only for understanding the material illustrated but also for illustrating best practices in the visualization of demographic information. In the appendixes on visualization of demographic data and visualization rules of thumb, they capture the gist of how to visualize demographic information.

Another major strength of this book is the choice of examples. The range of these examples is impressive, as noted above, but perhaps even more important, the examples are almost all interesting, stimulating, and thought provoking. Chapter 11 is a tour de force of 87 terse examples, ranging from survival pills to forensic entomology.

It was a great pleasure for me to read the draft of this book because it got better the more I read. The material covered in the first chapters is explained in a highly competent manner. As the basics are explained, it then becomes possible to introduce more innovative and more creative topics. So the first part of the book is satisfying in the knowledge conveyed; the second part of the book is stimulating in the new horizons opened up. The first part of the book, however, does much more than simply cover standard material any demographer should know: it introduces some basic concepts and methods that are powerful but generally not covered in overviews.

For example, all populations are heterogeneous and individuals of the same sex and age and at the same location may face radically different reproductive opportunities and mortality hazards. Surviving populations are transformed as those at greatest risk of some event drop out. Hence, underlying patterns of demographic events for individuals differ qualitatively from observed patterns for those remaining in the population. Demographers are interested in how the chance of reproduction or the risk of death changes with age for individuals, but they can only observe the changes for changing populations. This is a fundamental problem of demographic analysis that Carey and Roach tackle.

They also cover other topics not found in most demography or population biology books. These include stage-structured populations, stochastic rates of growth, two-sex models, kinship models, and thanatological demography.

A treasure chest of gems enrich all the chapters of this book—short discussions that capture the essence of an important concept or method. An excellent example is the discussion at the onset of chapter 3 of age-specific survival (which is the measure currently used by almost all biologists studying patterns of mortality over age) versus age-specific mortality (which is the measure demographers use). Carey and Roach persuasively explain why biologists should stop using age-specific survival and start using age-specific mortality. They make their case in a few cogent sentences that every biologist who uses age-specific survival should read.

In sum, this book is impressive. The authors aim to "enlighten and inspire" and they succeed. They do so in a highly original way—the book is reliable but unconventional. Standard material is covered but in new ways. Important material not found in other demography or population biology books is innovated. Originality of thought, of mode of explanation, of example, of graphic illustration sparkles on almost every page.

Perspective

Demography is the taproot of an interdisciplinary tree containing multiple branches whose demographic topics range from health, disease, marriage, and fertility to anthropology, population biology, paleontology, history, and education. Our book now adds a new branch to this tree—biodemography.

We have two major goals for this book. The first is to collate, merge, integrate, and synthesize the key literature from two epistemological domains concerned with demography. The first domain is the taproot itself; that is, mainstream human demography. This domain constitutes a field that straddles the social and formal sciences and serves as the primary pedagogical foundation for our book. The second domain is the emerging branch of demography that makes our book biodemography. That is the literature concerned with demographic concepts on or related to biology, including ecology and population biology of plants and animals, biogerontology, epidemiology, and biomedicine as well as the more applied areas such as wildlife biology, invasion biology, pest management, and conservation.

Our second goal is revealed in the book's subtitle, *Introduction to Concepts and Methods*, where our aim is to include the *conceptual* underpinnings for all demographic *methods*. Whereas the conceptual aspects serve as the basic ideas and then situate these ideas into demographic contexts, the methodological components provide a complementary operational framework. For example, life tables are useful for summarizing the actuarial properties of cohorts. But without a deeper appreciation of the underlying concepts of mortality, survival, risk, life-years, and life expectancy, life tables themselves are dry accounting tools at best and simple "cookbooks" at worst. Likewise, observing the ergodic dynamics of stable population theory using Leslie matrix population projections deepens an understanding of the formal mathematical properties that underlie these projections.

One goal we do not have for this book is to make any attempt to match, much less supersede, the scope or mathematical sophistication of any of the number of truly remarkable textbooks, reference books, handbooks, or book series in mainstream (human) demography. Rather, we strongly recommend that readers seek out these sources in order to both deepen and broaden their knowledge of demographic concepts and methods. A particularly deep source for demographic information is the awe-inspiring 140-chapter, four-volume series titled *Demography: Analysis and Synthesis—A Treatise in Population* edited by Graziella Caselli, Jacques Vallin, and Guillaume Wunsch (2006a). Other excellent sources of demographic concepts, models, methods, and techniques include two textbooks, one by Samuel Preston, Patrick Heuveline, and Michel Guillot titled *Demography: Measuring and Modeling Population Processes* (2001) and the other by Dudley Poston and Leon Bouvier titled *Population and Society: An Introduction to Demography* (2010). The book by Kenneth Wachter (2014) titled *Essential Demographic Methods* is an excellent source for clear, succinct presentations of basic demographic concepts. Nathan Keyfitz's book *Applied Mathematical Demography*, the third edition of which was coauthored with Hal Caswell, is a treasure trove of demographic models and ideas (2010). Excellent edited books on demography include the 22-chapter book edited by Jacob Siegel and David Swanson titled *The Methods and Material of Demography* (2004) and the 28-chapter *Handbook of Population* edited by Dudley Poston and Michael Micklin (2005). The book by Frans Willekens titled *Multistate Analysis of Life Histories with R* (2014) is one of a number of excellent more specialized books in the Springer Series on Demographic Methods and Population Analysis, which is edited by Kenneth Land.

Organization of the Book

There are 11 chapters that we separate into two parts. The first part (chapters 1–7) consists of chapters covering more traditional (albeit biologically oriented) demographic content that is considered standard fare among mainstream demographers. Chapter 1 contains the demographic basics and foundational concepts, such as life course, rates, and Lexis diagrams. Chapter 2 contains life table concepts and methods, including construction of period, cohort, and abridged tables. Chapter 3 covers the basics of mortality, including its importance in demography, derivation of the force of mortality, mortality metrics, the main mortality models, demographic heterogeneity, and selection. Chapter 4 contains basic information on reproduction, ranging from age-specific and lifetime birth metrics to reproductive heterogeneity, parity progression, and fertility models. Chapters 5, 6, and 7 cover population concepts and methods. The first of these introduces stable population theory, including Lotka and Leslie models; the next introduces stage-based matrix models, including the Lefkovitch model, and considers life table–stable theory congruency; and the last considers extensions of stable theory, such as two-sex and stochastic models.

In the second part (chapters 8–11) we move from the foundational concepts of demography to topics focused on areas that are either more specialized, more biologically relevant, more novel, or more idiosyncratic. This content includes chapter 8 on human demography, which summarizes various aspects of human demography as well as includes topics that are often not covered in books on general human demography, such as family demography and kinship. This is followed by chapters 9 and 10 on applied demography. Chapter 11 then contains descriptions of, and solutions to, demographic questions that we refer to as "demographic shorts." These narratives include a wide range of demographic concepts that are novel, heuristic, inherently interesting, thought provoking, or too mathematically complex for other chapters, but they are nonetheless relevant to our larger goals.

We complete the book with four appendixes, including appendix I on data visualization for demographic data, appendix II on demographic storytelling, appendix III that presents ten visualization rules of thumb, and appendix IV on data management.

Editorial Strategies

The editorial challenges in producing this book involved technical, philosophical, and epistemological components concerning decisions on (1) which material to include from each of the two primary literature domains of biology and demography, (2) how best to collate, integrate, and synthesize this information, and (3) determining the best practices for illustrating concepts and methods. The purpose of this preface is to provide context and background for meeting these challenges.

Data selection

Because the richness of demographic concepts, methods, and models is best illustrated with the use of data, the paucity of high-quality databases for nonhuman species posed one of our greatest challenges. What is taken for granted by mainstream demographers, who have access to databases consisting of detailed demographic information on tens of thousands if not tens of millions of people, is only a fantasy for scientists studying the demography of nonhuman species. No databases exist in biology that are even remotely close to the quality and quantity of those available for human studies.

A related challenge is an outcome of the fact that most demographic models were inspired by, developed for, and applied to humans. Indeed, the contours of the life history of *Homo sapiens* have framed the demographic questions and shaped the mathematical models where the human life history consists of, for example, sexual mating (e.g., no hermaphroditism, self-fertilization, or sterile castes), continuous development (e.g., no dormant stages, metamorphosis, or strong seasonality), potentially 100 or more age classes (i.e., an extremely long potential life span), and mostly singleton births (i.e., not litters, clutches, broods, clones, or seed set). Application of these models, originally developed for humans, for use in biodemography requires that, to greater and lesser degrees, they be adapted to vastly different life histories across the Tree of Life, including for species from within and between different kingdoms, phyla, classes, families, and genera.

In light of the challenges of model applicability and data availability for biodemography, and given our overarching goals for this book, we devised a data use strategy in which we selected demographic information based on three purposes.

(1) *Concept elucidation.* We believe that in order to understand the basics of biodemography, which involves life tables, mortality, reproduction, and stable population theory, we need to use databases consisting of detailed information on thousands, if not tens of thousands, of individuals that were both maintained and monitored under carefully controlled conditions. Although there are scores and perhaps hundreds of studies that meet most of these criteria, many studies lack the scale and raw data that are needed, such as age-specific births for each individual. We thus draw heavily on the fruit fly databases of coauthor Carey to elucidate the concepts and methods contained in the first five chapters.

(2) *Human demography.* We felt that including human demography was important since it provides us with ways to broaden the reader's demographic

perspective as well as, and especially, add demographic depth and breadth. There are a number of concepts and models, such as multiple-decrement life tables, family demography models, and Lexis diagrams, that are regularly used in human demography but not in biodemographic contexts with nonhuman species.

(3) *Applied biodemography.* Population biologists and applied ecologists have been developing and using models for well over a century, and many models and databases on nonhuman species are available in the literature that we could have used to parameterize the models presented. Whereas we illustrate many of the basic concepts using databases on either fruit flies or humans, we illustrate most of the applied content using other species, including birds, mammals, reptiles, and plants.

Tabulation and visualization

As strong believers in the power of visualization and illustration in teaching and learning, another editorial strategy we have used is to make liberal use of tables, graphs, and schematics to enhance conceptual and methodological clarity. Different content calls for different categories of each of these. For example, the different types of tables and figures we develop and use in our book include

(1) *Heuristic models.* Small, simplified tables of hypothetical data used to illustrate a mathematical concept (similar to what statisticians refer to as *toy models*).

(2) *Complete data.* Information for all age classes for species where the details contained in the patterns and quantities in all ages classes is important to illustrate the concept or method.

(3) *Abridged data.* Usually, tables containing age-specific data for models and methods where it is less important to have complete age information.

(4) *Summary information.* Tables typically containing information not age related.

Our visualization strategies include the abundant use of a wide variety of data graphs and schematics throughout the book, including appendixes for outlining and illustrating best practices in the visualization of demographic information. As Few (2013) notes, information flows to the viewer in the clearest and most efficient way when a chart is presented properly. Indeed, best practices in information design and typography matter because they help conserve the most valuable resource people have as writers and presenters—*reader and viewer attention* (Butterick 2015).

Concepts and models

As long-time instructors of a wide range of university subjects, including classes on evolution, aging, and demography, it is difficult for us to overstate the importance of using clear descriptions of the underlying concepts, including their mathematical foundation(s), and of using either hypothetical or real-world examples. The life table, for example, is essentially a tool for ordering actuarial information by age.

However, this tool can be transformed into an overarching concept when mortality is considered in actuarial and dynamic contexts, and into a mathematical model when various formulas are brought to bear on the data.

Epistemological

Unlike interdisciplinary fields such as biochemistry, historical economics, and mathematical statistics, which straddle two or more disciplines *within* larger scientific domains, the interdisciplinary field of biodemography bridges three scientific domains: the natural sciences (biology), the social sciences (demography), and the formal sciences (mathematical modeling). Whereas virtually all disciplines within the social sciences (demography) generate new knowledge primarily through observation, nearly all disciplines in the natural sciences (biology) generate new knowledge primarily through experimentation. Remarkably, integration of the formal sciences (i.e., mathematic/statistical modeling) into biology and especially into demography set the stage for the emergence of biodemography as a subfield of each of these parent disciplines. For example, the multistate models used widely in demography are virtually the same as the age/stage models used widely in ecology and population biology; the life-lived and -left concept discovered in formal demography is identical to the age structure and death rate identity discovered in biodemography; and the multiregional models used in demography are similar in concept to the metapopulation models used in population biology. Identifying these and many other interdisciplinary connections is important because the common mathematical models point to similar interdisciplinary concepts, which then lead toward interdisciplinary syntheses. Moreover, cross-references to similar mathematical models developed independently by scientists in different fields synergize science by making the scientists in the respective fields aware of the generality of the original models and thus of the underlying concepts themselves.

Human Demography in the Tree of Life Context

Human biology and the shaping of mainstream demography

For perspective on the challenges of using mainstream demographic concepts and tools in biological contexts, consider the entire corpus of formal demography, not only in the context of human demography but also in the context of the demography of species across the entire Tree of Life (Jones et al. 2013). As we discuss below, it becomes manifestly evident that classical demography is framed almost exclusively by *Homo sapiens*' life history traits, enabled by demographic databases on humans that are as large as they are detailed. Approaches are modeled using the birth, death, and migration processes that are specific to our species and that shape our populations. Our comparisons between the demographic characteristics of humans and the characteristics of species across the Tree of Life spectrum both highlight and underscore the conceptual constraints, methodological limitations, and empirical inadequacies for off-the-shelf application of demographic methods that were both informed by and created for use in human demography. The comparisons also provide insights into the species-specific demographic requirements for the unmodified use of classical demographic tools such as the life table and stable theory, as well as the underlying motivation and need for biologists to develop new concepts and tools that are relevant to the great many species whose demography cannot be easily described and modeled using conventional tools.

Demographic characteristics

Humans exist as clearly defined individuals—single entities that are unequivocally, unambiguously, and unmistakably discrete. The existence of species as separate, identifiable individuals is an implicit requirement for virtually all types of the more conventional demographic analyses of cohorts and populations. However, clearly defined individuals do not exist for many species of plants (e.g., aspens) and animals (e.g., corals) whose populations consist of expanding clones. Different levels of individualism may also exist for nonhuman species. The best example here is that of social insects whose individualism can be considered at two levels—the individual as the single insect itself, the honeybee worker, and the individual as a superorganism representing the second tier of the hierarchy, the honeybee colony.

Whereas humans and most other species of animals can move, this is not the case for species of plants and for many species of animals, such as barnacles after they enter their sessile (anchored) stage. Humans are warm blooded so that the chronological age of each person essentially defines his or her stage in life. This is not the case for the vast majority of organisms because, with the exception of birds and mammals (homeotherms), members of all other groups of organisms cannot control their temperature (poikilotherms). This means that, to greater and lesser extents, their demographic traits are driven by temperature and other factors that affect their biological age. Humans do not hibernate, estivate, quiesce, diapause, or otherwise become dormant. This is not the case for a large number of species who often enter dormancy-related stages ranging from tardigrades, insects, and other arthropod species to bears, rodents, and small birds. Humans also do not undergo metamorphosis like species of holometabolous insects, amphibians, or even some plants. Unlike species that reproduce by cloning or self-fertilization, humans reproduce sexually. Humans pair-bond, breed throughout the season, produce singleton offspring, and create families that consist of offspring of overlapping ages, each of which is cared for by both parents for 15 to 20 years. Human family units may last 50 or more years. The model for human families is unique to *H. sapiens* even in broad outline much less in fine detail. Relative to preadult survival for the vast majority of nonhuman species, which produce hundreds and even tens of thousands of offspring, human survival is extremely high. Whereas humans can live more than a century and thus have over 100 age classes (years), the life spans of the vast majority of bird and mammal species are much shorter and therefore have many fewer age classes. This high survival and long life in humans has profoundly affected the demographic methods and models that have been developed for use in mainstream, formal demography.

Databases

Demographic databases on humans often contain precise records on the sex, age, marital and family status, and medical/health conditions for individuals as well as on their manners, causes, and ages of death. The size of these databases, and the individual-level details they contain, enable demographers to estimate a wide variety of demographic rates for humans (e.g., reproduction; death; marriage; divorce) with great accuracy and precision for all but the most extreme ages. The databases on human demography stand in glaring contrast to the databases for all other species, those for model species (worms, flies, rodents) notwithstanding. No database exists on nonhuman species that is even remotely similar to databases for human demography with respect to numbers, cradle-to-grave details, and historical duration.

Acknowledgments

We thank the following colleagues, students, friends, and associates for their help, suggestions, and encouragement during various stages of the book's gestation or for information (including data) they provided for us to use. In alphabetical order, they are Melissa Aikens, Susan Alberts, Jeanne Altman, Martin Aluja, Robert Arking, Goshia Arlet, Annette Baudisch, Carolyn Beans, David Berrigan, Emily Bick, Nathanial Boyden, Nicolas Brouard, Thomas Burch, Hal Caswell, Ed Caswell-Chen, Dalia Conde, Alexandros Diamantidis, William Dow, Jeff Dudycha, Pierre-François Duyck, Michal Engelman, Erin Fegley, Caleb Finch, David Foote, Rachel Long, Jutta Gampe, Josh Goldstein, Jean-Michel Guillard, John Haaga, Rachid Hanna, James Harwood, Stephanie Held, Donald Ingram, Lionel Jouvet, Deborah Judge, Byron Katsoyannos, Robert Kimsey, David Krainacker, Gene Kritsky, Dick Lindgren, Freerk Molleman, Amy Morice, David Nestel, Vassili Novoseltzev, James Oeppen, Robert Peterson, Dudley Poston, Nick Priest, Daniel Promislow, Arni S. R. Srinivasa Rao, Roland Rau, Tim Riffe, Jean-Marie Robine, Blanka Rogina, Olav Rueppel, Alex Scheuerlein, Richard Shefferson, Sarah Silverman, Ana Rita Da Silva, Rahel Sollmann, Uli Steiner, Richard Suzman (deceased), Mark Tatar, Roger Vargas (deceased), Robert Venette, Elizabeth Vasile, Francisco Villavicencio, Maxine Weinstein, Frans Willekens, Nan Wishner, Pingjun Yang, Zihua Zhao, and Sige Zou.

We are indebted to a number of persons at Princeton University Press including editorial director Alison Kalett for her interest in our ideas and encouragement for writing this book, and especially for her patience in waiting for us to deliver the manuscript. We are also grateful to PUP editorial associates Kristin Zodrow, Abigail Johnson, and Lauren Bucca, production manager Jacqueline Poirier, production editor Mark Bellis, promotions associate Matthew Taylor, and publicist Julia Hall. We also thank Jennifer McClain for line editing and Julie Shawvan for indexing. Lastly, we wish to acknowledge the two anonymous reviewers for their extraordinarily constructive and insightful comments on earlier drafts of the manuscript. All figures were produced de novo by coauthor Carey.

Carey expresses his gratitude to the demographers and administrators associated with both the UC Berkeley Department of Demography and the Center for the Economics and Demography of Aging (CEDA), particularly to Ronald Lee, Kenneth Wachter, and Eugene Hammel, for their many years of support. He thanks members of the Department of Entomology at UC Davis for their nearly 40 years of support of biodemographic research; the undergraduate students who enrolled in his biodemography-based courses at UC Davis for their direct and indirect (e.g., course evaluations) feedback on demography models and concepts; the PhD students in the European Doctoral School of Demography for their feedback on best practices in visualizing demographic data; and the graduate students at UC Davis, who many years ago enrolled in his insect demography course, for their engagement and constructive critiques at the early stages of his biodemographic career.

Roach has been doing demography since she was an undergraduate, and she thanks the many graduate and undergraduate students and technicians who have spent long hours collecting demographic data with her in the field. She thanks the Monticello Foundation for use of her Shadwell field site, and the University of Virginia Foundation for use of her site at Morven Farms. She also thanks the UVA Department of Biology, including Henry Wilbur, Laura Galloway, and her other EEBio colleagues, and her demographic collaborator Jutta Gampe.

A major portion of the research, for both Roach and Carey, was supported through a series of NIH-NIA P01grants, and they both thank the other PI's on this grant, including Kaare Christensen, Tim Coulson, James Curtsinger, Michael Gurven, Lawrence Harshman, Carol Horvitz, Thomas Johnson, Hillard Kaplan, Nikos Kouloussis, Ronald Lee, Pablo Liedo, Valter Longo, Kenneth Manton, Hans Müller, Dina Orozco, Cindy Owens, Rob Page Jr., Nikos Papadopoulos, Linda Partridge, Patrick Phillips, Leslie Sandberg, Eric Stallard, Shripad Tuljapurkar, James Vaupel, Nancy Vaupel (deceased), Kenneth Wachter, Jane-Ling Wang, Anatoli Yashin, and Yi Zeng for many engaging discussions about the foundations of biodemography across species.

We express our deep gratitude to our respective spouses, Patricia Carey and Dennis Proffitt, for their patience and support over the several years we spent writing. Special thanks go to James W. Vaupel for his vision, inspiration, leadership, support, advice, and friendship, and for inviting both of us to become part of a series of research programs at Duke University on the Oldest-Old, which was funded by the National Institute on Aging. We also thank Jim for writing the foreword for this book.

James R. Carey
Davis, California

Deborah Roach
Charlottesville, Virginia

Permissions

A portion of this material is published by permission of Oxford University Press and is based on content from J. R. Carey's *Applied Demography for Biologists* (Oxford University Press, 1993). Specifically, this includes rephrased material found here from the following chapters of that book: chapter 1 (Introduction), including the elementary characteristics of populations and demographic rates; chapter 2 (Life Tables), the cohort and multiple-decrement life table general concepts; chapter 3 (Reproduction), the per capita reproductive rates; chapter 4 (Population I), parts of the discussion of stable population models and fundamental properties; chapter 5 (Population II), the material from sections on two-sex, stochastic, multiregional, and hierarchical models; and chapter 6 (Applications), the selected prose on stage duration and harvesting models. For permission to reuse this material, please visit http://global.oup.com/academic/rights.

BIODEMOGRAPHY

Introduction

||||||||||||||||||||||||||||||||||||||||||||||||||

> The relation of our disciplines has not been symmetric. Biology textbooks
> incorporate short courses in demography, an attention that is not reciprocated. But
> that . . . by no means forecloses work at the boundary of population and biology.
>
> Nathan Keyfitz (1984, 7)

Biodemography is an emerging interdisciplinary science concerned with identifying
a universal set of population principles, integrating biological concepts into demographic approaches, and bringing demographic methods to bear on population problems in different biological disciplines (Carey and Vaupel 2005). It is also an interdisciplinary science in the sense that it uses theories and analytical methods from
classical (human) demography and population biology to study biological systems
at levels of organization from the individual, to the cohort, to populations. In so doing,
biodemography provides quantitative answers to questions at the whole-organism
level concerned with birth, death, health, and migration.

Biodemography does not have university-level departments of its own, but it has
presence across departments within the fields of demography, economics, sociology,
gerontology, entomology, wildlife and fisheries biology, ecology, behavior, and evolution. Research efforts in biodemography are often initiated by scientists who
were traditional ecologists, demographers, economists, and gerontologists by train-
ing. It is concerned with the study of populations of organisms, especially the regulation of populations, life history traits, and extinction. Depending on the exact
definition of the terms used, biodemography can be thought of as a small, specialized branch of classical demography, or as a tool with which to investigate and
study ecology, evolution, and population biology.

Historical Perspectives on Biodemography

Demography began as the study of human populations and literally means "description of the people." The word is derived from the Greek root *demos*, meaning "the
people," and was coined by a Belgian, Achille Guillard, in 1855 as "demographie"—
elements of human statistics or comparative demography (Siegel and Swanson 2004).
He defined demography as the natural and social history of the human species or the
mathematical knowledge of populations, of their general changes, and of their physical, civil, intellectual, and moral condition.

Biology and demography

The field has had multiple points of contact with biology, as well as mathematics,
statistics, the social sciences, and policy analysis. Population biology and demography share common ancestors in both T. R. Malthus (1798) (i.e., populations grow

exponentially but resources do not) and Charles Darwin (1859) (i.e., differential birth and death rates resulting from variation in traits). The biology-demography interface also served as the research foundation for two distinguished demographers in the early decades of the twentieth century—Alfred J. Lotka (1880–1949) and Raymond Pearl (1879–1940). Lotka developed concepts and methods that are still of fundamental importance in biological demography, and his two most significant books are *Elements of Physical Biology* (1924) and *Theorie Analytique des Associations Biologiques* (1934). Pearl (1924, 1925) pioneered biological-demographic research on several species, including flatworms, the aquatic plant *Ceratophyllum demersum*, the fruit fly *Drosophila melanogaster*, and humans. He founded two major journals, the *Quarterly Journal of Biology* and *Human Biology*, and helped found both the Population Association of America (PAA) and the International Union for the Scientific Investigation of Population Problems (which later became IUSSP—the International Union for Scientific Study of Population).

Following the pioneering work of Lotka and Pearl in the 1920s and 1930s, there was very little interest among demographers in integrating biology into any part of the discipline until the 1970s. There were a few chapter entries on population studies in crosscutting disciplines such as demography and ecology (Frank 2007), demography and anthropology (Spuhler 1959), and genetics and demography (Kallmann and Rainer 1959), all of which are in the seminal book *The Study of Populations* by Hauser and Duncan (1959). These and other similar chapters served more as illustrations of how demographic methods were used by different disciplines than as sources of knowledge for demography.

Early developments

In the early 1970s a group of population biologists and demographers, including Nathan Keyfitz, launched the journal *Theoretical Population Biology* (TPB). The journal was intended to be a forum for interdisciplinary discussion of "the theoretical aspects of the biology of populations, particularly in the areas of ecology, genetics, demography, and epidemiology." This description is still used by the publisher to describe the journal, but the publisher describes the audience of the journal as "population biologists, ecologists, evolutionary ecologists," with no mention of demographers (or epidemiologists). In the late 1970s IUSSP members expressed concern that demography was at risk of isolating itself and becoming more a technique than a science. Demographer Nathan Keyfitz (1984b, 1) lamented that *"demography has withdrawn from its borders and left a no man's land which other disciplines have infiltrated."* Hence in 1981 a workshop titled "Population and Biology" was organized at the Harvard University Center for Population Studies (Keyfitz 1984a) to explore the possible impact of biological "laws" on social science (Jacquard 1984; Lewontin 1984; Wilson 1984), the selective effects of marriage and fertility (Leridon 1984), the autoregulating mechanisms in human populations (Livi-Bacci 1984), and the concepts of morbidity and mortality (Cohen 1984). That no notable papers or concepts emerged from this meeting between biologists and demographers, many of whom were among the most prominent scientists in their respective fields, was itself significant—the good intentions of top scientists are not enough to integrate two fields with fundamentally different disciplinary histories, professional cultures, and epistemological frameworks.

Traction

In the mid-1980s two separate meetings were organized that brought scientists together to address more circumscribed and focused questions that lie at the interface between biology and demography. The first workshop that brought biologists and demographers together during this period was organized in 1987 by Sheila Ryan Johannson and Kenneth Wachter at the University of California, Berkeley, titled "Upper Limits to Human Life Span," and supported by the National Institute on Aging (NIA). Although there were no publications and/or proceedings from this workshop, it was important historically because it was the first meeting to bring biologists and demographers together to focus expressly on a circumscribed topic of great importance to demographers, biologists, and policy makers—aging and longevity. This workshop set the stage for virtually all the subsequent research developments in the biological demography of longevity and aging.

The second workshop that helped frame biological demography was organized in 1988 at the University of Michigan by Julian Adams, Albert Hermalin, David Lam, and Peter Smouse, titled "Convergent Issues in Genetics and Demography" (Adams 1990). This resulted in an edited volume that included sections on the use of historical information, such as pedigree and genealogical data in genetics and demography, on the treatment and analysis of variation in the fields of genetics and demography, on epidemiology as common ground for the convergence of demography and genetics, and on issues in genetics and demography that have attracted the attention of scientists in both fields, such as two-sex models, minimum viable population size, and sources of variation in vital rates. This workshop on genetics and demography was significant because it revealed the importance of organizing research at the interface between biology and demography around a circumscribed topic, in this case genetics.

Coalescence

The Berkeley and Ann Arbor workshops set the conceptual stage for the organization of a cluster of three highly successful workshops held between 1996 and 2002. The first of these was a workshop titled "Biodemography of Longevity," organized and chaired by Ronald Lee of the Committee on Population of the US National Research Council, and held in Washington, DC (April 1996). This meeting fostered an interchange of demographic and biological ideas and was one of the seminal developments in biological demography because of the new insights and perspectives that emerged on the nature of aging and life span. The workshop led to the book *Between Zeus and the Salmon: The Biodemography of Longevity* edited by Kenneth Wachter and Caleb Finch (1997). This volume includes papers on the empirical demography of survival, evolutionary theory and senescence, the elderly in nature, post-reproduction, the human life course, intergenerational relations, the potential of population surveys in genetic studies, and synthetic views on the plasticity of human aging and life span.

The second workshop concerned with biological demography was organized by James Carey and Shripad Tuljapurkar. Titled "Life Span: Evolutionary, Ecological, and Demographic Perspectives," it was held on the Greek Island of Santorini in 2001. This workshop was a follow-up to the 1996 meeting on biological demography but with a greater emphasis on life span rather than aging per se. The edited volume from this workshop (Carey and Tuljapurkar 2003) included papers on conceptual and/or theoretical perspectives on life span and its evolution, ecological and life history correlates, and genetic and population studies of life span in both in humans and non-human species.

The third workshop, held at the National Academies in Washington, DC (June 2002) and organized and chaired by Kenneth Wachter and Rodolfo Bulatao, focused on fertility and was designed to complement the workshop on the biological demography of longevity. Like the others preceding it, this workshop brought together demographers, evolutionary biologists, geneticists, and biologists to consider questions at the interface between the social sciences and the life sciences. Topics in the resulting volume (Wachter and Bulatao 2003) included the biodemography of fertility and family formation and the genetic, ecological, and evolutionary influences on human reproduction.

At the beginning of the twenty-first century, biological demography is reemerging as the locus of cutting-edge demographic research. It is clearly accepted that fertility, mortality, morbidity, and other processes of profound interest to demographers have a basic biological component. Moreover, biology is fundamentally a population science and there is growing recognition that biological studies can benefit greatly from demographic concepts and methods. From a biologist's perspective, biological demography envelops demography because it embraces research pertaining to any nonhuman species, to populations of genotypes, and to biological measurements related to age, health, physical functioning, and fertility. Within this vast territory, several research foci are noteworthy and are briefly described in the next section.

Classical Demography

Classical demography is concerned with basically four aspects of populations (Siegel and Swanson 2004; Poston and Bouvier 2010). These are 1) *size*—the number of units (organisms) in the population; 2) *distribution*—the arrangement of the population in space at a given time; 3) *structure*—the distribution of the population among its sex and age groupings; and 4) *change*—the growth or decline of the total population or one of its structural units. The first three (size, distribution, structure) are referred to as population statics while the last (change) is referred to as the population dynamics. Hauser and Duncan (1953) regard the field of demography as consisting of two parts: *formal demography*—a narrow scope confined to the study of components of population variation and change (i.e., births, deaths, and migration); and *population studies*—a broader scope concerned with population variables as well as other variables, which may include genetics, behavior, and other aspects of an organism's biology. The methodology of demographic studies includes data collection, demographic analysis, and data interpretation.

Demographers conceive the population as the singular object for scientific analysis and research. However, as Pressat (1970, 4) notes, "population" is everywhere and nowhere in the sense that many aspects of demography can be studied simply as component parts of the disciplines considered. He states, "But to bring together all the theories on population considered as a collection of individuals subject to process of evolution, has the advantage of throwing into relief the many interactions which activate a population and the varied characteristics of that population." This is what demography is about, particularly mathematical demography.

Usefulness of Demography

Conceptual unification

Demography can be thought of in two ways. First, as a large collection of mathematical models that can be reduced to a small number of mathematical relationships. Or, as a small collection of metaphors that can be conceptually extended to a large number of biological problems. Both of these ways of thinking about demography provide conceptual as well as functional unification to ecology and population biology. In principle all life history events can be reduced to a series of transition probabilities and all events are interconnected in several ways. Demography provides the tools to connect these events, and most events reduce to one of two things—birth or death. Metaphorically, birth and death can represent a wide range of phenomena. In human demography divorce can be viewed as the death of a marriage, or in epidemiology hospital entry can be viewed as the birth of a case. In insect ecology metamorphosis can be viewed as the "death" of a larva and the "birth" of a pupa. It will become evident later that these perspectives extend beyond the rhetorical.

Projection and prediction

The terms *projection* and *prediction* are often used interchangeably in other disciplines. In demography, however, these terms apply to two distinctly different activities. *Population prediction* is a forecast of the future population. Because things are interconnected, we thus cannot know the future of one variable (population) without knowing the future of every other variable. *Population projection* refers to the consequences of a particular set of assumptions with no intention of accounting for the future population of a specific case (Keyfitz 1985). All predictions are also projections, but the reverse is not necessarily true.

Control, conservation, and exploitation

Caughley (1977) points out that the uses of demography in applied ecology fall into one of three categories. The first is control, where the objective is to reduce population number and growth rate. This obviously applies to the management of plants and animals. The second is conservation, where the goal is to increase growth rate to the point where the number of individuals are no longer threatened by extinction. The third is exploitation, where the purpose is to maintain a breeding stock of fixed size in order to harvest a fraction of their offspring (e.g., insect mass rearing) or gather products that they produce (e.g., honey). All three cases are concerned with conferring a predetermined population size or growth rate by manipulating life history traits. All involve demography.

Demographic Abstractions

Many early demographers tended to view the components of population change and the processes of population change separately, but more recently the trend has been to abstract and extend many of the components and mechanisms. These perspectives are important because they establish a unity among concepts and methods and they permit an easy extension to analysis of life history characteristics that may not currently have a protocol. Three useful abstractions include age, process, and flow.

Age

In conventional demography many events are measured with respect to the progression of age, but age is not the only progression in the life course. By viewing birth as the starting point and life progression as distance, then age becomes distance in time and total births become the distance in births. The general point is that all individuals that live to age 10 must also have lived to age 9, age 8, age 7, and so forth. Likewise, all individuals that live to produce the tenth offspring must also have lived to produce offspring number 9, number 8, number 7, down to number 1. This concept applies to any repeatable life history event.

Process

Demographic processes in which constituent events cannot be repeated are referred to as *nonrenewable processes,* and those events that can be repeated are *renewable processes*. Clearly, attainment of reproductive maturity and mortality are nonrenewable processes, and giving birth and mating are renewable in most species. By specifying the order of events in a renewable process, it is possible to examine the constituent nonrenewable parts using life table methods of analysis.

Flow

Demography provides a methodology for biological accounting. Gathering data in many respects is the measurement of current inventory that describes changes in stocks (individuals) that have occurred over two or more points in time. Changes arise as a consequence of increments and decrements associated with events such as births and deaths and with flows of individuals between ages or between cross-classifications. Hence net changes in birth and death account for changes in numbers, but interstate transitions or flows from what is considered the origin state to the destination state account for population structure.

These abstractions form the core of the biodemographic models and analysis that are presented in the chapters that follow.

Demography Basics

On every life-line there now lies only one death point, but with a
high density of births there would tend to be in each small interval
of births dying points of all the ages from 0 to ω.
         Wilhel Lexis (1875), *Einleitung in die Theorie der Bevolkerungs-Statistik*

Models and methods of demography describe variation and predict changes in populations over time and space. These models are used in two contexts, the first of which is *formal demography*, in which the characteristics of populations and population change (i.e., births, deaths, and migration) are described. In the second context, a *population study*, demographic analysis is used to evaluate a group of individuals from a general population that shares a common characteristic, such as physiology, genetics, behavior, or other aspects of an organism's biology. In this chapter we establish the foundation of demographic analysis first with a description of demographic levels and then with definitions of different aspects of the characteristics of populations and population change. We then present several examples of how to visualize data that we expand upon in appendices I and II.

Basic Formalization

Demographic levels and traits

Demography describes populations. But the basic unit of all populations, and the starting point for demographic analysis, is the individual—a single organism that is a carrier of demographic attributes (Willekens 2005). The individual is a natural unit and the basic attributes of an individual include all of its life history components, such as gender, development, and size as well as age-specific rates of mating, birth, and death. Depending on context, individual-level attributes in humans may also include cultural attributes, ranging from ethnicity and religion to education and profession. The next demographic level is the cohort, which is defined as *"a group of same-aged individuals"* or, more generally, *"a group who experience the same significant event in a particular time period, and who can thus be identified as a group for subsequent analysis"* (Pressat and Wilson 1987). For example, all individuals born in 2010 are considered a birth cohort, whereas all individuals married that year are considered a marital cohort. Finally, at the demographic level of the population, traits emerge from the interplay of cohorts and individuals. The traits at all levels are

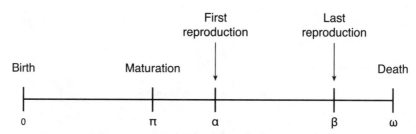

FIGURE 1.1. Generic diagram of life course with key states of birth (age 0), sexual maturity (π), first (α) and last (β) ages of reproduction, and age of death (ω).

considered both micro- and macrodynamic since the passage of time is identified for both individuals and cohorts with age and for the total population with time.

Age and the life course

Age is the characteristic, central variable in almost all demographic analysis and serves as a surrogate for more fundamental measures (e.g., physiological state; biological age) and duration of exposure to risk. Chronological age is defined as the exact difference between the time that a measurement is made and the time of an individual's birth. This difference is typically termed *exact age* and is to be contrasted with *age class*, which combines exact ages into *periods*. As individuals age and develop, they pass through various stages, the duration of which is typically greater than a single age class. The passage from one stage to another is defined as an *event*. The sequence of events and the duration of intervening stages throughout the life of the organism are defined as an individual's life course (fig. 1.1).

Types of data

Feeney (2013) defines *data* as "systematic information about the *entities* in a precisely specified group." He refers to entities as anything to which information can be associated, for example, an individual worm, fly, mouse, or person. The various kinds of information we associate with entities are referred to as *attributes*, each of which has a *name* (e.g., sex; age), a *meaning* (e.g., the gender; the time from birth), and a *range of values* (e.g., female; age 35 years). Like all numerical data, demographic data can be generated either by counting (i.e., 1, 2, 3 . . . n) or by measuring (i.e., process whereby a feature is evaluated), and they fall into one of three categories: discrete, continuous, or categorical. *Discrete demographic data* are numerical data that have a finite number of possibilities, for example, numbers of people, births, and deaths. These data are considered discrete for two reasons: the numerical values are the result of counting and not of measuring; and fractions of each do not exist (e.g., half a person or 90% of a death). *Continuous demographic data* are numerical data that are measured and have values within a range. For example, heights, weights, blood pressure, and walking speed are all considered continuous data because each can vary within a continuous range (e.g., person's height of 2.16 m and weight of 39.21 kg). *Categorical demographic data* are discrete data that can be sorted according to a group or a category, for example, numbers of males and females, Asians and Indians, or married and unmarried.

Population Characteristics

Demography is concerned with four aspects of populations: 1) *size*—the number of individuals in the population; 2) *distribution*—the arrangement of the population in space at a given time; 3) *structure*—the distribution of the population with respect to age, size, gender, and the like; and 4) *change*—the variation in the size, distribution, or structure of the population over time (Carey 1993; Siegel and Swanson 2004). The first three (size, distribution, structure) are referred to as population statics while the last (change) is referred to as population dynamics. Information on populations is obtained either through a census or a survey, the distinction of which is far from clear-cut. A complete canvass of an area is typically thought of as a census, where the intent is to enumerate every individual in the population by direct counting and, further, to cross-classify by age (stage), sex, and so forth. The intent of a survey is to estimate population characteristics based on a subsample of all possible individuals.

Population size

Population size refers to the total number of individuals within a population. Demographers concerned with human populations make a distinction between *de facto* enumeration, which records where each individual is located at the time of the census, and *de jure* enumeration, which records usual residence. This concept can be generalized in biology to distinguish between, for example, transient populations, such as those in temporary residence due to migration (e.g., migratory birds), and more permanent residents of a region (e.g., resident birds).

Population distribution

The spatial distribution of the number of individuals within a population can be characterized using one of three measures: number by spatial location, central location, or standard distance. The *number by spatial subdivision* is given as the percentage of the total population size within a particular location, or as a rank order of the subdivisions from the highest to lowest size. Depending upon which method is used, comparisons of two census times reveal the change in percent, or the change in rank, by spatial location. The *measure of central location* identifies the mean point of a population distributed over a particular area and is defined as the center of population gravity or population mass. The formula for the coordinates of the population center are given by

$$x = \sum_{i=1}^{n} \frac{p_i x_i}{p_i} \qquad (1.1)$$

$$y = \sum_{i=1}^{n} \frac{p_i y_i}{p_i} \qquad (1.2)$$

where n denotes the number of subdivisions, p_i the number in the population at point i, and x_i and y_i its horizontal and vertical coordinates, respectively. Population center can also be defined in three-dimensional space (e.g., vertical distribution on plants) by adding a z-coordinate and computing z.

The third measure of the degree of dispersion of a population in the xy-plane is known as the *standard distance*, which is a measure of number by both area and location. This measure bears the same kind of relationship to the center of the population that the standard deviation of any frequency distribution bears to the arithmetic mean. If x and y are the coordinates of the population center, then the standard distance, D, is given by

$$D = \sqrt{\frac{\sum f_i(x_i - x)^2}{n} + \frac{\sum f_i(y_i - y)^2}{n}} \tag{1.3}$$

where f_i denotes the number of organisms in a particular area and $n = \sum f_i$.

Population structure

The structure of a population is the relative frequency of any enumerable or measurable characteristic, quality, trait, attribute, or variable observed for individuals. These items include age, sex, weight, length, shape, color, biotype, genetic constitution, birth origin, and spatial distribution. Only age and sex are covered here since they are the most common traits by which individuals within populations are decomposed. An age pyramid is often used to illustrate the age-by-sex distribution of a population (fig. 1.2).

The age and sex data in the population pyramid can then be summarized further to measure, for example, the *sex ratio* (SR), which is defined as the number of males per female:

$$SR = \frac{N_m}{N_f} \tag{1.4}$$

where N_m and N_f denote the number of males and the number of females, respectively. Alternatively, the *proportion* of males (PM) in a population, as a fraction of the total population, is given by

$$PM = \frac{N_m}{N_m + N_f} \tag{1.5}$$

Additional measures of sex composition include the primary sex ratio (sex ratio at conception or birth); secondary sex ratio (sex ratio at adulthood or at the end of parental care); tertiary sex ratio (newly independent, nonbreeding, animals); quaternary sex ratio (older breeding adults in the population); and functional sex ratio (sexually active males to receptive females).

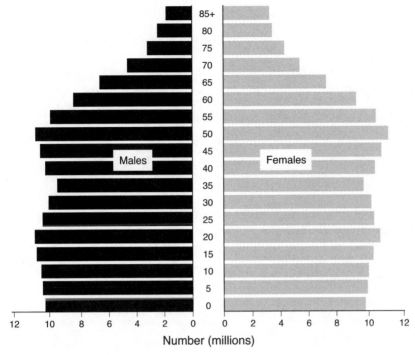

FIGURE 1.2. Population of the United States in 2012 by age and sex. Vertical numbers in center indicate age class (*Source*: US Census Bureau, Current Population Survey, Annual Social and Economic Supplement 2012)

The simplest kind of analysis of age, or stage, data is the *age frequency* distribution of the total population by age, or

$$f_x = \frac{N_x}{N_{total}} \tag{1.6}$$

where f_x is the frequency of individuals age x, and N_{total} denotes the total number in the population. Similarly, the *age-sex frequency* distribution by can be expressed as

$$f_x^m = \frac{N_x^m}{N_{total}^m} \tag{1.7}$$

$$f_x^f = \frac{N_x^f}{N_{total}^f} \tag{1.8}$$

where the superscripts m and f denote male and female, respectively.

Population change (size)

Population change in size can be expressed in the following three ways:

$$\text{Linear (change amount):} \quad N_t - N_{t+1} \tag{1.9}$$

$$\text{Ratio (change factor):} \quad \frac{N_{t+1}}{N_t} \tag{1.10}$$

$$\text{Fractional (proportional change):} \quad \frac{N_{t+1} - N_t}{N_t} \tag{1.11}$$

A linear population change occurs when the difference between the numbers in a population at two time periods is a constant. For example, the sequences 2, 4, 6, and 8 and 10, 20, 30, and 40 are both linear series because in the former series the constant difference is 2 and in the latter series the constant difference is 10. The recursive (iterative) formula is thus

$$N_{t+1} = N_t + c \tag{1.12}$$

where c is the constant difference.

A *geometric change* occurs when the ratio of the numbers in a population at two time periods is a constant. For example, the sequences 1, 2, 4, and 8 and 10, 100, 1,000, and 10,000 are both geometric series because in the former the *constant ratio* is 2 and in the latter it is 10. The recursive formula for the geometric model is

$$N_{t+1} = cN_t \tag{1.13}$$

where c is the constant ratio. The general formula for the geometric model is

$$N_{t+1} = N_0 c^t \tag{1.14}$$

where N_0 is the initial number in the series. An exponential change is a type of geometric change when the time step tends to zero. A geometric change can be converted to an exponential change as

$$N(t) = e^{at} \tag{1.15}$$

where $c = e^a$. For example, $\ln(10) = 2.303$ and therefore $a = 2.303$; so $N(4) = e^{(2.303*4)} = 10,000$.

Population change (space)

Within a spatial location, the number of individuals in a population is a function of fertility and mortality, but the dynamics of populations may vary if a population is spread across geographically defined spatial units and migration occurs from one area to another (Siegel and Swanson 2004). Every move is an out-migration with respect to the area of origin and an in-migration with respect to the area of destination, the balance of which is termed *net migration*. The sum total of migrants moving one direction or the other is termed *gross in-migration* or *gross out-migration*, and the sum total of both in- and out-migration is termed *turnover*. A group of migrants having a common origin and destination is termed a *migration stream*. The difference between a stream and its counterstream is the *net stream* or *net interchange* between two areas, and the sum of the stream and the counterstream is called the *gross interchange* between the two areas.

Various rates of migration can be expressed as

$$\text{Mobility rate} = \text{MR} = \frac{M}{P} \tag{1.16}$$

where M denotes the number of movers and P denotes the population at risk of moving. Other formulas for movement include

$$\text{In-migration rate} = M_1 = \frac{I}{P} \tag{1.17}$$

$$\text{Out-migration rate} = M_o = \frac{O}{P} \tag{1.18}$$

$$\text{Net migration rate} = \frac{I - O}{P} \tag{1.19}$$

where I and O denote the number of in-migrants and out-migrants, respectively.

Basic Demographic Data

Large variation is an inherent component of virtually all demography data sets. For example, with respect to life span and across all species, many individuals die shortly after birth. Thus, across the tree of life the minimum longevity is measured in days. In sharp contrast to this, the units used for the maximum longevity is species specific and highly variable. For example, maximum longevity of individuals can range from months for wasps and flies, years for mice and rats, decades for cats and dogs, centuries for quahogs and tortoises, and millennia for redwoods and bristlecone pines (Carey and Judge 2000). Equally large spreads in lifetime reproductive output occur with per capita offspring production, which can range from zero for individuals that die before maturing or are infertile (e.g., worker honeybees) to maxima in the single digits for cetaceans, in the scores for ungulates, in the hundreds for mice, in the thousands for moths, in the millions for ant queens, and in the tens of millions for large fish. Whereas variation is considered problematic in many branches of science, variability is an integral and essential part of most demographic data sets. Thus, decisions about data analysis and the figures used to support key findings in scientific publications are critically important (Weissgerber et al. 2015). In this section we outline a number of summary techniques for visualizing, analyzing, and graphing basic demographic data. We use data collected on the lifetime egg production in three species of fruit flies to illustrate these techniques. A more formal treatment of data graphics and data visualization is presented in appendices I and II.

Exploratory data analysis and summarization

Visual inspection of raw data

Visual inspection of raw data is the first step to understanding numerical information collected in demographic studies. Engaging with the numbers at this early stage helps to identify patterns, ascertain nuances, spot anomalies, compare trends, anticipate statistical requirements, conceive mathematical models, and envision appropriate graphics (Tufte 2001). This early process of data analysis, even while data are still being collected, is often used to either redirect studies with what appear to be dead-end approaches or fine-tune those in which the experimental results reveal new discoveries that need to be replicated or validated.

Visual inspection of the raw data for the lifetime reproduction in 1,000 individuals in each of three fruit fly species (fig. 1.3) reveals a number of patterns. First, there is a broad pattern that shows that the overall range of lifetime egg production is similar for the Medfly and the Drosophila but nearly twofold greater in the Mexfly. Second, with respect to infertility, there were very few Drosophila females that laid few or zero eggs, but there were a relatively large number of Medfly and Mexfly females that laid minimal eggs. With respect to high reproduction, the drop-off in numbers of females laying large numbers of eggs was abrupt in the Medfly but more gradual in the Mexfly and the Drosophila. Finally, the data show that, whereas in the Mexfly there was a relatively uniform distribution of individuals across the total range of lifetime egg production, the distribution was more concentrated in both the Medfly and the Drosophila with drop-offs at both high and low values of lifetime egg production.

Standard deviation

The standard deviation (SD) is a measure that is used to quantify the amount of variation of a set of data values. It is computed as $SD = \sqrt{\dfrac{\Sigma(x_i - \bar{x})^2}{n-1}}$, where $\bar{x}$ denotes the mean value computed as $\bar{x} = \dfrac{1}{n}\Sigma x_i$. Standard deviation has three properties that are useful in demographic analyses: (1) one, two, and three standard deviations provide statistical first approximation values of the data that lie within 68.27%, 95.45%, and 99.73% of the mean, respectively; (2) unlike the variance, it is expressed in the same units as the data; and (3) it is commonly used to measure confidence in statistical

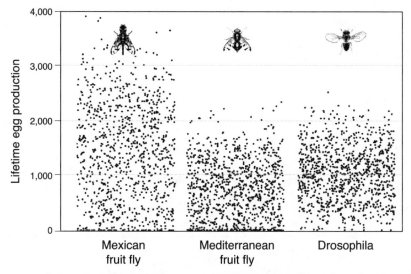

FIGURE 1.3. Lifetime egg production for 1,000 females in each of three fruit fly species—the Mexican fruit fly (*Anastrepha ludens*), the Mediterranean fruit fly (*Ceratitis capitata*), and the vinegar fly (*Drosophila melanogaster*). The horizontal axis for each species is a random number between 0 and 1 assigned to each individual used to spread the points and thus aid in data visualization.

conclusions with the reported margin of error typically twice the standard deviation (i.e., half-width of a 95% confidence interval).

Because different data sets can yield the same SD, a shortcoming of this statistic is that it sheds light on relative spread but little else (Weissgerber et al. 2015). This is illustrated by considering the SD for each of the data sets on lifetime reproduction in fruit flies (see fig. 1.4). Although the larger value of SD for the Mexfly shows that there is larger variation in the Mexfly relative to the other species, the similarity of SD's between the Medfly and the Drosophila does not tell us anything about the details of why they are similar. In fact, as shown below (see fig. 1.5), the distribution of lifetime egg-laying patterns between these two species is quite different. For additional perspective see Anscombe's quartet in chapter 11 (S35).

Histograms

Histograms are graphical representations of the distribution of numerical data and provide estimates of the probability distribution of continuous variables (e.g., birth and death distributions). They differ from bar charts that are used as graphical tools for categorical data (e.g., gender; ethnicity) with the bars being separated spatially as in fig. 1.4 for average egg production data from different species. Histograms for egg laying for the data shown in fig. 1.3 are given in fig. 1.5 and provide more detailed information than previous depictions or summaries. For example, the histograms show a flatter and wider distribution of egg laying in the Mexican fruit fly relative to the other two species. Also, the large number of females who laid zero or few eggs in the Mexfly and Medfly is readily apparent.

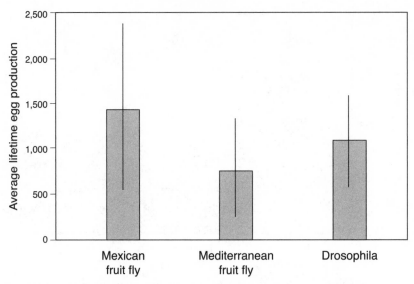

FIGURE 1.4. Means and standard deviations for the egg production in three fruit fly species. Mean and SD for the Mexfly are 1,406.9 and 917.4, respectively, for the Medfly are 749.6 and 527.9, respectively, and for the Drosophila are 1,030.6 and 503.8, respectively. (*Sources*: J. R. Carey, unpublished; R. Arking, unpublished)

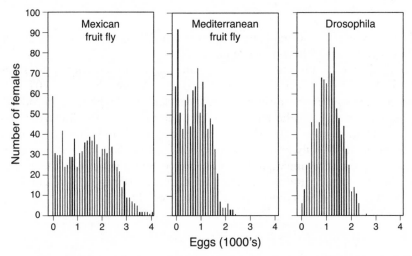

FIGURE 1.5. Histograms for lifetime egg production in three species of fruit flies. Bars depict 100-egg intervals.

Box plots

Box plots are graphic depictions of numerical data in quartiles, one form (box-and-whisker) of which is shown in fig. 1.6 for the fruit fly reproductive data. These plots provide a summary of the data and their variation in five numbers, including the minimum and maximum, and three quartiles, including Q_1 (25%), Q_2 (50%), and Q_3 (75%). These metrics are used to compute range (max-min), interquartile range

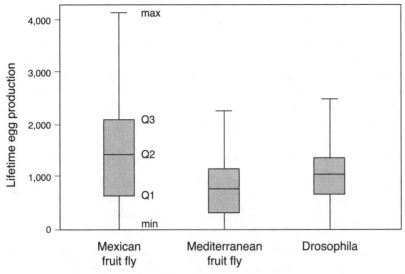

FIGURE 1.6. Box plots for lifetime egg production in 1,000 females in each of three species of fruit flies. Values for Q1, Q2, Q3, and max for the Mexfly are 625, 1,409, 2,111, and 4,034, respectively; for the Medfly are 300, 753, 1,145, and 2,349, respectively; and for the Drosophila are 663, 1,031, 1,371, and 2,537, respectively. Mins are zero eggs (infertile females) for all three species. The IQR, midhinge, and midrange for the Mexfly are 1,486, 1,368.0 and 2,017.0, respectively; for the Medfly are 845, 722.5, and 1,174.5, respectively; and for the Drosophila are 708, 1,017 and 1,268.5, respectively.

$(IQR = Q_3 - Q_1)$, midhinge (average of Q_1 and Q_3), and midrange (middle of the range) and can be computed as

$$\text{Interquartile range (IQR)} = Q_3 - Q_1 \qquad (1.20)$$

$$\text{Midhinge} = \frac{Q_3 + Q_1}{2} \qquad (1.21)$$

$$\text{Midrange} = \frac{\text{max} + \text{min}}{2} \qquad (1.22)$$

The box plots of the fruit fly reproductive data reveal a number of important relationships and patterns. First, a quarter of all Medfly females produce 300 eggs or fewer, and this first quartile is over half that for the other two species. An underlying reason for this low value is that a large fraction of Medfly females are infertile (zero egg layers). The box plots also show that the third quartile of Mexfly reproduction approaches the maximum of the other two species. In other words, the lifetime egg production of one in four Mexfly females is greater than the most fecund females in both the Medfly and the Drosophila. Additionally, the plots show that the Mexfly IQR of nearly 1,500 eggs/female is over 100% greater than that for the Drosophila (≈ 700) and around 75% greater than for the Medfly (≈ 850).

Lorenz curves and Gini coefficients

The Gini coefficient (GC) was originally developed to show the income distribution among residents of a population, but it can be used to represent the degree of inequality in the distribution of any quantitative variable among a group of individuals (Siegel and Swanson 2004). The GC is mathematically defined based on the Lorenz curve, which plots the proportion of the total of a given metric (e.g., income; offspring) possessed by the bottom x% of the population. The 45-degree line represents perfect equality of this metric.

The Gini index is computed from the equation

$$\text{Gini index} = \frac{A}{A+B} \qquad (1.23)$$

where the numerator is the area between the 45-degree line of equality and the Lorenz curve and the denominator is the total area under the line of equality.

Fig. 1.7 shows the Lorenz curve for the distribution of egg production in 1,000 Drosophila females. The area occupied by A in this figure is 0.139. Because the area of the right triangle (total of both shaded zones) in this figure is 0.5, the Gini coefficient for this Lorenz curve is computed as 0.139/0.5, or GC = 0.278, which is a moderately low value thus indicating moderate skew.

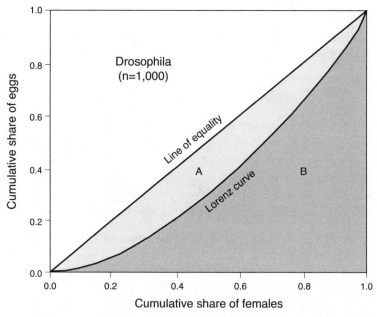

FIGURE I.7. Lorenz curve for *D. melanogaster* egg production. The Lorenz curve would be superimposed on the line of equality if all females had the same lifetime egg production. The departure reveals that around 20% and 50% of females produce 6% and 30% of all eggs, respectively. Or alternatively, the top-producing 6% and 30% of females produce 12% and 47% of all eggs, respectively.

At the extremes, 0 and 1 bound the Gini coefficient, the former indicating complete uniformity (e.g., of income) and the latter indicating the maximum skew (e.g., one single individual accounts for all income in the population). In the current case involving fruit fly reproduction, low GC values indicate greater uniformity of egg production, and higher values indicate smaller numbers of individuals in the population accounting for a larger fraction of total egg production. The strength of the Gini coefficient is that it is simple to compute, easy to interpret, and can be used to compare patterns of inequality across cohorts, subpopulations, or whole populations. It can thus serve as a comparative index for historical trends (e.g., human populations) or comparisons between treatments and/or species (e.g., in biological contexts). A limitation of this coefficient is that similar values can represent very different distributions. This is because the Lorenz curve can have different shapes and still yield the same GC. For example, consider two populations with identical coefficients in which the first population has half of its members who produce no income and the other half who have identical income. In the second population one member generates 50% of the total income and all the remaining members have the same income, but in total they generate the other 50% of total income. In both populations the GC is 0.5.

Lexis diagrams

Concept

Lexis diagrams are graphical representations of the relationships between demographic events in time and persons at risk (Feeney 2003) and are named after the German statistician Wilhelm Lexis (1837–1914). Two numbers characterize every demographic event: the time (e.g., year) at which it occurs and the age of the person to whom the event occurs. The Lexis diagram can be used to depict two types of demographic information: an individual's life-line and cohort effects. An individual's life can be depicted as a *life-line* where the line begins on the time axis at the time of the individual's birth and ends at the age/time point representing the individual's death. The sum of all the life-line lengths in a particular area of the diagram represents life-years lived or exposure in that area.

The Lexis diagram is a powerful visual, conceptual, and heuristic tool for visualizing and understanding the concept of *age-period-cohort effects* (Hobcraft et al. 1982). The effects of age on mortality risk and behavior are complex because they include cohort membership and history (Exter 1986). The biological origin of age effects usually endows them with a generality and regularity absent in cohort and period effects (Wilson 1985). The birth cohort reflects generation and, in turn, affects the value and habits of individuals within it. Cohorts (e.g., baby boomers; millennials) all carry with them habits unique to their generation that affect their mortality risk (e.g., food; exercise; smoking). Likewise, plants that germinate, or animals that are born, at particular times all possess biological and behavioral qualities unique to that particular time period during which their life began. History affects all age groups at once though different ages often react differently. For example, the response to epidemics, famines, droughts, and wars of the very young and the very old are quite different from the response of individuals of middle age.

The Lexis diagram consists of five different elements (fig. 1.8a, b), each of which depicts a different age-period-cohort relationship. The *upper Lexis triangle* depicts

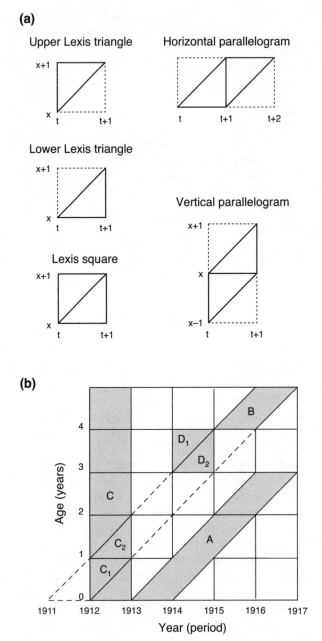

FIGURE 1.8. (a) Five basic Lexis elements. Squares are most frequently used for period analysis; parallelograms are most useful for the cohort analysis (vertical can also be used for period); triangles may serve to compute rates for both period and cohort analysis (data classified by calendar year and birth cohort and age in completed years). (*Source*: Caselli and Vallin 2006) (b) Lexis diagram applied to cohorts for illustration (see text).

the life-lines of individuals who turned age x in the previous year (time $t-1$) and will turn age $x+1$ the current year (time t). For example, the triangle C_1 in fig. 1.8b depicts person-years in 1912 of persons (babies) born in 1911. The *lower Lexis triangle* depicts the life-lines of individuals who will turn age x in the current year (time t) and will turn age $x+1$ the next year (time $t+1$). For example, the triangle C_2 in fig. 1.8b depicts person-years in 1911 of persons (babies) born that year. *The Lexis square* depicts the life-years lived for individuals age x to $x+1$ from time t to time $t+1$. For example, D1 and D2 show the life-years lived in 1914 to all individuals age 3. The Lexis *horizontal parallelogram* depicts the life-years lived from age x to $x+1$ from times t to $t+1$. This is shown in the area shaded B in fig. 1.8b for the life-years lived by a single birth cohort at age 4 from 1915 to 1917. The *Lexis vertical parallelogram* depicts life-years lived by a birth cohort from ages $x-1$ to $x+1$ from time t to time $t+1$. For example, the parallelogram formed by C_1 and C_2 in fig. 1.8b show life-years lived by the 1911 birth cohort between 1912 and 1913, all of whom turned one year old that year.

Lexis diagram: Example applications

As an example, a Lexis diagram depicting the age-period relationship of eminent biologists and others is presented in fig. 1.9. The naturalist Alfred Russel Wallace was an 8-year-old boy in 1831 when the 22-year-old Charles Darwin set off on his around-the-world voyage, he was a 36-year-old explorer in 1859 when the 50-year-old Darwin published his seminal *On the Origin of Species*, and he was a 59-year-old scholar in 1882 when, as a pallbearer, he carried the casket of the 73-year-old newly deceased Charles Darwin at Westminster Abbey. Wallace died in 1913 at age 90, the birth year of demographer Nathan Keyfitz. This was also the birth year of movie star Mary Martin,

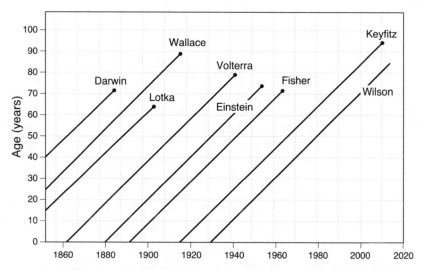

FIGURE 1.9. Lexis diagram with life-lines of eminent scientists. Charles Darwin (1809–1882; 73); Alfred Russel Wallace (1823–1913; 90); Alfred Lotka (1841–1904; 63); Vito Volterra (1860–1940; 80); Albert Einstein (1879–1955; 76); Ronald Fisher (1890–1962; 72); Nathan Keyfitz (1913–2010; 97); E. O. Wilson (b. 1929).

Israeli prime minister Menachem Begin, anthropologist Mary Leakey, and US president Gerald Ford. All five of these individuals, who were born in the same year, aged lock-step as part of this birth cohort, dying at ages 76, 78, 83, and 93, respectively.

Demographic time: Extensions beyond age-period-cohort

As we noted in the last section, demographer Nathan Keyfitz and anthropologist Mary Leakey were both in the 1913 birth cohort, and, therefore, their life-lines are in the same age-period plane. Both Keyfitz and another famous scientist, evolutionary biologist George Williams, died in 2010. They were ages 97 and 84, respectively. Both men died the same year, and, therefore, when Williams was born they became members of the same death cohort, each having 84 years remaining. They possessed identical thanatological ages (time-to-death) by virtue of their identical year of death.

Based on the six time measures contained in the description above, including chronological age (A), period (P), birth cohort (C), thanatological age (T), death cohort (D), and life span (L), Tim Riffe and his colleagues developed a unified framework of demographic time using identities and subidentities that connect all measures (Riffe et al. 2017). These demographers used this terminology and the concept of "Lexis measures" to create triad identities where any two pieces of information will give the third. For example, given that Nathan Keyfitz was 50 (A) on June 29, 1963 (P), then it is possible to derive that he was born on June 29, 1913 (C). Or given that Keyfitz died in 2010 (D) at age 97 (L), it is possible to derive that he had one year to live (T) in 2009.

Riffe and his colleagues noted that there are four informative triad identities formed by the various combinations of three measures, the relationships of which are presented in Lexis-like diagrams in fig. 1.10, including APC, TPD, TAL, and LCD. We use their approach, below, applied to biographical and historical information on eminent biologists and demographers, to illustrate these identities by giving brief descriptions of three permutations within each of the four triads.

Variants of age-period-cohort (APC)

The AP(C) temporal plane constitutes the classical Lexis diagram. If Nathan Keyfitz was age 10 (A) in 1923 (P), then he must have been born in 1913: $(C) = (P - A)$ (fig. 1.10a). The life-lines are run diagonally to the A and P axes. The AC(P) temporal plane is equivalent to the Lexis diagram except birth cohort is given and period is derived rather than the other way around. Given that Charles Darwin was born in 1809 (C), and he was 22 (A) when he departed on his epic around-the-world expedition on the HMS *Beagle*, then the year he departed must have been 1831: $(P) = (C + A)$ [fig. 1.10b]. The CP(A) temporal plane is equivalent to the Lexis diagram except birth cohorts are given and age is derived rather than the other way around. Plant geneticist Barbara McClintock was born in 1902 (C), and, therefore, in 1983 (P) when she received the Nobel Prize, she was age 81: $(A) = (P - C)$ [fig. 1.10c].

Variants of thanatological age–period–death cohort (TPD)

Wilhelm Lexis had 39 years of life left (T) in 1875 (P) when he published his original Lexis diagram; he thus belonged to the 1914 death cohort: $(D) = (P + T)$ [fig. 1.10d]. These lines are orthogonal (sloped downward) to life-lines because as time-to-death (T) decreases, calendar date (P) increases. Benjamin Gompertz died in 1865 (D); thus in 1825 (P) when he published his mortality model, he had 40 years left to live: (T) = (D – P) [fig. 1.10e]. Thomas Malthus died in 1834 (D); when he had 36 years of life left (T), he published his *Essay on the Principle of Population*. Therefore, he published this seminal work in 1798: (P) = (D – T) [fig. 1.10f].

Variants of thanatological age–chronological age–life span (TAL)

The time already lived and the time still left to live sum to total life span. Alfred Lotka was 45 years old (A) when he published his classic article in 1925, at which time he had 24 more years to live (T); thus he lived to age 69: (L) = (T + A) [fig. 1.10g]. John Graunt lived to the age of 54 (L); when he had 12 years left (T), he published *Bills of Mortality*, which meant he was 42 years old at the time of publication: (A) = (L – T) [fig. 1.10h]. The British plant population biologist John Harper was 52 years old (A) when he published his seminal book *Population Biology of Plants*; he lived to age 84 (L) and he thus had 32 years to witness the book's impact on plant science: (T) = (L – A) [fig. 1.10i].

Variants of life span–chronological age–death year (LCD)

Biologist W. D. Hamilton was born in 1936 (C) and lived to be only 64 (L), implying an untimely death in 2000: (D) = (C + L) (fig. 1.10j). Francis Crick was born in 1916 (C) and died in 2004 (D), thus he had a life span of 88 years: (L) = (D – C) [fig. 1.10k].

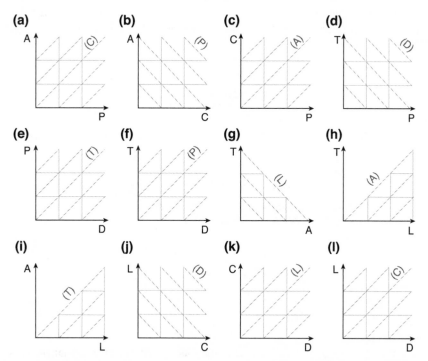

FIGURE 1.10. Didactic juxtapositions of the six measures of demographic time (Riffe et al. 2017). Variants of APC: (a) AP(C); (b) AC(P); (c) CP(A). Variants of TPD: (d) TP(D); (e) PD(T); (f) TD(P). Variants of TAL: (g) TA(L); (h) TL(A); (i) AL(T). Variants of LCD: (j) LC(D); (k) CD(L); (l) LD(C).

Isaac Newton died in 1727 (D) with a completed life span of 84 (L), putting his birth year in 1643: (C) = (D – L) [fig. 1.10l].

Cohort Concepts

A cohort of individual plants, animals, or humans either is born at the same time or all began a treatment at the same time. Multiple cohorts can be followed either *cross-sectionally*, where demographic data are collected by observing subjects across all or a subset of cohorts at the same point in time, or *longitudinally*, where demographic data are collected by observing the same subjects over time (fig. 1.11). The main advantage of a cross-sectional study is that the information can be collected on all age groups over a short period. The disadvantage is that it confounds age differences with cohort or period effects. The main advantage of longitudinal studies is that they track long-term trends as cohorts/individuals age. The main disadvantage is that longitudinal studies require much more time and typically many more resources.

Cohort analysis is the study of dated events as they occur from the time of the event that initiated the cohort (e.g., birth)—for example, a comparison of marriages and mortality occurring in the 1950 and 2000 birth cohorts. Cohort analysis is contrasted with *period analysis*, the study of events occurring in multiple cohorts at a particular time. Cohorts may differ because they experience certain key historical events at different times and critical ages (Ryder 1965). For example, all individuals whose mothers experienced a famine while they were in utero, or when they were young, may be collectively subject to a higher risk of death at older ages (Barker 1994).

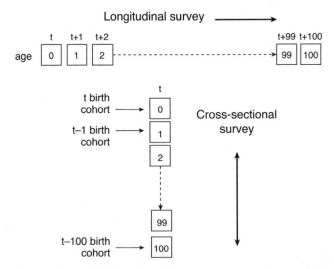

FIGURE 1.11. Schematic showing difference in the concept between longitudinal and cross-sectional surveys or sampling programs. In the former the traits (e.g., fertility; mortality; marriage) of a cohort are tracked as the cohort ages. In the latter the traits are measured in different cohorts over a single period, such as a year.

Ratios, Proportions, and Rates

Most quantities in demography can be expressed as one of three kinds of measures. The first is a *ratio*, which compares frequencies of two mutually exclusive quantities; the second is a *proportion*, which is a special type of ratio in which the numerator is part of the denominator [e.g., $p = a/(a + b)$]; and the third is a *rate*, which is defined as a measure of change in one quantity (e.g., number) per unit of another quantity (e.g., time) on which the first quantity depends. Rates are usually associated with dynamics of populations or cohorts, such as growth. In contrast, ratios and proportions are often associated with population statics, such as sex ratio or age structure.

Demographic rates can be grouped into five categories according to either the kind of population trait counted in the denominator or the kind of events counted in the numerator (Ross 1982). These include the following:

(1) *Crude rates.* This category uses the total population as the denominator. These include crude birth rate (number of births per number in the population), crude death rate (number of deaths per number in the population), and crude rate of natural increase (difference between crude birth and death rates). The results are "crude" in that they consider all individuals rather than individuals grouped by age or sex.

(2) *Age-specific rates.* These are the same as crude rates except with age restrictions for both numerator and denominator. For example, the age-specific rate for fecundity for an individual age 20 counts only offspring produced by females in that age group and counts only females of that age in the denominator.

(3) *Restricted rates.* These rates apply to any special subgroup. For example, in human demography births to married women (rather than to all women) is termed "marital age-specific fertility rate." In nonhuman species a restricted rate could be fecundity rates of females that produce at least one offspring, which would exclude all infertile individuals.

(4) *Rates by topic.* These rates apply to each specialized topic in demography. For example, rates by topic could include total, gross, and net rates for topics ranging from mating and fertility to marriage and migration.

(5) *Intrinsic rates.* The rates that prevail in a stable population are referred to as intrinsic rates in that they cannot reflect any accidental or transient short-term feature of the age distribution. In other words, they are considered "intrinsic" or "true" rates.

2

Life Tables

From Thurfday at Noon the 21ft of December, Anno 1592, unto the 20th Day of
December, being Thurfday at Noon, Anno 1593

Buried	17844
Whereof have died of the Plague	10662
Chriftened this whole Year	4021
Parifhes clear of the Plague	0

"Collection of the Yearly Bills of Mortality," London

Life tables are among the most important tools in demography because they provide
an organizational framework for mortality data. Within this framework they can provide detailed and transparent descriptions of the actuarial properties of a cohort,
generate simple summary statistics that are useful for comparisons (e.g., life expectancy; cohort survival), disaggregate mortality data by cause of death, provide information useful in mortality forecasting, and serve as models of stationary (zero growth)
populations (Carey 1993). In an even broader context where death can be generalized as a change of state, life table techniques can be brought to bear on the analysis
of a wide range of processes involving the exit from a current state. These processes
may include states such as reproduction, migration, divorce, unemployment, or
incarceration.

There are two types of life tables used in demography, and the type used is conditional on the purpose and the data available (i.e., longitudinal or cross-sectional). A
cohort life table provides a longitudinal perspective from the moment of birth through
consecutive ages until the death of all individuals in the original cohort. A period life
table is cross-sectional and assumes a hypothetical (*synthetic*) cohort that is subject,
throughout its lifetime, to the age-specific mortality rates prevailing for the actual
population over a specified period. The construction of the elements of both types of
life tables are discussed in this chapter. We also compare complete and abridged life
tables and other actuarial concepts that can be derived from the results of life table
analyses.

The Basic Life Table

For simplicity, we present a single-decrement, complete cohort life table, which refers
to lumping all forms of death, inclusion of all age classes, and the use of longitudinal
data, respectively. We start with a brief description of the life table radix as a congru-

ency and structural concept, and follow this with parameter definitions and computational examples.

Life table radix

In mathematics a *radix* is a number that is arbitrarily made the fundamental number of a system of numbers. The radix of a life table can be considered its root, which in a cohort life table will be the initial size of the cohort. In human demography a typical radix is assigned a number, such as 100,000. Thus any number remaining at successive ages can be conveniently expressed as the number of survivors out of 100,000. The life table radix is often assigned the value of unity (1.0000) so that subsequent survivors are expressed as a fraction of the original number. The life table radix is the single most important number in the life table because all numbers can be traced to it; in other words, the normalized births initialize the cohort. The radix initializes the organizational stage for the life table.

Life table rate functions

The most basic and widely used life table function is cohort survival, which is defined as the proportion of a newborn cohort surviving from birth ($x = 0$) to age x. Cohort survival, designated l_x, is computed using the formula

$$l_x = \frac{N_x}{N_0} \tag{2.1}$$

where N_0 and N_x denote the number of individuals alive at age 0 (initial number) and age x, respectively. Whereas the denominator, N_0, in this equation is the numerical radix, l_0 denotes the radix for the l_x schedule at age 0.

The second life table parameter, denoted p_x, is age-specific (or period) survival probability, defined as the probability of surviving from age x to $x + 1$. It is computed from the l_x schedule as

$$p_x = \frac{l_{x+1}}{l_x} \tag{2.2}$$

There is an important distinction between survival concepts l_x and p_x. Whereas l_x, the cohort survival, denotes the probability of a *newborn* surviving to age x, p_x, period survival, denotes the probability of an individual that is alive at age x survives to $x + 1$, that is, the next age interval.

The complement of the period survival schedule is the age-specific (period) mortality schedule, denoted q_x, and defined as the probability of an individual dying in the interval x to $x + 1$. It is the complement of p_x such that one minus the probability of living from one age to the next is equal to the probability of dying in this same interval. The parameter q_x is computed using the formula

$$q_x = \frac{l_x - l_{x+1}}{l_x} \tag{2.3}$$

$$= 1 - \frac{l_{x+1}}{l_x} \tag{2.4}$$

$$= 1 - p_x \tag{2.5}$$

The sum of p_x and q_x must equal unity (i.e., $p_x + q_x = 1.0$).

The fourth life table parameter, denoted d_x, is the fraction of the cohort that dies in the interval x to x + 1, which is computed with the formula

$$d_x = l_x - l_{x+1} \tag{2.6}$$

This parameter represents the incremental decrease in l_x due to mortality within the age interval.

Both d_x and q_x are concerned with deaths in the interval x to x + 1. The former refers to the probability that a newborn individual will die in this interval during its lifetime, while the latter refers to the probability that an individual alive at age x will die before age x + 1. The distinction between these parameters can be illustrated if we consider an example of extreme age classes in humans where the fraction of a birth cohort dying in the interval 105 to 106 years (i.e., d_{105}) is extraordinarily small (i.e., $\ll 0.01$) because of the low likelihood of surviving to that age. However, if an individual survives to age 105, their probability of dying before their 106th birthday (i.e., q_{105}) is extraordinarily high (i.e., ≈ 0.5).

At this point we have derived all the life table parameters using the l_x function, and it should be noted that these other parameters can also be used to derive l_x. For example, in terms of p_x,

$$l_x = p_0 p_1 p_2 \cdots p_{x-1} \tag{2.7}$$

$$l_x = \prod_{y=0}^{x-1} p_y \tag{2.8}$$

In terms of q_x,

$$l_x = (1 - q_0)(1 - q_1)(1 - q_2) \cdots (1 - q_{x-1}) \tag{2.9}$$

$$l_x = \prod_{y=0}^{x-1} (1 - q_y) \tag{2.10}$$

And in terms of d_x,

$$l_x = d_0 + d_1 + d_3 \cdots + d_{x-1} \tag{2.11}$$

$$l_x = \sum_{y=0}^{x-1} d_y \tag{2.12}$$

These formulas, used for computing l_x from the other life table parameters, are important because actuarial data frequently involve cohort deaths rather than cohort survival.

The last life table parameter we present in this section is age-specific expectation of life, denoted e_x, defined as the number of years (days) remaining for the average individual age x. The value of e_x is computed as the sum of life-years remaining for a cohort age x normalized by the fraction of those alive at age x. This computation can be demonstrated using the number of life-years remaining to a newborn (i.e., e_0), determined using the l_x schedule as

$$e_0 = \frac{(l_0 + l_1)}{2} + \frac{(l_1 + l_2)}{2} \cdots + \frac{(l_{\omega-1} + l_\omega)}{2} \tag{2.13}$$

The fractions in the right-hand expression give the midpoint values for survival between l_x and l_{x+1}. These midpoints represent the height of an imaginary rectangle centered at this midpoint (fig. 2.1). Given that the area of a rectangle is the product of its height and width, where the width is 1.0 (i.e., 1 year), the midpoint value thus equals the area. In other words, the midpoint value equals the number of life-years lived between x and x+1. It then follows that the sum of the midpoint values for all the imaginary rectangles under the l_x curve gives the remaining number of life-years beyond age x (i.e., the life expectancy).

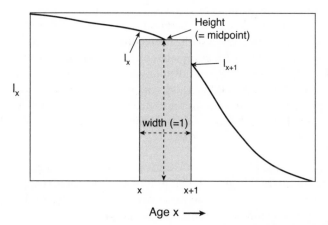

FIGURE 2.1. Depiction of area estimation for life-years lived under the l_x curve. Note that the rectangle area representing life-years equals the product of its height (= midpoint) and width (= single year).

The computational formula for e_x is derived from eqn. (2.13) by bringing the denominator for all of the fractions outside the expression and summing the duplicate l_x's. Thus

$$e_0 = \frac{1}{2}(l_0 + 2l_1 + 2l_2 \cdots + 2l_\omega) \tag{2.14}$$

Given that l_0 always equals unity, then e_0, the expectation of life at age 0, can be computed using the shorthand formula

$$e_0 = \frac{1}{2} + \Sigma_{x=1}^\omega l_x \tag{2.15}$$

The general formula for expectation of life at age x, e_x, is given as

$$e_x = \frac{1}{2} + \frac{1}{l_x}\Sigma_{y=x+1}^\omega l_y \tag{2.16}$$

In eqn. (2.16), the inverse of l_x normalizes the accumulated life-days beyond age x. It should also be noted that the age index outside of the summation symbol begins at x, whereas the initial (lower) index for the summation begins at $x + 1$.

Construction of cohort life table

A complete, single-decrement, cohort life table is constructed by starting with a group of same-aged individuals and recording the number of deaths in the group at each point in time until all individuals have died. The raw data are organized in columns starting with the number alive at the beginning of each age interval. This and other information is arranged in a table with seven columns, as specified below, with rows for age classes. An example of a complete cohort life table is given in table 2.1 for the fruit fly, *Drosophila melanogaster*. By convention, a cohort life table for the Drosophila begins at eclosion (adult "birth"), when a new adult emerges from the pupa stage. In other words, the life stages for a fly from egg to emerging adult are not generally included for logistical reasons. In an equivalent sense, complete life tables for humans begin at birth.

> Column 1: *Age class*, x. This is the age index and designates the exact age that an interval begins (with a unit of days for the Drosophila). Age class x refers to the interval from exact age x to exact age $x + 1$ (e.g., age class 0 specifies the interval from age 0 to age 1).
>
> Column 2: *Number alive at age* x, N_x. The data in this column contain the number of individuals alive at age x. For example, from table 2.1, $N_0 = 666$ and $N_{80} = 2$, which in words means that the cohort started at age 0 with 666 individuals, and at age 80 only 2 individuals remained.

Column 3: *Cohort survival to age* x, l_x. The first fraction in this column, l_0, is the radix, and each successive number represents the fraction of survivors at the exact age x from the cohort of size l_0 (normalized to 1.0). For example, from table 2.1,

$$l_{10} = \frac{N_{10}}{N_0} \qquad\qquad l_{25} = \frac{N_{25}}{N_0}$$

$$= \frac{656}{666} \qquad\qquad = \frac{516}{666}$$

$$= 0.9851 \qquad\qquad = 0.7741$$

In words, the fraction of the initial cohort that survives to ages 10 and 25 days are 0.9851 and 0.7741, respectively. Fewer than 2% were dead after 10 days, but nearly a quarter were dead 15 days later.

Column 4: *Period survival*, p_x. This parameter is defined as the proportion of those alive at age x that survive through the interval x to x + 1. For example,

$$p_{10} = \frac{l_{11}}{l_{10}} \qquad\qquad p_{25} = \frac{l_{26}}{l_{25}}$$

$$= \frac{0.9806}{0.9851} \qquad\qquad = \frac{0.7527}{0.7741}$$

$$= 0.9954 \qquad\qquad = 0.9724$$

The probability of a 10-day-old female surviving to day 11 is 0.9954. Fifteen days later the probability of this fly, now 25 days old, surviving for one more day is reduced to 0.9724, or by over 2%. Although this difference in one-day survival probability appears small, when reexpressed as a probability of dying, the proportional difference is relatively large as will be seen with the next parameter, q_x.

Column 5: *Period mortality*, q_x. This parameter is defined as the probability of individuals alive at age x dying in the interval x to x + 1. For example,

$$q_{10} = 1 - \frac{l_{11}}{l_{10}} \qquad\qquad q_{25} = 1 - \frac{l_{26}}{l_{25}}$$

$$= 1 - \frac{0.9806}{0.9851} \qquad\qquad = 1 - \frac{0.7527}{0.7741}$$

$$= 0.0046 \qquad\qquad = 0.0276$$

The probability of a 10-day-old female dying before she is age 11 days is 0.0046. Fifteen days later the probability of this fly, now 25 days old, dying the next day is increased to 0.0276—a sixfold increase. Indeed, a 75-day-old fly has over a fiftyfold greater risk of dying in the next 1-day period relative to a 10-day-old fly (see table 2.1).

Column 6: *Frequency distribution of deaths*, d_x. This column contains the fraction of the original cohort, l_0, that die within the age interval x to x + 1. The d_x column thus represents the frequency distribution of deaths in the cohort, and the sum of d_x across all ages is unity. In table 2.1, d_x is computed for ages 10 and 25 days as

$$d_{10} = l_{10} - l_{11} \qquad\qquad d_{25} = l_{25} - l_{26}$$
$$= 0.9851 - 0.9806 \qquad\qquad = 0.7741 - 0.7527$$
$$= 0.0045 \qquad\qquad = 0.0214$$

In words, a new adult fly has a likelihood of dying in the intervals 10 to 11 days and 25 to 26 days of 0.0045 and 0.0214, respectively.

Column 7: *Expectation of life at age* x, e_x. This gives the average remaining lifetime for an individual who survives to the beginning of the indicated age interval. For example,

$$e_{10} = \frac{1}{2} + \frac{1}{l_x} \sum_{y=11}^{\omega} l_y$$
$$= \frac{1}{2} + \frac{l_{11} + l_{12} + l_{13} + \cdots + l_{81} + l_{82}}{l_{10}}$$
$$= \frac{1}{2} + \frac{0.9806 + 0.9751 + 0.9678 \cdots + 0.0018 + 0.0000}{0.9851}$$
$$= 29.5 \text{ days}$$

$$e_{25} = \frac{1}{2} + \frac{1}{l_x} \sum_{y=26}^{\omega} l_y$$
$$= \frac{1}{2} + \frac{l_{26} + l_{27} + l_{28} + \cdots + l_{81} + l_{82}}{l_{25}}$$
$$= \frac{1}{2} + \frac{0.7527 + 0.7317 + 0.7108 \cdots + 0.0018 + 0.0000}{0.7741}$$
$$= 20.0 \text{ days}$$

In words, the average 10-day-old fly has 29.5 days of life remaining and the average 25-day-old fly still has 20.0 days of life remaining.

Table 2.1. Life table for the fruit fly, *Drosophila melanogaster*

Age (days)	Number alive at age x	Fraction of initial cohort alive at age x	Fraction surviving from age x to x+1	Fraction dying in age interval x to x+1	Fraction of initial cohort dying in interval x to x+1	Remaining life at age x
x	N_x	l_x	p_x	q_x	d_x	e_x
(1)	(2)	(3)	(4)	(5)	(6)	(7)
0	666	1.0000	1.0000	0.0000	0.0000	39.0
1	666	1.0000	1.0000	0.0000	0.0000	38.0
2	666	1.0000	0.9998	0.0002	0.0002	37.0
3	666	0.9998	0.9994	0.0006	0.0006	36.0
4	665	0.9992	0.9987	0.0013	0.0013	35.0
5	665	0.9979	0.9982	0.0018	0.0018	34.1
6	663	0.9961	0.9982	0.0018	0.0018	33.1
7	662	0.9943	0.9978	0.0022	0.0022	32.2
8	661	0.9921	0.9967	0.0033	0.0033	31.3
9	659	0.9888	0.9963	0.0037	0.0037	30.4
10	656	0.9851	0.9954	0.0046	0.0045	29.5
11	653	0.9806	0.9944	0.0056	0.0055	28.6
12	649	0.9751	0.9925	0.0075	0.0073	27.8
13	645	0.9678	0.9912	0.0088	0.0086	27.0
14	639	0.9592	0.9896	0.0104	0.0100	26.2
15	632	0.9492	0.9885	0.0115	0.0110	25.5
16	625	0.9383	0.9867	0.0133	0.0125	24.8
17	617	0.9258	0.9847	0.0153	0.0142	24.1
18	607	0.9116	0.9822	0.0178	0.0163	23.5
19	596	0.8953	0.9812	0.0188	0.0168	22.9
20	585	0.8785	0.9792	0.0208	0.0183	22.3
21	573	0.8603	0.9772	0.0228	0.0196	21.8
22	560	0.8406	0.9741	0.0259	0.0218	21.3
23	545	0.8189	0.9733	0.0267	0.0219	20.8
24	531	0.7970	0.9712	0.0288	0.0230	20.4
25	516	0.7741	0.9724	0.0276	0.0214	20.0
26	501	0.7527	0.9721	0.0279	0.0210	19.5
27	487	0.7317	0.9715	0.0285	0.0209	19.1
28	473	0.7108	0.9698	0.0302	0.0215	18.6
29	459	0.6893	0.9705	0.0295	0.0204	18.2
30	446	0.6690	0.9675	0.0325	0.0217	17.7
31	431	0.6472	0.9680	0.0320	0.0207	17.3
32	417	0.6265	0.9667	0.0333	0.0209	16.9
33	403	0.6056	0.9681	0.0319	0.0193	16.4
34	390	0.5863	0.9661	0.0339	0.0199	15.9
35	377	0.5664	0.9653	0.0347	0.0197	15.5
36	364	0.5468	0.9629	0.0371	0.0203	15.0
37	351	0.5265	0.9592	0.0408	0.0215	14.6
38	336	0.5050	0.9582	0.0418	0.0211	14.2
39	322	0.4839	0.9566	0.0434	0.0210	13.8
40	308	0.4629	0.9536	0.0464	0.0215	13.4
41	294	0.4414	0.9530	0.0470	0.0207	13.0

(*continued*)

Table 2.1. (*continued*)

Age (days)	Number alive at age x	Fraction of initial cohort alive at age x	Fraction surviving from age x to x+1	Fraction dying in age interval x to x+1	Fraction of initial cohort dying in interval x to x+1	Remaining life at age x
x	N_x	l_x	p_x	q_x	d_x	e_x
(1)	(2)	(3)	(4)	(5)	(6)	(7)
42	280	0.4207	0.9482	0.0518	0.0218	12.6
43	266	0.3989	0.9482	0.0518	0.0207	12.3
44	252	0.3782	0.9412	0.0588	0.0222	11.9
45	237	0.3560	0.9421	0.0579	0.0206	11.7
46	223	0.3354	0.9397	0.0603	0.0202	11.3
47	210	0.3151	0.9361	0.0639	0.0201	11.0
48	196	0.2950	0.9312	0.0688	0.0203	10.8
49	183	0.2747	0.9294	0.0706	0.0194	10.5
50	170	0.2553	0.9275	0.0725	0.0185	10.3
51	158	0.2368	0.9250	0.0750	0.0178	10.0
52	146	0.2190	0.9240	0.0760	0.0167	9.8
53	135	0.2024	0.9226	0.0774	0.0157	9.6
54	124	0.1867	0.9212	0.0788	0.0147	9.3
55	115	0.1720	0.9201	0.0799	0.0137	9.1
56	105	0.1583	0.9212	0.0788	0.0125	8.8
57	97	0.1458	0.9209	0.0791	0.0115	8.6
58	89	0.1343	0.9170	0.0830	0.0111	8.3
59	82	0.1231	0.9241	0.0759	0.0093	8.0
60	76	0.1138	0.9194	0.0806	0.0092	7.6
61	70	0.1046	0.9125	0.0875	0.0092	7.2
62	64	0.0954	0.9107	0.0893	0.0085	6.8
63	58	0.0869	0.9066	0.0934	0.0081	6.4
64	52	0.0788	0.8966	0.1034	0.0081	6.1
65	47	0.0707	0.8833	0.1167	0.0082	5.7
66	42	0.0624	0.8797	0.1203	0.0075	5.4
67	37	0.0549	0.8653	0.1347	0.0074	5.1
68	32	0.0475	0.8584	0.1416	0.0067	4.8
69	27	0.0408	0.8524	0.1476	0.0060	4.5
70	23	0.0348	0.8256	0.1744	0.0061	4.1
71	19	0.0287	0.8211	0.1789	0.0051	3.9
72	16	0.0236	0.8171	0.1829	0.0043	3.7
73	13	0.0193	0.7900	0.2100	0.0040	3.4
74	10	0.0152	0.7700	0.2300	0.0035	3.1
75	8	0.0117	0.7600	0.2400	0.0028	2.9
76	6	0.0089	0.7500	0.2500	0.0022	2.7
77	4	0.0067	0.7400	0.2600	0.0017	2.4
78	3	0.0049	0.7240	0.2760	0.0014	2.1
79	2	0.0036	0.7118	0.2882	0.0010	1.7
80	2	0.0025	0.7000	0.3000	0.0008	1.2
81	1	0.0018	0.0000	1.0000	0.0018	0.5
82	0	0.0000	—	—	0.0000	0.0

Source: Unpublished data from Robert Arking, by permission.

Parameter visualization

Each parameter of the life table reveals different information from the same basic data (fig. 2.2). For example, the survival curve (l_x) shows a gentle slope at younger ages, followed by a moderately rapid drop in survival at middle and late ages. The fine details of daily survival and mortality are difficult to discern from the survival curve, but they are clearer from the mortality curve. The age-specific mortality curve (q_x) shows a low risk of death at young ages followed by the ever-increasing risk of death through the remaining life course. Although the death distribution can be inferred from the l_x schedule, the d_x schedule shows this distribution directly and clearly highlights the pattern showing that the vast majority of deaths occur between 20 and 60 days. The remaining expectation of life (e_x) is a stand-alone metric that graphically shows the actuarial outcome of present and future mortality at each age. Life expectancy (e_x) cannot be inferred directly from any of the other parameters.

Life table censoring

Cohort life tables present data for a large group of individuals followed over time, but information on some individuals may be unknown if subjects drop out of a study or are otherwise removed from a study for reasons other than death. Individuals who are lost from a study are considered "censored," and a framework for computing period-specific survival rates with censoring was first published by statisticians Kaplan and Meier (1958). The computation is expressed as

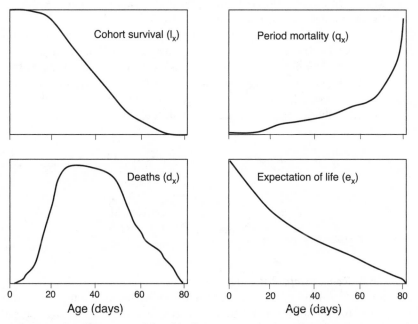

FIGURE 2.2. Graphs of four main life table functions in the fruit fly, *D. melanogaster* (from table 2.1).

$$f_x = \prod_{x=0}^{\omega} \frac{N_x - D_x}{N_x} \qquad\qquad (2.17)$$

where N_x is the number alive at age x, D_x is the number dying in the interval x to x + 1, and f_x is the fraction surviving from birth to age x.

Consider a life table study in which the survival of 100 individuals is monitored from birth through age 6 (table 2.2). After one week, 9 individuals had died, but censoring was required because an additional 5 individuals had escaped. The probability of surviving that period must be computed using only the individuals that were observed to be at risk throughout the period. In this case 86 remained alive by the end of the period, and the probability of surviving from age 0 to age 1 is computed as 86/95, or 0.905. These survival probabilities, corrected for censoring, are computed for each period, and the cumulative probabilities of survival from birth to age x are, in turn, computed from these.

A step-by-step method for constructing a survival schedule for a cohort subject to censoring is described as follows (table 2.2):

Step 1. *Organize basic data.* Arrange original data by age (col. 1), by number at risk at the beginning of each age class (col. 2), by the number censored in each age class (col. 3), and by the number dying at the end of each interval (col. 5).

Step 2. *Number at risk throughout interval* (col. 4). Correct the at-risk number of individuals in each interval by subtracting the values in col. 3 (number censored in interval) from col. 2 (number at risk at start of interval) and insert in col. 4 (at risk throughout interval). For example, 100 − 5 = 95 gives the number at risk from ages 0 to 1 (col. 4).

Step 3. *Number surviving interval.* This number is computed as the number at risk (col. 4) minus the number who died (col. 5). For example, 95 − 9 = 86 gives the number of at-risk individuals surviving the interval 0 to 1 (col. 2).

Step 4. *Fraction surviving interval.* This number is computed as the difference between the number at risk (col. 4) and the number who died (col. 5) divided by the original number at risk (col. 4):

95 − 9 = 86 (i.e., number of at-risk individuals surviving the interval 0 to 1)

86/95 = 0.905 (i.e., fraction of at-risk individuals surviving the interval 0 to 1)

Step 5. *Repeat steps 2 through 4 to complete col. 6.*

Step 6. *Compute censoring-corrected cohort survival* (col. 7). Components for this column are computed iteratively by first inserting the radix of 1.000 at age 0, then multiplying this number by the fraction surviving the interval. For example:

Table 2.2. Example of Kaplan-Meier estimator to correct for censoring (e.g., killed; escaped; lost tag)

Age (x)	Number alive at start of interval	Number censored during interval	At risk throughout interval	Number who died at end of interval	Fraction surviving this interval	Censoring-corrected survival
(1)	(2)	(3)	(4)	(5)	(6)	(7)
0–1	100	5	95	9	0.905	1.000
1–2	86	4	82	25	0.695	0.905
2–3	57	1	56	14	0.750	0.629
3–4	42	7	35	39	0.143	0.472
4–5	5	1	4	4	0.000	0.067
5–6	0	0	0	—	—	0.000

Iteration #1: $1.000 \times 0.905 = 0.905$

Iteration #2 $0.905 \times 0.695 = 0.629$

Iteration #3 $0.629 \times 0.750 = 0.472$

Iteration #4 $0.472 \times 0.143 = 0.067$

Iteration #5 $0.067 \times 0.000 = 0.000$

Period life tables

Background

Period life tables, also referred to as current life tables, are cross-sectional and thus represent the patterns of survival and mortality during a specific, limited time period. A period life table assumes a hypothetical, synthetic cohort subject throughout its lifetime to the age-specific mortality rates prevailing for the actual population over a specified period. It is constructed by using the cross-sectional mortality data for each age to compute age-specific survival, p_x, and, in turn, to iteratively calculate the survival schedule, l_x. This schedule is then used to compute all the remaining life table parameters, including d_x, L_x, T_x, and e_x. A period life table can be abridged, in the same way that a cohort life table can be, but here we only present the construction of a complete period table.

Construction

A period life table is constructed by creating a hypothetical birth cohort of 100,000. Survival of these individuals is then projected forward to create a synthetic cohort using the prevailing age-specific survival rates applied to the iterative equation

$$N_{x+1} = {}_np_xN_x \qquad\qquad (2.18)$$

where n denotes the age interval.

Example computations for values of N_x in table 2.3 include

$$N_5 = {}_5p_0N_0$$
$$= 0.99482 \times 100,000$$
$$= 99,482$$

$$N_{10} = {}_5p_5N_5$$
$$= 0.99944 \times 99,482$$
$$= 99,426$$

$$N_{15} = {}_5p_{10}N_{10}$$
$$= 0.99926 \times 99,426$$
$$= 99,352$$

Once the N_x column is complete, all additional columns of the life table can be completed according to the formulas presented earlier.

Primacy of life expectancy

Life expectancy is widely used as an indicator of mortality conditions because it is based on the mortality experience at all ages and is independent of age structure. It provides a projection for the average number of years remaining to an individual of a specified age with the assumption that the mortality conditions will be invariant. Life expectancy at birth (e_0) is the most widely used value, and in situations where life expectancy is referred to without qualification, the value at birth is normally

Table 2.3. Example of period life table

Age class	$_np_x$	N_x
0–5	0.99482	100,000
5–10	0.99944	99,482
10–15	0.99926	99,426
15–20	0.99809	99,352
20–25	—	99,162

Note: Survival probabilities from one age class to the next are derived from prevailing rates during a specified period (i.e., cross-sectional data) rather than from an actual cohort. These rates are applied to a hypothetical cohort of 100,000 individuals.

assumed. At the earliest life stages, high levels of infant mortality may mean that those who survive this high-risk year may have higher life expectancies than the newborns themselves. The inverse of life expectancy at birth $(1/e_0)$ gives the per capita death rate in a population. For example, the 80-year life expectancy at birth of contemporary US females can also be presented as a per capita annual death rate of $0.0125 \, (= 1/80)$.

The Abridged Life Table

Thus far we have been considering complete cohort life tables where values are given across time units of single days or years. Alternatively, an abridged life table, which is an abbreviated version of a complete life table, can be constructed with longer age intervals. An abridged table retains all the basic functions, properties, and concepts of the basic life table while at the same time removing two disadvantages of the complete life table. First, it is often either extremely expensive or virtually impossible to monitor mortality in a cohort over their entire life course in the short intervals required to construct a complete cohort life table. This is especially true for organisms that develop in stages (e.g., eggs; larvae; pupae in insects) in which the only realistic solution for determining survival is to record the number entering and exiting a stage of several days' duration. Second, the many hundreds of values in a table may contain details that are not of interest. By grouping deaths into age increments larger than a single unit (i.e., day; week; year), it is possible to summarize the information more concisely, with minimal compromising of accuracy and precision, while still retaining the basic life table format.

Notation, concepts, and columns

In a complete life table, the basic parameters apply to a single age interval; for example, the parameter p_{10} implies survival from age 10 to 11. In an abridged table where the interval of concern is greater than a single age class, a subscript n is used to designate its length beyond age x. For example, $_np_x$, $_nq_x$, and $_nd_x$ denote period survival, period mortality, and death frequency, respectively, in the interval x to x+n. This notation is also used to denote the number of life-years lived by the cohort in this interval as $_nL_x$, where

$$_nL_x = \frac{n(l_x + l_{x+n})}{2} \qquad (2.19)$$

For the computation of expectation of life at age x, which is based on the remaining life-years to be lived by the average individual alive at age x, an additional parameter, T_x, is needed, which is the sum total of remaining life-years:

$$T_x = \sum_{y=x}^{\omega} {}_nL_x \qquad (2.20)$$

This column is then used to compute remaining expectation of life in an abridged life table given by the equation

$$e_x = \frac{T_x}{l_x} \qquad (2.21)$$

In words, eqn. (2.21) states that the expectation of life at age x is the dividend of the total number of life-years remaining to a cohort at age x and the fraction of the cohort surviving to age x. Similar to the complete life table computations, the fraction, l_x, normalizes the life-days remaining.

Construction of abridged cohort life table

Here we provide example computations for constructing an abridged life table using the data presented in table 2.1 on *D. melanogaster*, where we lump the information into 10-day age groups (table 2.4):

Column 1: *Age class*, x.
Column 2: *Number alive at age* x, N_x.
Column 3: *Survival to age* x, l_x. This is computed the same as in the complete life table, $l_x = \dfrac{N_x}{N_0}$.
Column 4: *Fraction surviving in an age interval*, $_np_x$. This is the proportion of flies alive at age x that survive to x+n. In the example, $_{10}p_x$ is the proportion of flies that survive from age x to age x+10. For example,

$$_{10}P_0 = \frac{l_{10}}{l_0} \qquad\qquad\qquad _{10}P_{50} = \frac{l_{60}}{l_{50}}$$
$$= \frac{0.9851}{1.000} \qquad\qquad\qquad = \frac{0.1138}{0.2553}$$
$$= 0.9851 \qquad\qquad\qquad\qquad = 0.4457$$

In words, 0.9851 of all flies 0 days old survived to age 10 and 0.4457 50-day-old flies survived to age 60.

Column 5: *Fraction dying in an age interval*, $_nq_x$. This parameter is the proportion of those alive at age x that die in the interval x to x+n. For example,

$$_{10}q_0 = 1 - \frac{l_{10}}{l_0} \qquad\qquad\qquad _{10}q_{50} = 1 - \frac{l_{60}}{l_{50}}$$
$$= 1 - \frac{0.9851}{1.000} \qquad\qquad\qquad = 1 - \frac{0.1138}{0.2553}$$
$$= 0.01491 \qquad\qquad\qquad\qquad = 0.5543$$

In words, only 0.0149 of the newly eclosed flies died before they reached age 10, but 0.5543 of the 50-day-old flies died before reaching age 60. Comparing age intervals, flies were 37 times more likely to die in the 10-day interval from 50 to 60 days than in the interval from 0 to 10 days.

Column 6: *Fraction of initial cohort dying during an age interval*, $_nd_x$. This column contains the fraction of the original cohort that die in the age interval x to x + n. For example,

$$_{10}d_0 = l_0 - l_{10}$$
$$= 1.000 - 0.9851$$
$$= 0.0149$$

$$_{10}d_{50} = l_{50} - l_{60}$$
$$= 0.2553 - 0.1138$$
$$= 0.1415$$

In words, 0.0149 of the entire cohort died between age 0 and age 10 days, but between age 50 and age 60 days 0.1415, or nearly 15%, of the entire cohort died.

Column 7: *Life-days in an interval*, $_nL_x$. This is computed for the age interval x to x + n, for example,

$$_{10}L_0 = \frac{10(l_0 + l_{10})}{2}$$
$$= \frac{10(1.000 + 0.9851)}{2}$$
$$= 9.9255$$

$$_{10}L_{50} = \frac{10(l_{50} + l_{60})}{2}$$
$$= \frac{10(0.2553 + 0.1138)}{2}$$
$$= 0.1415$$

Column 8: *Cumulative days lived beyond age* x, T_x. For example,

$$T_0 = \sum_{y=0}^{\omega} {}_{10}L_y$$
$$= 9.9255 + \cdots + 0.0127$$
$$= 39.0$$

$$T_{50} = \sum_{y=50}^{\omega} {}_{10}L_y$$
$$= 1.8454 + \cdots + 0.0127$$
$$= 10.9$$

Column 9: *Expectation of life at age* x, e_x. This is the expected number of additional days the average individual age x will live, e_x. For example,

$$e_0 = \frac{T_0}{l_0}$$
$$= \frac{39.0189}{1.0000}$$
$$= 39.0$$

$$e_{50} = \frac{T_{50}}{l_{50}}$$
$$= \frac{2.7873}{0.2553}$$
$$= 10.9$$

Table 2.4. Abridged life table for *D. melanogaster*

Age (days) x	Number alive at age x N_x	Survival to age x l_x	Fraction surviving from age x to x+10 $_{10}P_x$	Fraction dying in interval x to x+10 $_{10}q_x$	Fraction of cohort dying x to x+10 $_{10}d_x$	Life-days in interval x to x+10 $_{10}L_x$	Cumulative life-days from x to x+10 T_x	Expectation of life at age x e_x
(1)	(2)	(3)	(4)	(5)	(6)	(7)	(8)	(9)
0	666	1.0000	0.9851	0.0149	0.0149	9.9255	39.0189	39.0
10	656	0.9851	0.8918	0.1082	0.1066	9.3182	29.0934	29.5
20	585	0.8785	0.7615	0.2385	0.2096	7.7375	19.7752	22.5
30	446	0.6690	0.6920	0.3080	0.2061	5.6593	12.0377	18.0
40	308	0.4629	0.5515	0.4485	0.2076	3.5910	6.3783	13.8
50	170	0.2553	0.4457	0.5543	0.1415	1.8454	2.7873	10.9
60	76	0.1138	0.3055	0.6945	0.0790	0.7427	0.9419	8.3
70	23	0.0348	0.0732	0.9268	0.0322	0.1865	0.1993	5.7
80	2	0.0025	0.0000	1.0000	0.0025	0.0127	0.0127	5.1
90	0	0.0000			0.0000	0.0000	0.0000	

Note: See table 2.1 for the original complete life table data.

Table 2.5. Comparison of *D. melanogaster* life expectancy estimates

	Life expectancy at age x (days)		Estimate comparison	
	Abridged	Complete	Absolute (difference) (abridged/complete)	Relative (ratio) (abridged/complete)
0	39.0	39.0	0.0	1.00
10	29.5	29.5	0.0	1.00
20	22.5	22.3	0.2	1.01
30	18.0	17.7	0.3	1.02
40	13.8	13.4	0.4	1.03
50	10.9	10.3	0.6	1.06
60	8.3	7.6	0.7	1.09
70	5.7	4.1	1.6	1.39
80	5.0	1.2	3.8	4.17

Note: Comparison based on complete versus abridged life tables (i.e., results from tables 2.1 and 2.2). Absolute differences in days.

In words, average newly eclosed flies and 50-day-old flies have, respectively, 39 and 10.9 days of life remaining.

Comparison of abridged and complete life table

There are a few differences between a complete and an abridged life table. The reduction of the number of age classes from 82 in the complete life table (table 2.1) to 10 in the abridged life table (table 2.4) preempts the study of some of the actuarial details, such as the mortality trajectory at older ages. However, in many applications researchers are primarily concerned with life expectancy comparisons, and it is thus useful to understand the extent to which accuracy of this metric is traded off by lumping actuarial information into broad age classes, in our case, 10-day age classes. Differences in estimates between the complete and abridged life table (table 2.5) show that the estimates of remaining life expectancy at ages 0, 10, 20, 30, and 40 days for the abridged life table are identical with or differ by less than 3% from the calculations from the complete life table at these ages. At days 50 and beyond, the absolute differences between the two types of life tables are small but the relative differences are moderate to large. In addition to the progressive increase in the differences in remaining life expectancy, one should also note that the estimates for the abridged model exceeded those for the complete life table at all ages. The reason for this discrepancy is that, because of the rapid decrease in survival (fig. 2.3), the rectangle method of life-day estimation overcompensates to the right of the midpoint.

Period and cohort life table concepts

Gaps and lags

Whereas cohort life tables are constructed from longitudinal data of an actual cohort, period life tables are constructed from cross-sectional data and, therefore, are synthetic and hypothetical. However, period life expectancy is not as hypothetical as is often believed (Goldstein and Wachter 2006; Wachter 2014). These authors provide

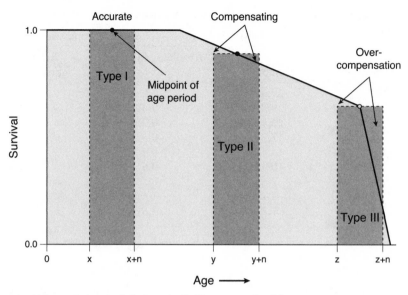

FIGURE 2.3. Schematic l_x schedule (survival) illustrates why life-year (area) estimates at older ages using the midpoint in survival at two ages are often inaccurate because the survival rates are dropping rapidly. The midpoint of the l_x schedule within the age interval determines the rectangle height, and the interval duration (n) is the width. The product of these (height × width) yields the area estimate.

an empirical and analytical foundation for regarding period life expectancy as a lagged indicator such that under contemporary conditions the expectation of life is approximately equal to cohort life expectancy for the cohort born 40–50 years earlier. The vertical distance between the period and cohort curves represents the gap. The magnitude of the gap is the "bonus" (years) that a real cohort receives as it experiences the benefits of future improvements in mortality. The horizontal distance between cohort and period curves represents the lag. This metric gives the number of years in the past that life expectancy was equivalent to period life expectancy in a given year. The authors' analysis also evaluates changes in these gaps and lags. First, as mortality has fallen, the lag between periods and cohorts has increased. This is due largely to the greater ages at which deaths are occurring and thus the greater ages at which mortality improvement is taking place. Second, the absolute difference between period and cohort life expectancy (i.e., the gap) has risen and then fallen over time. The pace of change in period mortality has flattened in recent years. Finally, although the pace of mortality decline plays an important role in determining the magnitude of the gap, this rate of reduction has less of an effect on the size of the lag.

A hypothetical illustration of the gap between survival estimates for cohort versus period life tables is given in table 2.6. Mortality for ages 0 to 3 is improving for periods 1 through 6, and then mortality is constant for the subsequent periods 7 through 10. An illustration of how the survival "gap" (=0.10) arose between period survival (=0.85) and cohort survival (=0.75) at period 4 is shown in the following discussion.

Table 2.6. Illustration of period and cohort life table concepts

	Period									
	Mortality improving						Mortality unchanging			
Age	1	2	3	4	5	6	7	8	9	10
3	0.751	0.834	0.900	0.934	0.945	0.956	0.967	0.967	0.967	0.967
2	0.833	0.889	0.933	0.955	0.963	0.970	0.978	0.978	0.978	0.978
1	0.888	0.925	0.955	0.970	0.975	0.980	0.985	0.985	0.985	0.985
0	0.925	0.950	0.970	0.980	0.983	0.987	0.990	0.990	0.990	0.990
Period	0.51	0.65	0.78	0.85	0.87	0.90	0.92	0.92	0.92	0.92
Cohort	0.75	0.82	0.87	0.90	0.91	0.92	0.92	0.92	0.92	0.92
Gap	0.23	0.17	0.09	0.05	0.04	0.02	0.00	0.00	0.00	0.00

Note: Each column contains the time-specific survival rates for age classes 0–3. Each row contains the age-specific rates for each of the periods with progressively higher survival through period 7 at which time the rates become fixed. Period survival rates through three age classes are shown at the bottom, the former the product of the prevailing rates during that period and the latter the product of the diagonals. For example, the cohort rates for period 4 are the product of the values in the left diagonal sequence. The middle and right diagonal sequences show, respectively, cohort survival probabilities for the period 4 birth cohort when they complete period 7 and for the period 7 birth cohort when they complete period 10. See text for discussion.

Period (cross-sectional) survival for period 4:

$$(0.980) \times (0.970) \times (0.955) \times (0.934) = 0.85$$

Note for this case the computation is based on the survival rates prevailing in all age classes in period 4.

Cohort (longitudinal) survival for period 4:

$$(0.980) \times (0.975) \times (0.970) \times (0.967) = 0.90$$

Note for this case the computation is based on the survival rates that the period 4 birth cohort actually experienced each year from period 4, when they were newborn, to period 7, when they were age 3. The advantage of 0.05 in the cohort survival over the period survival is derived from the fact that mortality improved in each age class as the period 4 birth cohort survived forward. The period and cohort survival rates converge when mortality is no longer improving each year. This is evident in periods 7 through 10 when the gap is 0.

Period life expectancy can be interpreted as a lagged measure of underlying cohort experience. For age 65, for example, the lag for the period life expectancy in the current life table corresponds to the life expectancy for the cohort that reached age 65 around 15 years ago. On one hand, period mortality has become more out of date, but its divergence from cohort mortality has shrunk because the pace of change in period mortality has itself flattened. As Goldstein and Wachter (2006, 268) note, "It

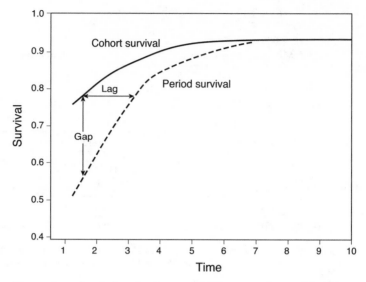

FIGURE 2.4. Illustration of period and cohort survival patterns from table 2.6.

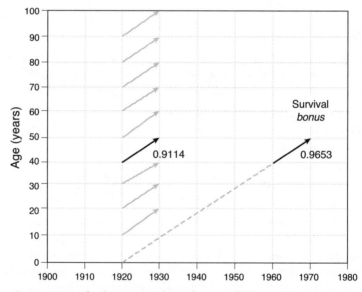

FIGURE 2.5. Comparison of cohort versus period survival illustrating the 10-year survival rate bonus resulting from a survival increase of 0.0539 (from 0.9114 to 0.9653) between 1920 and 1970 for 40-year-olds born in 1920.

takes more years to cover less ground." The gap and lag concepts for the US are illustrated in fig. 2.5, which shows the survival "bonus" resulting from the increase in survival rates as a cohort gets older. These overall trends are shown for the US population through the twentieth century in fig. 2.6.

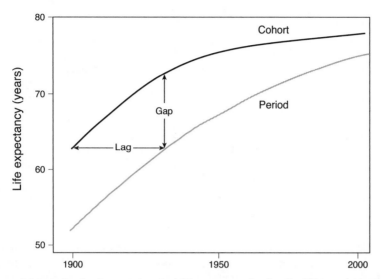

FIGURE 2.6. Schematic of cohort and period life expectancies for the US population through the twentieth century illustrating the lag and gap concepts (smoothed approximations from fig. 2 (US population) in Goldstein and Wachter 2006). Note the narrowing of the gap and the lengthening of the lag over the 100-year period.

Cross-sectional average length of life (CAL)

A measure known as the cross-sectional average length of life [CAL(t)] was first introduced by Brouard (1986) as a measure of the average of a population's mortality experience. As Guillot (2003) notes, this metric can be defined in two additional ways, including (1) the average number of person-years lived between time t and t + dt in a closed population where a constant number of annual births is exposed to actual cohort mortality conditions, and (2) the life expectancy at birth of a cohort aged a at time t (i.e., born at time t − a), whose survival advantage relative to the population's younger or older cohorts is equal to its survival disadvantage. This basic concept can be understood when it is related to the two standard measures of survival, including cohort survival derived from longitudinal survival data and period survival derived from cross-sectional single-year survival data, both of which are depicted in fig. 2.7 (see caption). The CAL(t) concept is based not on the single-year survival of the respective cohorts, as used in the period life table, but rather as the survival of each birth cohort to time t. Note that CAL(t) refers to general time t, whereas CAL(T) refers to specific time T as shown in the figure.

Canudas-Romo and his colleagues (2018) note that the truncated versions of CAL, denoted TCAL, are a novel way of comparing mortality between populations. It has the advantage over the standard cross-sectional period life table where only a single piece of information, the age-specific mortality, is used from each cohort. In contrast, CAL and TCAL use the aggregate survival probabilities of entire cohorts from their birth to the year in question. Therefore, CAL and TCAL capture longitudinal information as a cross-sectional composite metric:

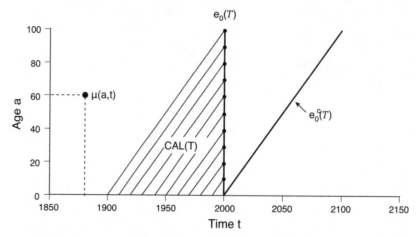

FIGURE 2.7. Lexis diagram showing the location of death rates used in the computation of cross-sectional average length of life, CAL(T), period life expectancy at birth, $e_0(T)$, and cohort life expectancy at birth, $e_0^c(T)$ (redrawn from Guillot 2003). Each node on the vertical line at T = 2000 represents two cohort metrics: (1) age-specific probability of age cohorts surviving from 2000 and 2001 that are used to compute $e_0^W(2000)$ and (2) survival of each cohort from birth to the year 2000 used to compute CAL (2000).

$$CAL(t) = \sum_0^\omega l^c(x, t - x) \qquad (2.22)$$

where $l^c(x, t - x)$ denotes cohort survival to age x at time $t - x$. The formula for TCAL is the same except truncated at year Y_1.

$$TCAL(t, Y_1) = \sum_0^\omega l^c(x, t - x) \qquad (2.23)$$

Thus the TCAL(2015) for ages 0 through 80 is computed as

$$
\begin{aligned}
TCAL(2015) &= l^c(80,1935) + l^c(79,1936) + \cdots + l^c(1,2014) + l^c(1,2014) \\
&= 0.2474 + 0.2592 + \cdots + 0.9947 + 1.000 \\
&= 69.7 \text{ years}
\end{aligned}
$$

This value can be contrasted with the period (cross-sectional) life expectancies for 1935 and 2015 of 62.1 and 75.7 years, respectively.

Metrics of Life Tables

A number of different actuarial concepts, analytics, and metrics can be derived from life tables; we discuss these here. We also provide a breadth of examples for the use of life tables with single-decrement processes.

Measures of central tendency

There are several metrics that describe age at death. *Life expectancy* at birth is a life table measure that refers to the average number of years remaining to a newborn, or the average age of death in the cohort. *Median survival*, or the probable length of life, refers to the age at which half of all deaths have occurred. And *mode of survival*, or the modal age at death, refers to the age corresponding to the highest frequency of deaths. The convergence of the ages for the mean, median, and modal ages of death for US females in the year 2000, relative to the differences in the ages at which these respective metrics occur, is shown in fig. 2.8. In every population on record, the mode of the length of life is several years higher than the mean, which is pulled down by premature (early) deaths.

Central death rate

The central death rate, or age-specific death rate for a cohort, denoted m_x, measures the number of deaths during a period divided by the average number of individuals who were living during that period. (Note that the m_x notation is also often used to

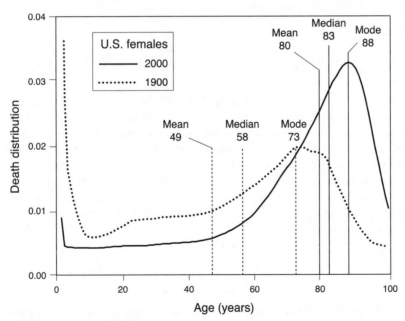

FIGURE 2.8. Distribution of deaths (d_x schedules) for US females in 1900 and 2000. (*Source*: HumanMortalityDatabase 2018)

denote age-specific fecundity.) The central death rate is not a probability but rather an observed rate—the number of individuals that die relative to the number at risk. In other words, it is essentially a weighted average of the force of mortality between ages x and x + 1. The relationship between m_x and q_x is

$$m_x = \frac{q_x}{1 - \frac{1}{2}q_x} \qquad (2.24)$$

$$q_x = \frac{m_x}{1 - \frac{1}{2}m_x} \qquad (2.25)$$

For example, a value of $m_{10} = 0.03027$ gives the *per capita number dying* in the interval 10 to 11, and the value of $q_{10} = 0.02982$ gives the *probability of an individual dying* in the age interval 10 to 11. The central death rate is used to compute another important actuarial parameter discussed in the next section—the life table aging rate.

Life table aging rate

Horiuchi and Coale (1990) introduced the parameter life table aging rate (LAR), denoted k_x, which is defined as the rate of change in age-specific mortality with age. This measure is based on *relative* rather than *absolute* rate of change in mortality with age. The formula is given as

$$k_x = \ln(m_{x+1}) - \ln(m_x) \qquad (2.26)$$

where m_x denotes the central death rate. The life table aging rate is a measure of the slope of mortality with respect to age and is thus an age-specific analog of the Gompertz parameter *b* (see chapter 3). Unlike the Gompertz parameter, which typically assumes constancy of the mortality slope over a large age interval, LAR is used to measure the slope of mortality over short intervals. As an illustration, computations of k_x from the *D. melanogaster* data (see table 2.1) for ages 5 and 40 are $k_{10} = 0.1972$ and $k_{70} = 0.0279$, which indicates that mortality is changing nearly 20% per day at day 10 but less than 3% per day at day 70. Additional perspectives on LAR applied to the Medfly and to the bean beetle, *Callosobruchus maculatus*, can be found elsewhere (Tatar and Carey 1994; Carey 1995).

Life table entropy

Life table entropy provides a metric to evaluate improvements in mortality and survival in a population. If all individuals die at exactly the same age, the l_x schedule is rectangular, whereas if all individuals have exactly the same probability of dying at each age (i.e., all p_x's are identical), the l_x schedule decreases geometrically. The distribution of deaths by age varies greatly between these two patterns. A measure of this heterogeneity known as *entropy*, H (Demetrius 1978), can be computed as

$$H = \sum_{x=0}^{\omega} \frac{e_x d_x}{e_0} \qquad (2.27)$$

The sum of products $e_x d_x$ in the numerator can be viewed as either the weighted average of life expectancies at age x, the average days of future life that are lost by the observed deaths, or the average number of days an individual could expect to live, given a second chance on life (Goldman and Lord 1986). The denominator is the expectation of life at birth, e_0, which converts an absolute effect into a relative effect.

Entropy, H, can be interpreted as (1) the proportional increase in life expectancy at birth if every individual's first death were averted; (2) the percentage change in life expectancy produced by a reduction of 1% in the force of mortality at all ages; and (3) the number of days lost due to death per number of days lived (Vaupel 1986). This last interpretation provides a quantitative measure of the survival pattern that, in turn, is an index of variation or heterogeneity. If $H=0$, then all deaths occur at exactly the same age, and if $H=1$, then the l_x schedule is exponentially declining. The intermediate value, $H=0.5$, suggests that the l_x schedule is linear.

As an example, consider entropy for the *D. melanogaster* life table given in table 2.1:

$$\begin{aligned}
H &= \frac{e_0 d_0 + e_1 d_1 + e_2 d_2 \cdots + e_\omega d_\omega}{e_0} \\[2mm]
&= \frac{(39.0 * 0.0000) + (38.0 * 0.0000) + (37.0 * 0.0002) \cdots + (0.500 * 0.0018)}{39.0} \\[2mm]
&= \frac{15.017}{39.0} \\[2mm]
&= 0.385
\end{aligned}$$

Note that for reference when $H=0$, all individuals die at once, thus heterogeneity in the death rate is 0. When $H=1$, the number of days lost in the cohort due to death equals the average number of days lived by a newborn. The case of $H=0.5$ is intermediate between these two extremes.

Sensitivity analysis

In the analysis of survivorship, it is possible to evaluate how a small change in survival at a specified age would change the expectation of life at birth, e_0. The summation of all l_x that is used for calculating e_0,

$$e_0 = \sum_0^\infty l_x \qquad (2.28)$$

can be reexpressed as

$$e_0 = \sum_0^\omega l_0 \prod_{y=0}^{x-1} p_y \qquad (2.29)$$

or

$$= 1 + p_0 + p_0 p_1 + p_0 p_1 p_2 + \cdots p_{\omega-3} p_{\omega-2} p_{\omega-1} \qquad (2.30)$$

In a four-age-class cohort we can compute the effect on e_0 of a small change in p_1, for example,

$$e_0 = 1 + p_0 + p_0 p_1 + p_0 p_1 p_2 \qquad (2.31)$$

and

$$\frac{de_0}{dp_1} = p_0 + p_0 p_2 \qquad (2.32)$$

The derivative can also be expressed as

$$\frac{de_0}{dp_1} = \frac{1}{p_1} \sum_{x=2}^3 l_x \qquad (2.33)$$

Since

$$p_0 + p_0 p_2 = \left(\frac{p_0 p_1}{p_1}\right) + \left(\frac{p_0 p_1 p_2}{p_1}\right) \qquad (2.34)$$

the term $(1/p_1)$ can be factored out of the right-hand side, yielding

$$\frac{1}{p_1}(p_0 p_1 + p_0 p_1 p_2) \qquad (2.35)$$

or

$$\frac{1}{p_1}(l_2 + l_3) \qquad (2.36)$$

Finally, the general form for computing the sensitivity of a change in p_x on e_0 is

$$\frac{de_0}{dp_x} = \frac{1}{p_x}\sum_{y=x+1}^{\omega} l_y \qquad (2.37)$$

This expression illustrates two important aspects of the sensitivity of e_0 to a small change in age-specific survival. First, as x increases, the sum of the l_x's from x to ω continually decreases. Therefore, all else being equal, the effect of a change in survival on e_0 is always greater at younger ages than at older ones. Second, e_0 is most greatly affected by changes in period survival that are low rather than those that are high. This is evident by noting that the term outside the summation is an inverse of a fraction. For example, if $p_x = 0.9$, the factor by which the sum of l_x's is multiplied is 1.1 (i.e., 1/0.9), which can be contrasted with the situation where $p_x = 0.5$ and the sum is multiplied by 2.0 (i.e., 1/0.5).

Vaupel (1986) outlines a different type of analysis concerned with the effect of changes in mortality on life expectancy when he considered the hypothetical question, Which decade of life would be best to avert 100 deaths? The answer to this question is the first decade of life because children lose many years of life expectancy due to high early mortality. However, if this question is slightly modified and one asks, If deaths could be reduced by 1% during any decade of life, which decade would be best? As Vaupel notes, an intuitive reasonable guess is that a decade at young ages, such as 0 to 10 or 17 to 27, would be most beneficial. However, using the life tables for Swedish men and women, he shows that the correct answer is 67 to 77 for men and 74 to 84 for women because, although many infants die the first year of their lives, very few die over the next 9 years. This is in contrast to the very large numbers of men and women who die in their seventh and eighth decades of life. The mathematical details for this type of analysis can be found in his paper.

Example single-decrement processes

Life table concepts and analytics can be applied to a wide range of dichotomous processes, many examples of which are given in table 2.7. The general concept is applied to organisms that exist in an original state that can be dichotomized as an entry/exit concept. For example, being alive is the original state and exiting life by dying is the transition out of the live state, and being unmarried is an original state that is exited upon marriage. This can be extended to processes ranging from unemployment and incarceration to diapause and plant flowering.

Life Table as Stationary Population

A life table is a model for analyzing mortality data and summarizing their consequences on survival, life expectancy, and the age distribution of deaths, but it can also be considered a stationary population model with a zero growth rate. As a stationary population model, the radix, which in a cohort model is the initial number of newborn individuals, can also be considered the first age class of a population model if we make the assumption that this number, as well as age-specific survival, is fixed. Given that e_0 is the number of life-years lived by the average newborn, then, $1/e_0$ represents

Table 2.7. Examples of single-decrement processes that can be studied by means of a life table

| Process | State | | | Vertical dimension |
	Studied when	Entered	Left when	
Mortality	Being alive	Born	Die	Age
Nuptiality	Being unmarried	Born	Marry	Single life (age)
Migration	Living in place of birth	Born	Move	Duration of residence
Entering labor force	Never worked	Born	Entering labor force	Age
Becoming a mother	Having no births	Born	First birth	Age
Subsequent childbearing	Not having additional birth	Having a birth	Having an additional birth	Duration since having a birth
Marital survival	Being in intact marriage	Marrying	Marriage ending	Duration of marriage
Unemployment spells	Being unemployed	Becoming unemployed	Leaving state of employment	Duration of unemployment
Incarceration	Being in jail	Entering jail	Leaving jail	Duration of incarceration
Diapause/ hibernation	Prehibernation	Hibernation	Posthibernation	Duration of hibernation
Sexual maturation	Sexual immaturity	Sexual maturity	Sexually mature	Age
Parity	Reproductive	New birth	Next birth	Number of offspring
Flowering	Preflowering	Flowering	Postflowering	Duration of flowering

the per capita death rate in the population. In a stable population, this expression also equals the per capita birth rate because the radix equals unity, which equals the number of births in the life table population. The total number of births divided by the number of life-days in the population (e_0) yields the per capita birth rate, thus

$$b = d = \frac{1}{e_0} \tag{2.38}$$

Since L_x denotes the average number of life-days from x to $x+1$, the proportion of individuals in the population from x to $x+1$, denoted c_x, is given as

$$c_x = \frac{L_x}{e_0} \tag{2.39}$$

The life table stationary population model is useful because it provides explicit expressions that connect the demographic parameters—life expectancy, birth rates, death rates—to age structure (Preston et al. 2001). This model can be used to estimate one demographic parameter on the basis of another, for example, archaeologists using skeletons (Chamberlain 2006). More broadly, given that every human and nonhuman population has an underlying life table, this approach means that every population can form the basis of a model with a stationary population.

The relationship between age distribution and remaining lifetime distribution for a stationary population is the life table identity (Carey, Müller, et al. 2012). The source of this identity is a simple arithmetic relationship—the number of deaths in each time-to-death class equals the number of living in each population age class. The graphs presented in fig. 2.9 help visualize how these concepts are connected. The upper panel in this figure depicts the percentage of the hypothetical population in each of four age classes and corresponds to the N_x column in table 2.8. The lower panel shows the time-to-death distribution in this hypothetical population and corresponds to the d_x^* column in the table. The bar segments labeled a, b, c, and d in the top panel represent the respective percentages of each age class that die in time intervals 0 to 1, 1 to 2, 2 to 3, and beyond 3. All segments (a, b, c, and d) are represented in the age class 0 bar (top panel) because this bar depicts members of the youngest age class, and the members here will be subject to a risk of dying in every successive time interval. In contrast, all of the oldest members of the population in the top panel (age class 3) are contained in a single segment labeled a, and all of these members will die in the initial time interval. All members in each of the age classes will die according to the mortality rates associated with each of their ages. The total deaths in the initial time step correspond to the sum of the segments labeled a. This result illustrates the general concept that applies to all successive time steps—the cumulative percentage of deaths in the initial time interval (40%) equals the percentage of the original population in the first population age class (40%). Similarly, the cumulative percentage of deaths that occur over the subsequent time steps from each of the remaining age classes, and shown in the bottom panel (i.e., 30%, 25%, and 5%, respectively), equals the percentage of the original population in the corresponding age classes (i.e., 30%, 25%, and 5%, respectively), which is shown in the top panel.

Although hypothetical, these examples illustrate the relationship of population age structure and the death distribution of the individuals from which it is constituted. Aging and survival in the wild can be inferred using the life table identity, given the assumptions (1) that the captive cohort is randomly sampled from the wild cohort, so that each age group is sampled with probability corresponding to the frequency of this age group in the wild, and (2) that the survival schedule of the captive cohort remains the same as the survival schedule of the wild cohort.

The identity behind the captive cohort method described in the earlier papers by Müller, Carey, and their colleagues (Müller et al. 2004, 2007; Carey, Müller, et al. 2008, 2012) is the classical formula for residual renewal time in a stationary renewal process described in the early work on stationary population theory by both Lotka (Lotka 1907, 1928) and Feller (Feller 1950). More recent work involving stationary population theory relevant to this method includes the classic papers by Ryder (Ryder 1973, 1975) on replacement populations, the article by Kim and Aron (1989) showing the equivalency of the average age and average expectation of remaining life in a stationary population, the book section in Keyfitz (1985,74) containing a general formula for the average expectation of life in a stationary population, the demography

Table 2.8. Relationship of age distribution (c_x), in a stationary population with four age classes, to time-to-death distribution of its members (d_{x*})

Stationary population				Fraction captured at				Captive life table		
Age x	N_x	l_x	c_x	Age 0	Age 1	Age 2	Age 3	l_x*	d_x*	$x*$
0	40	1.000	0.40	0.40	0.30	0.25	0.05	1.00	0.40	0
1	30	0.750	0.30	0.30	0.25	0.05	—	0.60	0.30	1
2	25	0.625	0.25	0.25	0.05	—	—	0.30	0.25	2
3	5	0.125	0.05	0.05	—	—	—	0.05	0.05	3
4	0	0.000	0.00	—	—	—	—	0.00	0.00	4

Note: The left subtable contains the hypothetical population demographic data where N_x, l_x and c_x denote the number in the population, cohort survival, and fraction at age x, respectively (see Müller et al. 2004 for details). The middle subtable contains the fraction of the total number captured in the top row (i.e., in proportion to their field abundance of 40%, 30%, 25%, and 5%). These are survived downward with captive age, x* according to their age-specific survival rates in the left subtable. The right subtable contains the postcapture survival of all individuals by captive age. Note that the sums of the middle subtable rows equal the captive survival, $l*_x$, for each captive age. The d* column in this subtable equals the fraction of all deaths that occur in the age interval, the fractions of which reveal the identity (see all bold numbers in main table).

text Preston et al. (2001, 53–58) outlining the basic properties of a stationary population, and the paper by Goldstein (2009) showing the equivalency of life lived and life left in stationary populations. Vaupel (2009) rederived the basic model contained in Müller et al. (2004). He also shows that the captive cohort concept is a general model in which the proportion of individuals younger than age a equals the proportion whose remaining life span is less than age a. Or, conversely, the proportion of individuals age a or older is equal to the proportion of individuals who will still be alive in a years. Rao and Carey (2019) generalized this equality and introduced a theorem on stationary populations (for extension, see Rao and Carey 2019). Papers by French demographer Nicola Brouard (1986, 1989) and by American biodemographer James Carey and his colleagues (Carey et al. 2018; Carey 2019; Carey and Vaupel 2019) contain additional details of the life table population identity, including proofs, extensions, and applications.

Consider Further

Life table concepts and methods are foundational (literally) to demography, and therefore extensions and example applications are scattered throughout this book. The parity progression life tables in chapter 4 (Reproduction), the life table as a stationary population in chapter 5 (Population I: Basic Models), and the age/stage models in chapter 6 (Population II: Stage Models) are effectively life table models when populations are stationary. We include a baseline model life table of humans in chapter 8 (Human Demography) and also introduce the multiple-decrement life table in this chapter. We introduce the concept of active life expectancy using the healthy life expectancy life table in chapter 10 (Application II: Evaluating and Managing Populations). A number of life table-related questions are included in the demographic shorts in Chapter 11 including life overlap (S6), the probability of same-year deaths (S12) and demographic selection (S43).

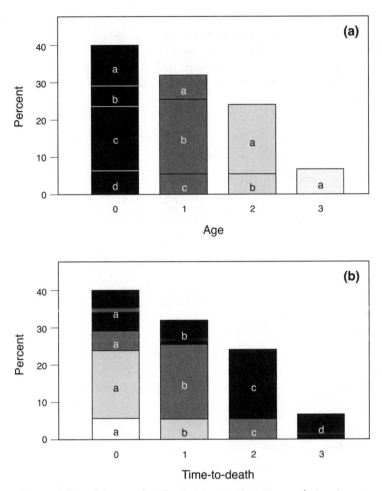

FIGURE 2.9. The equality of the age distribution in a stationary population (upper panel) and the distribution of time to death of its members (lower panel). The implied age-specific mortality rates (q_x) are 1/4, 1/6, 4/5, and 1.00 for age classes 0, 1, 2, and 3, respectively. Each segment labeled a, b, c, or d in the upper panel corresponds to the fraction of the age class of the population that dies in intervals 0, 1, 2, and 3, respectively (lower panel). For example, the percentage of the standing population that dies in each segment labeled a (upper panel) sums to 40% in the time-to-death class 0 (lower panel). Or, alternatively, as the population ages the number of individuals in segments a, b, c, and d in age class 0 (upper panel) die off at times 0, 1, 2, and 3, respectively, as shown in the time-to-death distribution (lower panel). (*Source*: Carey et al. 2012)

Scores of books have been published on life table concepts and methods, the best of which we believe are those by Namboodiri and Suchindran (1987) and Chiang (1984). Overviews and synopses of life table methods include the entry by Michel Guillot in the *Encyclopedia of Population* (Guillot 2005), a chapter in the *Handbook of Population* by Kenneth Land (Land et al. 2005), two chapters by Italian demographers Jacques Vallin and Graziella Caselli in *Demography: Analysis and Synthesis* (Vallin and Caselli 2006a, b), and a major chapter on life tables by Hallie J. Kintner in

Methods and Materials in Demography (Kintner 2004). Textbooks on general demography that include life table methods include those by Lundquist (2015), Rowland (2003), Preston et al. (2001), and Wachter (2014). Three excellent textbooks on life tables from the actuarial sciences include *Life Contingencies* by Jordan (1967), *Mortality Table Construction* by Batten (1978), and *Actuarial Mathematics* by Bowers et al. (1986). Harper (2018) gives a synopsis of demography in general and life tables in particular.

3

Mortality

> It is possible that death may be the consequence of two generally co-existing causes; the one, chance, without previous disposition to death or deterioration; the other, a deterioration, or an increased inability to withstand destruction.
>
> Benjamin Gompertz (1825, 513)

> who by water and who by fire
> who by sword and who by beast
> who by famine and who by thirst
> who by earthquake and who by plague
> who by strangling and who by stoning
> Unetanneh Tokef, ancient Jewish poem (Parsing death in the eleventh century)

Although cohort survival to age x (l_x) and age-specific survival (p_x) are the life table parameters most used in ecology and evolutionary literature, age-specific mortality (q_x) represents the most fundamental actuarial concept in the life table and is the statistical foundation for the actuarial sciences (Land et al. 2005; Roach and Carey 2014). Mortality modeling and the concepts of actuarial aging, relative risk, odds ratio, sex mortality differentials, cause-specific and all-cause mortality, and average lifetime mortality are all based on risk concepts. For example, the ratio of daily survival for two cohorts, one with a value of 99.9% and the other 99.0%, is slightly over 1.0. However, their relative risk of dying [i.e., $(0.01/0.001) = 10$] and their ratio of remaining life expectancies [i.e., $(1.0/0.01) = 100$; $(1.0/0.001) = 1,000$)] differ by 10 fold (Roach and Carey 2014). These and many other quantitative relationships between the two hypothetical cohorts are not evident without considering risk. Age-specific mortality is conceptually simple, easily measured, readily modeled, and applicable to virtually all species (Carey and Judge 2000).

The basic role of mortality is evident by considering that death is an *event* indicating a change of state from living to dead, a failure of the system. In contrast, survival is a *nonevent* inasmuch as it is a continuation of the current state. This orientation toward events rather than nonevents is fundamental to the analysis of risk. Mortality rates can be disaggregated by cause of death and thus can shed light on the biology, ecology, and epidemiology of deaths, the frequency distribution of causes, and the likelihood of dying of a particular cause by age and sex. An individual can die due to a number of causes, such as an accident, a predator, or a disease. This concept of *cause* does not apply to survival; in other words, there is no cause of survival.

Another aspect of age-specific mortality is that its value is mathematically (as distinct from biologically) independent of demographic events at other ages. In contrast, cohort survival (l_x) is conditional upon survival to each of the previous ages, life expectancy at age x (e_x) is a summary measure of the consequences of death rates over all ages greater than x, and the fraction of all deaths (d_x) that occur at young ages will determine how many individuals remain to die at older ages. This mathematical independence of mortality relative to events at other ages is important because age-specific rates can be directly compared among ages or between populations that, in turn, may shed light on differences in relative age-specific frailty or robustness.

Finally, a number of different mathematical models of mortality have been developed (e.g., Gompertz; Weibull; logistic) that provide simple and concise means for expressing the actuarial properties of cohorts with just a few parameters. These concise models facilitate comparisons of mortality and longevity variation across populations.

In this chapter we focus on three indexes of mortality that are of fundamental importance in demography and actuarial science. The first two, which were introduced in chapter 2, include the discrete functions period mortality, q_x, defined as the probability of dying in the age interval x to $x+1$, and the central death rate, m_x, defined as the number of deaths per person-years. The third mortality concept, one of the most important in demography, is the force of mortality, μ_x, defined as the instantaneous death rate. This latter function is continuous and is thus used primarily in actuarial calculus.

Discrete Mortality

As noted in the previous chapter on life tables, the q_x and m_x are related by the formula

$$m_x = \frac{q_x}{1 - \frac{1}{2} q_x} \tag{3.1}$$

The difference between them is evident in their algebraic definitions, where

$$q_x = \frac{\text{number dying in interval x to } x+1}{\text{number alive age at age x}} \tag{3.2}$$

$$m_x = \frac{\text{number dying in interval x to } x+1}{\text{number of years lived x to } x+1 \text{ by those alive at age x}} \tag{3.3}$$

Given that the numerators are identical, the distinction between these two measures can be seen in eqn. (3.2) and eqn. (3.3) in their denominators. The denominator in the q_x expression is the number of individuals that started the interval and were subject to the risk of dying throughout the interval; thus, q_x is a *probability* of dying between x and $x+1$. If 100 individuals are at risk and 88 survived, then the probability of dying in the age interval is 0.12.

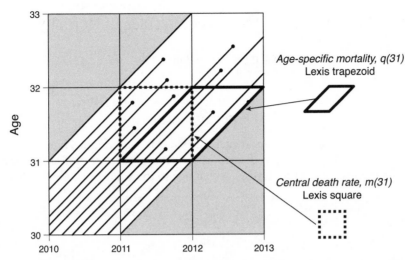

FIGURE 3.1. Visualization of the conceptual and analytical differences between age-specific mortality and the central death rate using the Lexis diagram. Whereas m(31) is computed as the number of deaths between 31 and 32 divided by the total life-years lived by individuals in the age interval 31 to 32 between 2011 and 2012, q(31) is the fraction of individuals who turned age 31 in 2011 who survived to turn age 32 in 2012.

The denominator in the m_x expression (eqn. 3.3) is the total number of life-years (life-days) lived in the interval. This includes the total life-years for individuals who survived the entire interval as well as the total life-years lived by individuals who died within the interval; thus, m_x is a death *rate* per life-year. In this case, if 100 individuals were alive at the start of the interval and the 12 individuals who died did so at a rate of one per month, then the average number of life-years lived in that interval is 88 for those that survived plus 6 for those that died, on average, at the midpoint of the interval. The central death rate for this hypothetical example is 12 deaths (numerator) divided by 106 life-years (denominator), or $m_x = 0.113$ deaths per life-year lived in the interval.

We can express the right-hand terms for both q_x and m_x using the life table parameters d_x and l_x from the life table, which yields

$$q_x = \frac{l_x - l_{x+1}}{l_x} = \frac{d_x}{l_x} \qquad (3.4)$$

$$m_x = \frac{d_x}{(l_x - d_x) + \frac{1}{2}d_x} \qquad (3.5)$$

Note that the left-hand expression in the denominator of eqn. (3.5) gives the life-years lived in the interval (l_x) minus the (uncorrected) life-years lost due to deaths (d_x). These uncorrected life-years lost are corrected by the right-hand expression in the denominator by assuming that the average individual died at the midpoint of the interval (i.e., $d_x/2$). The differences between m_x and q_x are illustrated in the Lexis diagram shown in fig. 3.1.

Continuous Mortality: The Calculus of Mortality

Mathematics, rightly viewed, possesses not only truth,
but supreme beauty—a beauty cold and austere . . .

Bertrand Russell

The Force of Mortality

Consider the number of survivors at each age x (l_x) as a continuous function of age x; then the force of mortality, μ_x, is defined as the ratio of the rate of decrease of l_x (i.e., the instantaneous effect of mortality) at that age to the value of l_x. Algebraically,

$$\mu_x = \frac{(l_{x+n} - l_x)}{l_x} \tag{3.6}$$

$$\text{or} \quad \mu_x = \frac{-\dfrac{d}{dx}(l_x)}{l_x} \tag{3.7}$$

where d/dx represents differentiation with respect to age x, which indicates the rate of change of l_x over an infinitesimally small increment of age. The minus sign is introduced to make μ_x positive since l_x is a decreasing function of age.

From basic calculus, the derivative of the natural log of a function of x, ln(f(x)), is

$$\frac{d}{dx}(\ln(f(x))) = \frac{1}{f(x)}\frac{d}{dx}(f(x)) \tag{3.8}$$

$$\text{Thus,} \quad \mu_x = \frac{d}{dx}(\ln(l_x)) \tag{3.9}$$

Integrating both sides over 0 to n yields

$$\int_0^n \mu_{x+t}dt = -\int_0^n \frac{d}{dt}(\ln(l_{x+t}))dt = [-\ln(l_{x+t})]_0^n \tag{3.10}$$

$$= -\ln(l_{x+n}) - \ln(l_x) = -\ln\left(\frac{l_{x+n}}{l_x}\right) \tag{3.11}$$

$$= {}_np_x = \exp\left(-\int_0^n \mu_{x+t}dt\right) \tag{3.12}$$

If n = 1, then this expression can be simplified to

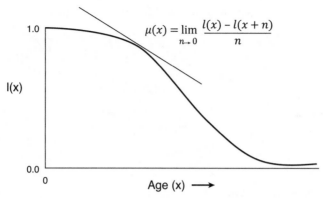

FIGURE 3.2. Illustration of the force of mortality as the first derivative of the l_x function.

$$p_x = \exp(-\mu_x) \tag{3.13}$$

Note that

$$-\ln(p_x) = \mu_x \tag{3.14}$$

and

$$q_x = 1 - \exp(-\mu_x) \tag{3.15}$$

Thus $l(x)$ equals the exponent of the summation of μ_x's from 0 through x

$$l(x) = l_0 \int_0^x \mu_a \, da \tag{3.16}$$

$$\text{or} \quad l(x) = l_0 \left[\exp\left(\int_0^x \mu_a \, da \right) \right] \tag{3.17}$$

The force of mortality can be thought of as either a measure of mortality at the precise moment of death or the first derivative of the l_x function (see fig. 3.2). In actuarial modeling, the force of mortality is preferred over age-specific mortality (q_x) because it is not bounded by unity, it is independent of the size of the age intervals, and it forms the argument of numerous parametric mortality functions.

Three formulas that are commonly used for computing μ_x include

$$\mu_x = -\ln p_x \tag{3.18}$$

$$\mu_x = -\frac{1}{2}(\ln p_{x-1} + \ln p_x) \tag{3.19}$$

$$\mu_x = -\frac{1}{2n}\ln_e\left(\frac{l_{x-n}}{l_{x+n}}\right) \tag{3.20}$$

where n denotes the band width (age interval).

An empirical model

We can create a simple model in which mortality at each subsequent age equals the product of the current age and the factor by which mortality changes at each age:

$$q_{x+1} = \lambda q_x \tag{3.21}$$

where λ denotes the rate of change in mortality with age. We can use the geometric mean of the rates of change from ages 70 through 75 as the constant, $\lambda = 1.0824$ (from fig. 3.3a) and project mortality from 70 through 75. Let $q_{70} = 0.020420$; then

$$q_{71} = \lambda q_{70} = (1.0824) \times (0.020420) = 0.022102$$

$$q_{72} = \lambda q_{71} = (1.0824) \times (0.022102) = 0.023923$$

$$q_{73} = \lambda q_{72} = (1.0824) \times (0.023923) = 0.025893$$

$$q_{74} = \lambda q_{73} = (1.0824) \times (0.025893) = 0.028026$$

$$q_{75} = \lambda q_{74} = (1.0824) \times (0.028026) = 0.030335$$

Note that since

$$q_{71} = \lambda q_{70}$$

and

$$q_{72} = \lambda q_{71}$$

then by substituting λq_{70} for q_{71}, we find

$$q_{72} = \lambda q_{70} \times \lambda q_{70}$$

or

$$q_{72} = q_{70}\lambda^2$$

By defining q_{70} as the initial level of mortality, denoted a, this model can be generalized as

$$q_x = a\lambda^x$$

which states that age-specific mortality at age x is the product of the initial level of mortality and the rate of change of mortality raised to the xth power. By setting $\lambda = e^b$, this model can be expressed in continuous form as

$$\mu(x) = ae^{bx}$$

This empirical example and associated graphs are shown in fig. 3.3a, b.

Smoothing age-specific mortality rates

Smoothing mortality rates for plotting is often useful because of the binomial noise present due to small numbers or environmental variation. A formula for computing the running geometric mean of an age-specific mortality schedule, denoted $\hat{q}_x$, is given as

$$\hat{q}_x = 1 - \left[\prod_{y=x-n}^{x+n} p_x \right]^{-(n+1)} \tag{3.22}$$

where $p_x = 1 - q_x$ and n denotes the "width" of the running geometric average. The analytical counterpart for the running mean of force of mortality, denoted $\hat{\mu}_x$, is

$$\hat{\mu}_x = \frac{1}{n+1} \sum_{y=x-n}^{x+n} \mu_y \tag{3.23}$$

More sophisticated methods for smoothing hazard rates using locally weighted least squares techniques are described in Müller et al. (1997) and Wang et al. (1998).

(a)

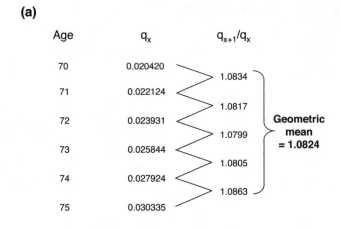

(b)

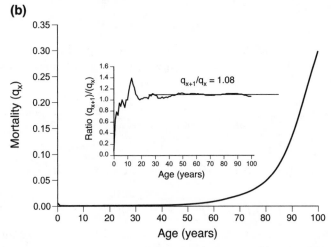

Figure 3.3. Illustration of constancy of change in mortality in US females (2000). (a) Rate of change in age-specific mortality from age 70 through 75. Note that the change is 1.08-fold to the second decimal place over all of these ages. (b) Rate of change in age-specific mortality viewed in the context of the life course. The main graph depicts age-specific mortality, and the inset shows the ratio of q_{x+1} to q_x from age 30 onward.

Mortality Models

There are three principal justifications for postulating an analytical form for mortality and survival functions (Bowers et al. 1986). First, many phenomena in the physical sciences can be explained efficiently by simple formulas, therefore some authors have suggested that animal survival is governed by simple laws (cited in Bowers et al. 1986). Second, it is easier to communicate a function with a few parameters than it is to communicate a life table with several hundred parameters and probabilities. Finally, a simple analytical survival function is easily estimated using a few parameters from the original determination of the mortality data. Six of the most frequently used mortality models in demographic and gerontological research are presented in

Table 3.1. Six of the major age-specific mortality [μ(x)] and survival [l(x)] models

Model	μ(x)	l(x)	Notes
de Moivre (1729)	$(\omega - x)^{-1}$	$1 - \dfrac{x}{\omega}$	ω = oldest age; survival can also be expressed as $l_x = a - bx$ where $a = 1.0$ (radix) and $b = 1/\omega$
Gompertz (1825)	ae^{bx}	$\exp\left[\left(\dfrac{a}{b}\right)(1 - e^{bx})\right]$	a = initial mortality rate; b = "Gompertz" parameter; linearized version: $\ln \mu(x) = a + bx$
Makeham (1860)	$ae^{bx} + c$	$\exp\left[\left(\dfrac{a}{b}\right)(1 - e^{bx}) - cx\right]$	c = age-independent (accidental) mortality
Exponential	c	$\exp[-cx]$	constant hazard rate, c
Weibull (1939)	ax^n	$\exp\left[-\left(\dfrac{a}{n+1}\right)x^{n+1}\right]$	a = location parameter; n = shape parameter; $n > 0$; linearized version: $\ln \mu(x) = a + n \ln x$
Logistic	$\dfrac{nx^{n-1}}{g^n + x^n}$	$\left(1 + \left(\dfrac{x}{g}\right)^n\right)^{-1}$	g and n are parameters to be fitted; both parameters control shape and location
Siler	$a_1 e^{-b_1 x} + c + a_2 e^{b_2 x}$		a_1 and b_1 control the scale and rate of change in infant mortality, c is age-independent (accidental) mortality and a_2 and b_2 control mortality scale and trajectory in young, middle and older age.

Note: Life expectancy at birth, e(0), is computed using the formula $e(0) = \int_0^\omega l(x)\,dx$ for all models (Carey 2001).

table 3.1, along with the corresponding expression for age-specific survival and associated parameters.

De Moivre model

Mortality rate in the de Moivre model equals the inverse of the difference between maximal and current age and tends to unity as age approaches a putative maximum, ω (Smith and Keyfitz 1977). The resulting survival schedule is a linearly decreasing function of age from 1.0 at age $x = 0$ to zero at age $x = \omega$. The advantage of this model is its simplicity—the model is transparent, easily understood, requires only a single parameter (ω), and produces a linear survival function. The assumption that survival is a linear function of age is often applied over short age intervals.

Gompertz model

The assumption of the Gompertz model (Gompertz 1825) is that mortality beyond the age of sexual maturity (or another predetermined age) is an exponentially increasing function of age. This model contains two parameters—the initial mortality rate, a, which denotes mortality at the youngest age class in the specified age interval, and the exponential rate of increase in death rate, b, which denotes the age-specific slope of the mortality function and is often referred to as the Gompertz parameter. Comparisons of how variation in these two Gompertz parameters influence both the survival curve and the death distribution are illustrated in figs. 3.4 and 3.5, respectively.

The mortality trajectory for the Gompertz model is exponential, and its survival trajectory is sigmoidal. The Gompertz model provides two useful formulas: (1) the mortality doubling time, denoted MDT, defined as the time required for the mortality to increase by twofold (MDT = ln(2)/b); and (2) the estimated maximum life span, denoted T_{max}, and defined by Finch (1990) as the age when a population subject to Gompertzian mortality rates has diminished to one survivor ($T_{max} = \ln[1 + b(\ln N)/a]/b$). Additional perspectives on the Gompertz model are given in Ricklefs and Scheuerlein (2002).

Makeham model

The Makeham model (Makeham 1867), also known as the Gompertz-Makeham model, represents an improvement in the Gompertz model. Makeham found that overall mortality levels could be better represented if a constant term was added to the Gompertz formula for μ(x) to account for causes of mortality not dependent on age (i.e., accidental deaths). There is no analogous transformation for the Gompertz-Makeham equation in which linear regression can be used to estimate the parameter c. Instead, it has been suggested that one can use trial and error, after first estimating parameters a and b from the Gompertz regression equation, and then adjust c until the closest approach to a straight line is attained (Elandt-Johnson 1980).

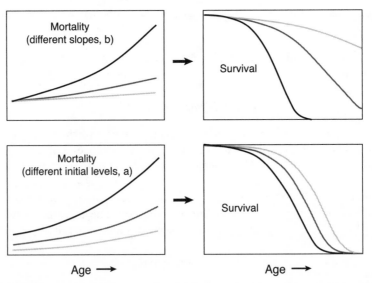

FIGURE 3.4. Comparison of how different values of a (initial mortality) and b (mortality slope) in the Gompertz model affect survival patterns.

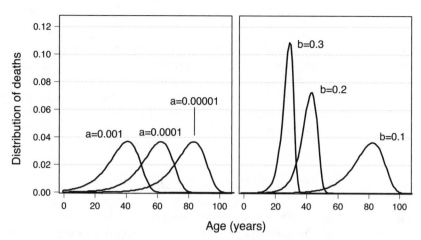

FIGURE 3.5. Comparison of how different values of a (initial mortality) and b (mortality slope) in the Gompertz model affect the death distribution (d_x) patterns. (*Source*: Canudas-Romo et al. 2018)

Exponential model

The exponential model is effectively the Gompertz-Makeham model minus the Gompertz component. In other words, it accounts for only the accidental deaths, c, and thus assumes age-independent mortality. Because mortality is constant with age in the exponential model, its plot is simply a horizontal line intercepting the y-axis at c and extending to the right. The survival function for this model decreases exponentially with age.

Weibull model

Whereas the de Moivre, Gompertz, and Makeham models were derived in an actuarial context, the Weibull model was developed in the context of reliability engineering (Weibull 1951). This model is a generalization of the exponential model but, unlike the exponential model, does not assume a constant hazard rate and thus has broader application. The Weibull model has two parameters, the value of n determines the shape of the distribution curve, and the value of a determines its scaling. The Weibull hazard function increases if $n > 0$, decreases if $n < 0$, and is constant if $n = 0$.

Logistic model

The logistic model was introduced to demography by Pearl and Reed (1920), who used this model to estimate the ceiling or asymptote of the US population. Wilson (1994) showed that the logistic model provided a good fit to the results of the large-scale Medfly study. Comparable to the logistic model is the Perks model (Perks 1932), which also exhibits leveling-off behavior at older ages.

Siler model

An important parametric model of mortality across the entire life span is the Siler competing hazards model (Siler 1979). Siler added a third component to the Gompertz-Makeham model to represent the early-life segment of humans and virtually all non-human species in which mortality is typically high. Siler called his model a competing hazards model precisely because he interpreted its three components as sets of risks that compete simultaneously throughout life.

Mortality Drives the Life Table

Mortality risk specifies the likelihood of transitioning from the live state to the dead state, and it thus underlies, and shapes, the life table. Fig. 3.6 illustrates how four different mortality patterns shape the survival (l_x) and death distribution (d_x) patterns, including (1) constant (type I) mortality, which produces geometrically decreasing survival and death curves; (2) exponential or Gompertzian (type II) mortality, which, for the mortality rate depicted, produces high survival through older ages, which, in turn, skews the deaths to older ages; (3) the Siler model with high early (infant) mortality followed by Gompertzian mortality (type III), which reduces survival at young ages and distributes deaths accordingly; and (4) slowing mortality at older ages (type IV), which produces long tails in both survival and death distributions.

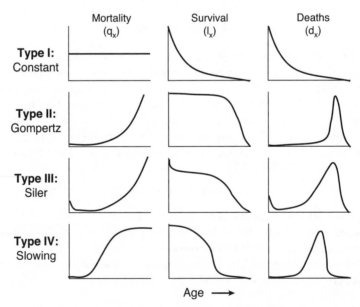

FIGURE 3.6. Age patterns of q_x, l_x, and d_x demonstrated by four different types of mortality curves.

Demographic Heterogeneity and Selection

One assumption in the mortality models presented earlier is that all individuals at birth experience the same probability of survival at each age. This assumption is virtually always violated because there are a wide variety of factors that can have profound effects on the risks of death, including within-group differences in genetic endowment or in developmental conditions. As populations age, the composition of individuals in the population changes, and they become more selected. Over time members of subcohorts with higher death rates will die out in greater numbers than will the members of subcohorts who experience lower death rates. The concept of subgroups endowed with different levels of frailty (see fig. 3.7) is referred to as *demographic heterogeneity*; the winnowing process as the cohort ages is referred to as *demographic selection*. The actuarial consequence of a cohort consisting of subcohorts, with different levels of frailty, is that the mortality trajectory of the whole cohort will change with the change in cohort heterogeneity. As a consequence, the whole cohort may depart substantially from Gompertz mortality rates even though each of the subcohorts is subject to Gompertz rates. Those individuals with the greatest frailty will die off first, leaving progressively fewer of them in the cohort, and thus altering the trajectory of mortality for the whole cohort. For example, if in a cohort there are two subgroups, one with a low and constant probability of death and the other with a rapidly rising mortality rate, the overall cohort mortality will rise to a peak and then fall off as the weak die off and the overall mortality then tracks the trajectory of the least frail (most robust) subgroup. Other examples of how heterogeneity can affect the mortality trajectory of a cohort are shown in fig. 3.8.

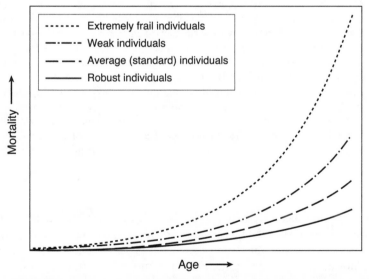

FIGURE 3.7. Mortality patterns of members of subcohorts within a larger cohort. Although mortality occurs in all subcohorts throughout the life course, their differential mortality rates, due to differences in frailty, shape the composition of the aging cohort and thus shape the composite mortality pattern.

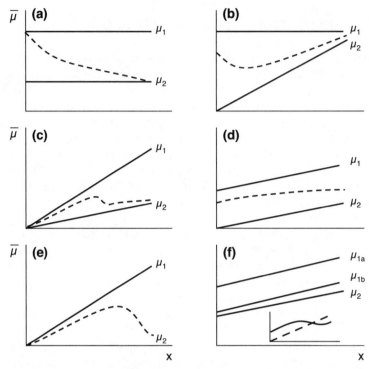

FIGURE 3.8. Examples of how heterogeneity can affect the mortality trajectory of a cohort (Vaupel and Yashin 1985): (a) the observed hazard rate may decline even though the hazard rates for the two subcohorts are constant; (b) the observed hazard rate may decline and then rise even though the hazard rate for one subcohort is rising steadily and that of the other is constant; (c) the observed hazard rate may rise steadily and then decline, and then rise again even though the hazard rates for the two subcohorts are steadily rising; (d) the observed hazard rate may increase more slowly than the hazard rates for the other two subcohorts; (e) the observed hazard rate may increase and then decline if the hazard rate for one subcohort is increasing and the other subcohort is immune; (f) a disadvantaged cohort may appear to suffer lower mortality rates than an advantaged cohort at older ages as their members die off. The frail subcohort of a disadvantaged population faces higher mortality chances than the frail subcohort of the advantaged population. Consequently, the frailer members of the disadvantaged population die off relatively quickly, leaving a surviving population that largely consists of the robust subcohort. If this selection effect is strong enough, a crossover may be observed for the two populations as shown in the inset (redrawn from figs. 1–6 in Vaupel and Yashin 1985).

The concept of subcohort variance in the risk of death and its actuarial consequences was introduced to the literature over four decades ago by James Vaupel and his colleagues (1979), and their model of heterogeneity yielded four major insights that were not apparent at the time. First, the mortality rates for individuals may increase faster with age than the observed mortality rates for cohorts because, as death selectively removes individuals with high frailty, the average frailty of a cohort decreases. This is because the cohort mortality rates increase less rapidly than the mortality rates for any individual in the cohort. Second, the life expectancy of those whose lives might be saved by some health intervention may be less than currently estimated because measures of progress consider only changes in cohort mortality and do not recognize that the proportion of cohorts that reach a particular age has been increasing. Third, past progress against mortality may be underestimated, and as a consequence predictions of future progress against mortality may be too low. Finally, heterogeneity in frailty may be a factor in observed declines and reversals with age of mortality differentials between pairs of populations.

Model derivation for demographic heterogeneity

Given that $\mu_i(x, y, z)$ is the force of mortality for individuals in population group I, at exact age x, at some instant in time y, and with frailty of z (Vaupel et al. 1979), the frailty parameter, z, is defined according to the relationship

$$\mu_i(x, y, z) = z * \mu_i(x, y, 1) \tag{3.24}$$

where an individual with the frailty parameter of $z = 1$ is a "standard" individual. This means that an individual with a frailty of 2 is twice as likely to die at any particular age and time, and an individual with a frailty ½ is half as likely to die.

For simplicity Vaupel et al. (1979) dropped the subscripts and arguments, so eqn. (3.24) is reduced to

$$\mu(z) = z * \mu \tag{3.25}$$

Let s be the probability that an individual will survive to age x and let H denote the cumulative hazard function through age x. Then

$$s = e^{-H} \tag{3.26}$$

And it follows that

$$s(z) = s^z \tag{3.27}$$

where $S = s(x, y, 1)$. In words, if a standard individual has a 0.50 chance of surviving to age x, then individuals with frailties of 1/3, 1/2, 1, 2, and 3 will have a 0.79, 0.71, 0.50, 0.25, and 0.125 chance of survival, respectively. And if there were equal initial numbers of individuals in each of these five subcohorts, the average survival for the cohort as a whole would be $s = 0.475$. If, alternatively, the initial numbers were $n = 5,000, 4,000, 3,000, 2,000$, and $1,000$ for the most to the least frail group, respectively, then $s = 0.356$. Finally, if the order of the initial numbers were reversed, from the least to the most frail group, then $s = 0.594$. These hypothetical examples demonstrate the importance of understanding not only the effects of the distribution of within-cohort frailty on the group survival, but also the effects on survival of the distribution of initial numbers for each subcohort.

Distribution of frailty

Vaupel and his colleagues (Vaupel, Manton, and Stallard 1979; Vaupel and Yashin 1985) assumed that frailty at birth is gamma distributed:

$$f_0(z) = \lambda^k z^{k-1} \frac{e^{-\lambda z}}{\Gamma(k)} \tag{3.28}$$

where λ and k are the parameters of the distribution. The mean and variance of a gamma variate are given by

$$\bar{z} = \frac{k}{\lambda} \tag{3.29}$$

and

$$\sigma^2 = \frac{k}{\lambda^2} \tag{3.30}$$

They chose this distribution because it is analytically tractable and readily computable. They note that it is a flexible distribution that takes on a variety of shapes as k varies; for example, when $k = 1$ it is identical to the exponential distribution, and when k is large it assumes a bell-shaped form.

Example of two-subcohort heterogeneity model

The concept of demographic selection resulting from demographic heterogeneity is illustrated with a hypothetical two-subcohort example applied to the fruit fly *D. melanogaster*. Using Gompertzian mortality rates for each subcohort with differences in the rate parameter, b, and starting with a 10-fold difference in initial numbers, the effect of selection is shown in table 3.2 and fig. 3.9. The overall cohort mortality rate closely tracks that for the subcohort with the highest mortality rate, subcohort A, and, because of its higher initial numbers, continues through 50 days. At this age the mortality of the whole, the dashed line in fig. 3.9, differs only slightly from mortality in subcohort A. However, by age 65–70 roughly the same number of survivors remain in each subcohort, and thus the composite mortality rate begins to bend toward that for subcohort B. With few remaining members of the high-mortality subcohort A still living, the overall mortality converges to the mortality of the low-mortality subcohort B. At this point the average mortality slows and begins to decrease even though mortality in both subcohorts continues to increase through all ages.

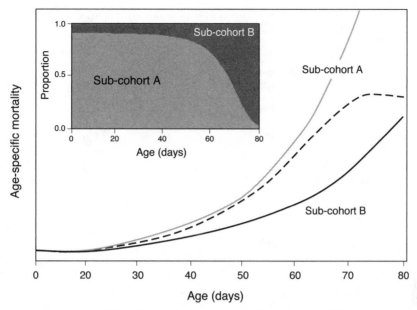

FIGURE 3.9. Illustration of how the composite mortality for two subcohorts, A and B, each subject to Gompertz mortality rates, departs from the Gompertz model (dashed curve). Inset shows the composition of the overall cohort starting with a 10-to-1 ratio of initial numbers of subcohort A relative to subcohort B.

Compositional interpretation of the Medfly mortality

Vaupel and Carey (1993) used mixtures of 12 different Gompertz curves to closely fit mortality patterns observed in the 1.2 million Medfly study showing slowing of mortality at older ages. To model the age trajectory of Medfly mortality, they assumed that

$$\mu(x, z) = z\mu^0(x) \tag{3.31}$$

where μ^0 is the baseline hazard function. Specifically, they assumed a baseline function of a Gompertz curve, where

$$\mu^0(x) = 0.003e^{0.3x} \tag{3.32}$$

The proportion of the population of 1.2 million flies that survived from eclosion (birth) to age 121 defined the baseline (average) survival with an expectation of life of 13.1 days for the "standard" individual (i.e., $z = 1$).

The z-values and initial proportions for each of 12 subcohorts are given in table 3.3. One should first note that the frailty levels for some cohorts differ by 10 orders of magnitude; for example, an extreme frailty of 3.2 with an expectation of life less than 10 days to a subcohort with the least frailty of 0.00000000073 and an expectation of life of over 100 days. The first three subcohorts in table 3.3 constitute 92% of the

Table 3.2. Hypothetical example of a *Drosophila melanogaster* cohort consisting of subcohorts A and B subject to Gompertzian rates of aging

Age x	Age-specific mortality, $\mu(x)$			Survival (l_x)	
	Subcohort A (low)	Subcohort B (high)	Composite	Subcohort A	Subcohort B
0	0.0050	0.0050	0.0051	100.0	1000.0
1	0.0052	0.0053	0.0054	99.5	994.9
2	0.0055	0.0056	0.0058	99.0	989.4
3	0.0057	0.0060	0.0061	98.4	983.7
4	0.0060	0.0064	0.0065	97.8	977.6
5	0.0063	0.0067	0.0069	97.2	971.3
20	0.0123	0.0166	0.0165	85.0	824.2
21	0.0129	0.0176	0.0175	84.0	810.2
22	0.0135	0.0187	0.0186	82.9	795.6
23	0.0141	0.0199	0.0197	81.7	780.4
24	0.0147	0.0211	0.0209	80.6	764.6
25	0.0154	0.0224	0.0221	79.4	748.2
40	0.0302	0.0551	0.0523	57.1	433.8
41	0.0316	0.0585	0.0554	55.3	409.8
42	0.0331	0.0621	0.0585	53.6	385.8
43	0.0346	0.0660	0.0619	51.8	361.9
44	0.0362	0.0701	0.0654	50.0	338.1
45	0.0379	0.0744	0.0691	48.2	314.5
60	0.0744	0.1830	0.1429	21.4	51.5
61	0.0778	0.1943	0.1482	19.8	42.6
62	0.0814	0.2063	0.1531	18.3	34.9
63	0.0852	0.2191	0.1577	16.8	28.2
64	0.0891	0.2326	0.1619	15.4	22.5
65	0.0932	0.2470	0.1654	14.1	17.7
70	0.1167	0.3334	0.1721	8.4	4.2
71	0.1221	0.3540	0.1713	7.4	3.0
72	0.1277	0.3759	0.1701	6.5	2.1
73	0.1335	0.3992	0.1688	5.7	1.4
74	0.1397	0.4239	0.1677	5.0	0.9
75	0.1461	0.4501	0.1670	4.3	0.6

Note: The Gompertz parameter a is identical for both (initial mortality, a = 0.005), but the Gompertz parameter b is different (b = 0.045 and 0.060 for subcohorts A and B, respectively). Note the 10-fold difference in initial numbers at age 0.

main cohort, and of this group only subcohort 1 had a frailty that was greater than the standard individual. Deaths in these three subcohorts were primarily responsible for shaping the overall trajectory of mortality through 50 days, the age at which mortality peaked. With respect to life span, only 0.7% (1 of 150) of the original Medfly cohort, and the modeled cohort, lived beyond 50 days. This means that only a tiny fraction of the whole cohort consisted of extremely long-lived individuals and was responsible for the deceleration and decline of cohort mortality at advanced ages. Inasmuch as each subcohort was experiencing ever-increasing probabilities of death with age, the decreasing overall mortality, in both the empirical and modeled data,

Table 3.3. Values of frailty, z, expectation of life at eclosion, and proportions, p, of flies at each level of frailty for Gompertz model described in text

Subcohort	Frailty, z	Expectation of life (days)	Proportion, p
1	3.7	4.4	0.41
2	0.75	16.4	0.38
3	0.17	47.6	0.13
4	0.03	93.3	0.46
5	0.0093	98.8	0.020
6	0.0020	100.4	0.0082
7	0.00036	100.7	0.0017
8	0.000074	100.8	0.00046
9	0.000011	100.8	0.00013
10	0.0000014	100.8	0.000053
11	0.000000058	100.8	0.000013
12	0.00000000073	100.8	0.000043

Source: Vaupel and Carey (1993). *Note:* The expectations of life for subcohorts 8–12 differ only by 2 to 4 decimal places (not given).

required a quick sequential cascade of cohort die-offs, which then dropped the age-specific mortality to the lower rate of the less frail (longer-lived) subgroup (fig. 3.10). This accounts for the "stair-step" appearance in the decreasing segment of the mortality curve after age 50. Finally, fine-tuning the heterogeneity model required for best fit revealed the importance of the interplay of subcohort levels of frailty (z) with their individual proportion (p) to achieve mortality deceleration and decline in Medfly mortality. It is thus evident that changing these two parameters either in the heterogeneity model or in empirical studies (i.e., by chance or by design) can have profound effects on the overall trajectory of mortality.

Mortality Metrics

Average lifetime mortality

The inverse of life expectancy at birth, e_0, is the average mortality experienced by the cohort, given as $\bar{\mu} = 1 / e_0$. More generally, the inverse of life expectancy at age x, e_x, is the average mortality experienced by the cohort beyond age x, given as $\bar{\mu}_x = 1 / e_x$. For example, the average yearly mortality probabilities for cohorts with 25-, 50-, and 75-year life expectancies are 0.04, 0.02, and 0.133, respectively.

Mortality change indicators

Kannisto (1996) describes several different mortality change indicators, some of which are included in table 3.4. These indicators reflect both the relative (proportional) and absolute (arithmetic) differences in three parameters—mortality, survival, and expectation of life. Expressing differences in mortality between two cohorts (table 3.4, 1a) as a proportion reflects relative changes with age and is particularly useful when mortality is low and thus absolute differences are small. Proportional differences at low

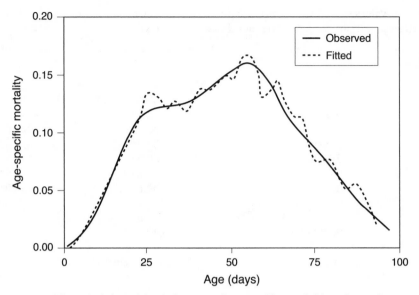

FIGURE 3.10. Daily probability of death for 1.2 million Medflies. Solid line shows the empirical data from Carey and his colleagues (1992). Dashed line is the output of the heterogeneity model, consisting of 12 different subcohorts with different initial proportions and frailty levels (Vaupel and Carey 1993).

mortality rates may be substantial, for example, the order-of-magnitude difference between $q_x = 0.0001$ and $q_x = 0.001$, and therefore they better highlight significant differences in the underlying biology. In contrast, absolute differences in mortality terms (table 3.4, 1b) are useful when absolute risks are high, for example, the small proportional differences between $q_x = 0.4$ and $q_x = 0.5$ have substantial absolute differences. Similar reasoning applies when comparing survival or life expectancy differences between two cohorts (table 3.4, 2a, b, and 3a, b). Relative differences are especially useful when comparing these properties at old ages when the probabilities of survival and the remaining life expectancies are both quite small, but absolute differences may be more useful at the younger ages when survival and life expectancies are greater.

An important use of both relative and absolute differences in age-specific mortality is in identifying mortality crossovers (Manton and Stallard 1984), which occur when the relative rate of change and level of age-specific mortality rates in two populations differ. One group is "advantaged"—in other words, it has a lower relative mortality rate—and the other group is "disadvantaged" with a higher relative mortality rate. The disadvantaged population must manifest age-specific mortality rates markedly higher than the advantaged population through middle age, at which time the rates change. An example of a male-female mortality crossover in the Medfly is presented in Carey et al. (1995).

Mortality scaling

Observed mortality (μ_x) can be scaled in one of two ways in order to create a modified mortality schedule ($\hat{\mu}_x$), including proportional scaling and age-specific scaling (table 3.4). Proportional scaling changes the level of observed mortality by a constant

Table 3.4. Selected mortality parameters and formulas, change indicators, and scaling models

Parameter/model	Description	Notation	Formula/model
General mortality parameters			
Force of mortality	Instantaneous mortality rate	μ_x	$-\dfrac{dl_x}{l_x dx}$
	Estimation formula (1)	μ_x	$-\ln p_x$
	Estimation formula (2)	μ_x	$-\dfrac{1}{2}(\ln p_{x-1} + \ln p_x)$
	Estimation formula (3)	μ_x	$\dfrac{1}{2n}\ln_e\left(\dfrac{l(x-n)}{l(x+n)}\right)$
Central death rate	Number dying at age x relative to number at risk	m_x	$\dfrac{q_x}{1-\frac{1}{2}q_x}$
Life table aging rate	Rate of change in age-specific mortality with age	k_x	$\ln(m_{x+1}) - \ln(m_x)$
Mortality smoothing	Smoothing for discrete form	$\hat{q}_x$	$1-\left[\displaystyle\prod_{y=x-n}^{x+n} p_x\right]^{-(n+1)}$
	Smoothing for continuous form	$\hat{\mu}_x$	$\dfrac{1}{n+1}\displaystyle\sum_{y=x-n}^{x+n} \mu_y$
Average daily mortality	Daily mortality given expectation of life, e_0	$\bar{\mu}$	$\dfrac{1}{e_0}$
Entropy	Days gained per averted death	H	$\dfrac{\displaystyle\sum_{x=0}^{\omega} e_x d_x}{e_0}$
Mortality change indicators (cohort A vs. cohort B)			
1a	Mortality increase/decrease (relative)	—	$\dfrac{\mu_x^A}{\mu_x^B}$
1b	Mortality increase/decrease (absolute)	—	$\mu_x^A - \mu_x^B$
2a	Survival increase/decrease (relative)	—	$\dfrac{l_x^A}{l_x^B}$
2b	Survival increase/decrease (absolute)	—	$l_x^A - l_x^B$
3a	Life-days gained/lost at age x (relative)	—	$\dfrac{e_x^A}{e_x^B}$
3b	Life-days gained/lost at age x (absolute)	—	$e_x^A - e_x^B$

Sources: Kannisto (1996); Carey (2001).

amount at all ages, whereas age-specific scaling changes mortality at each age by a specified proportion.

Threshold mortality

A major concern of any study designed to estimate the actuarial rate of aging in a cohort is the decrease in sample size at older ages due to attrition. However, this problem of insufficient sample size may also apply to the measurement of mortality at young ages even when the number of individuals at risk is at or near the initial number, n. This has been referred to as the *left-hand boundary problem* (Promislow et al. 1999), which occurs whenever the "actual" mortality rate is less than 1/n. For example, a mortality rate of $\mu = 0.001$ cannot be detected with a sample size of $n = 100$ since when a single individual dies the estimate will be $\mu = 1/100 = 0.01$. The main point is that even though the number of individuals at risk is highest at the youngest ages, a sample size constraint still exists inasmuch as mortality is often quite low at young ages and thus lower than 1/n. In general, at small sample sizes, where there are few individuals at risk, the observed death rates do not provide reliable estimates of the underlying distribution of the probability of death (fig. 3.11).

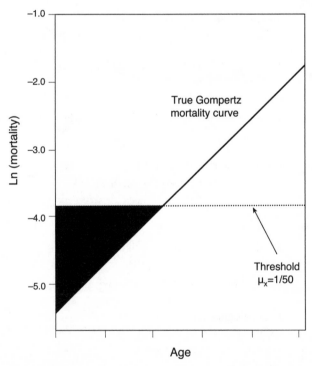

FIGURE 3.11. Illustration of threshold mortality concept. If sample size $n = 50$ individuals, then 0.02 (1/50) is the threshold beyond which mortality cannot be measured for that sample size.

Standard deviation of mortality

The equation for the 95% confidence interval for age-specific mortality is given as

$$CI_{95\%} = \hat{q}_x \pm 1.96 S_{\hat{q}x} \qquad (3.33)$$

where $S_{\hat{q}x}$ denotes the standard deviation of the death rate at age x. The formula for $S_{\hat{q}x}$ is

$$S_{\hat{q}x} = \hat{q}_x \sqrt{\frac{1}{D_x}(1-\hat{q}_x)} \qquad (3.34)$$

where $\hat{q}_x$ is the age-specific mortality probability at age x and D_x is the number of deaths at age x.

Mortality in the human population

The characteristics of the mortality patterns of contemporary humans living in developed societies vary with life-cycle stage, and age and gender account for most of the within-population differences in overall mortality (fig. 3.12). At the early life stages

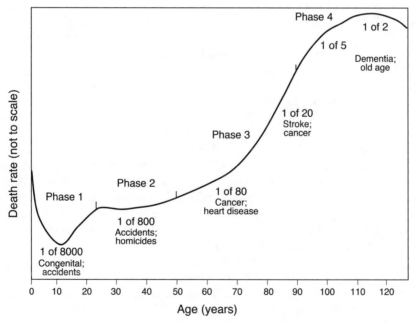

FIGURE 3.12. Anatomy of human mortality. This schematic diagram of the mortality schedule of modern humans (not to scale) shows various properties, including the general age pattern, rate of change with age, probability of dying, and principal causes of death (redrawn from Carey and Judge 2000).

newborns experience the highest death rates, and then after the first few months of life mortality drops quickly and reaches the lowest point of the life course at ages 10 or 11 years. An 11-year-old girl's probability of dying by her twelfth birthday is less than 1 in 8,000. Mortality then increases at a moderate rate through the early to mid 20s, at which point a slight change in trajectory occurs, the details of which depend on gender. Male mortality exceeds female mortality, and the rate of increase is greater for males, resulting in a slight male "mortality bump." These differences can be linked to the high-risk behavior more common in males in the late teens and early 20s than in females of the same age. Between ages 50 and 90, the rate of change in mortality with age first accelerates, then decelerates, forming a bell-shaped curve, quite regular for women but not for men. Mortality continues to decelerate from ages 90 to 110, causing a moderate leveling-off and slight decrease in mortality at the most advanced ages. The cause-of-death structure in developed countries like the United States progresses from congenital diseases, accidents, and homicides at younger ages to cancer, heart disease, stroke, and dementia at the oldest ages.

Consider Further

Inasmuch as mortality is central to demography at individual, cohort, and population levels, various concepts and metrics directly related to mortality are present throughout the book. These include concepts in the Lexis diagram in chapter 1 (Basics); the life table in chapter 2 (Life Tables); age-specific mortality (q_x) and the central death rate (m_x) in this chapter; the parity-progression life table in chapter 4 (Reproduction); and models throughout the three chapters on population, including chapter 5 (Population I) with the intrinsic death rate (d) and the relationship of fertility and mortality in shaping age structure, chapter 6 (Population II) with mortality used in the context of stage matrix modeling, and chapter 7 (Population III) with sex-specific mortality used in two-sex models, the stochastic mortality component in stochastic models, and the worker death rate in models of social insects. Mortality concepts are included in chapter 8 (Human Demography) with a depiction of the mortality curve and an example of a life table for humans. Mortality rates and related metrics are included in the two chapters on applied demography, including chapter 9 (Applied Demography I: Estimation) with mark-recapture and residual demography and chapter 10 (Applied Demography II: Evaluation) with comparative mortality dynamics, active life expectancy, irreplaceable mortality in biological control, and throughout the sections on harvesting and wildlife management. There are over a dozen of the short entries in chapter 11 (Biodemography Shorts) that are on or related to mortality, ranging from the theoretical consequences of a life course based on the ultralow mortality of 10-year-old girls (S7) to the impact on life expectancy of eliminating mortality at multiple ages (S10).

Books, chapters, and articles that provide more advanced treatments of mortality and related concepts include Kleinbaum and Klein (2012), Hosmer and coauthors (2008), Collett (2015), Liu (2012), Elandt-Johnson (1980), multiple chapters on mortality or closely related topics in the edited books by Skiadas and Skiadas (2018) and Vallin et al. (1990), trends in mortality analysis by Manton and Stallard (1984), and a review of mortality modeling by Anatoli Yashin and his colleagues (2000). General books that cover mortality include those by Carmichael (2016) and Preston et al.

(2001). The second volume of *Demography: Analysis and Synthesis* by Caselli, Vallin, and Wunsch (2006a) contains a total of 16 chapters on topics related to mortality, including chapters on the relationship between morbidity and mortality by cause (Egidi and Frova 2006), endogenous mortality to maximum human life span (Vallin and Berlinguer 2006), environmental factors of mortality (Sartor 2006), and mortality, sex, and gender (Vallin 2006).

4

Reproduction

In looking at Nature it is most necessary to ... never forget that every single organic being around us may be said to be striving to the utmost to increase in numbers.

Charles Darwin (1859, 65)

Background

Overview of terms and concepts

Under most demographic circumstances, reproduction is the most important determinant of population dynamics and growth. It follows that methods for analyzing, quantifying, and characterizing fertility are fundamental to the study of demography in general and, even more specifically, for life history concepts.

Reproduction can be characterized in one of two ways: physiological (i.e., the *process* that results in offspring) or demographic (i.e., the per capita *rate* of offspring production in a given period of time) (Carey 1993). The terms *reproductive rate, renewal rate, recruitment rate,* and *natality rate* all denote the same reproductive concept and are frequently used synonymously. In biology *fecundity* and *fertility* are sometimes used interchangeably with reproductive rate, but we use the convention from formal demography of defining fecundity as the total number of offspring a female is capable of producing in a specified age interval, and fertility as the total number of offspring a female actually produces in the interval. In human demography *fecundability* refers to the probability of becoming pregnant in a single menstrual cycle.

Our objective in this chapter is to provide an overview of the analysis of reproduction in biological populations that goes beyond what may be found in standard biology and ecology texts. We divide the subject into three general areas: reproductive averages (e.g., gross and net reproduction), reproductive heterogeneity (e.g., interindividual differences in reproductive rates), and generalizations of reproduction (e.g., clutch size; parity progression).

Patterns of Reproduction

One way to classify the reproductive diversity of organisms is by the number of reproductive events in an individual's lifetime (Roff 1992). Semelparous species reproduce once, and this pattern is often termed "big bang" reproduction because individuals die after reproduction. Examples of semelparous reproduction include

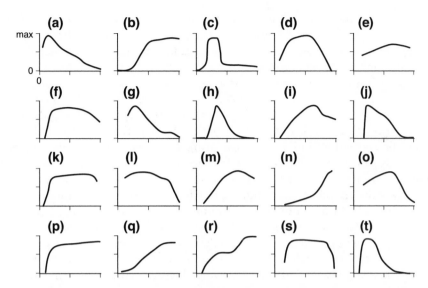

FIGURE 4.1. Shapes of age-specific reproductive schedules in selected taxonomic groups and species. Birds—(a) Japanese quail (Abplanalp and Woodard 1970); (b) South polar skua (Ainley et al. 1990); (c) white roman goose (Wang et al. 2002); (d) sparrow hawk (Newton and Rothery 1997); (e) magpie (Birkhead and Goodburn 1989); Mammals—(f) northern fur seal (Eberhardt and Siniff 1977); (g) Japanese monkey (Gouzoules et al. 1982); (h) killer whale (Olesiuk et al. 1990; Brault and Caswell 1993); (i) yellow-bellied marmot (Armitage and Downhower 1974); (j) chimpanzee (Teleki et al. 1976); (k) red deer (Clutton-Brock 1994); (l) vole (Leslie and Ransom 1940); (m) mountain goat (Côté and Festa-Bianchet 2001); (n) European rabbit (Rödel et al. 2004); (o) African lion (Packer et al. 1998); Amphibians—(p) Blue ridged two-lined salamander (Bruce 1988); (q) bullfrog (Howard 1988); (r) plains garter snake (Stanford and King 2004); Insects—(s) body louse (Evans and Smith 1952); (t) six-spotted thrips (Coville and Allen 1977).

Pacific salmon, American eels, octopods, and bamboo. Iteroparous species, on the other hand, breed more than once and thus exhibit age-specific patterns of reproduction; it is these patterns that are the focus of this chapter. Schedules of species that exhibit iterative reproduction are shown in fig. 4.1 for selected avian, mammalian, amphibian, and insect species. Roff (1992, 2002) divides the shape of the age schedule of reproduction into three classes: (1) uniform—an age-independent or constant level of offspring production with age; (2) asymptotic—an increasing reproductive output with age, a pattern that often occurs in plants and ectotherms that increase in size after reproductive maturity; and (3) triangular—an age-dependent reproductive pattern that starts low, increases to a peak typically in early middle age, and decreases with age.

Basic Concepts

A *reproductive age schedule* shows the per capita age-specific fecundity and fertility rates over specified age intervals or over a cohort's lifetime. A *gross schedule* is one in which mortality is not taken into account, and a *net schedule* weights reproduction by the proportion or number in the cohort that survive to each age class. Measures of gross and net reproduction give the mean number of offspring produced by the cohort, but they do not capture the reproductive heterogeneity within the cohort. In other words, they do not reflect the fact that not all individuals produce the same number of offspring. Parameters that express various aspects of this heterogeneity include the following: *reproductive interval*, which gives the frequency of offspring production by individuals (i.e., egg-producing days versus non-egg-producing days); *daily parity,* which denotes the fraction of the cohort that produce specified numbers of offspring at a given age; *cumulative parity,* which expresses the fraction of living individuals within a cohort that have had a specified sum total of offspring at given ages; and *concentration of reproduction,* which gives the frequency distribution of lifetime reproduction ranked by individual. Offspring produced by females are often produced as clutches (e.g., insects) or litters (e.g., rodents; canids), thus clutch or litter size designate offspring groupings that can be expressed as an age schedule. Most organisms produce more than one *offspring type* of which the most common type category is gender. Other types may include sexual/asexual offspring or winged/wingless offspring. All of the reproductive relations discussed here pertain to females and are thus *maternity functions*. Although only limited effort has been directed toward reproductive rates of males, a number of *paternity functions* can be patterned after these maternity concepts.

Birth Intervals and Rates

Variation in reproduction among individuals within a cohort can be quantified by comparing birth intervals and birth rates. Birth intervals provide order (i.e., age sequence), duration (width of age class) and timing (period), and birth rates can be expressed as age- and period-specific or as total reproductive rates, without adjustments for mortality (i.e., gross rate) or with adjustments for mortality (i.e., net rate).

Basic age-specific birth metrics

If all individuals in a cohort are (literally) exactly the same age, period effects on birth rates will, at least in theory and with respect to statistical analysis, have the same impact on all individuals. The schematic shown in fig. 4.2 illustrates how a number of reproductive metrics are measured and computed. The hypothetical raw data include the number of offspring by age for each individual in the rows (individuals A–D) and columns (ages 0–4). The summary data include total births by individual, exposure time (which refers to the duration of life for an individual), and average births per epoch (which are births divided by exposure time: $9/4.0 = 2.3$ for individual A). The summary metrics for the total group of individuals are computed from the sums of each of three columns, which shows a total of 24 offspring, 12 epochs exposure (average life span), and 7.3 total births. From these sums, averages can be computed, such as the average lifetime births (24 births/4 individuals = 6 lifetime births/individual), the average lifetime (12 epochs/3 individuals = 4 epochs), and the average reproduction per epoch (7.3 time units/4 individuals = 1.8).

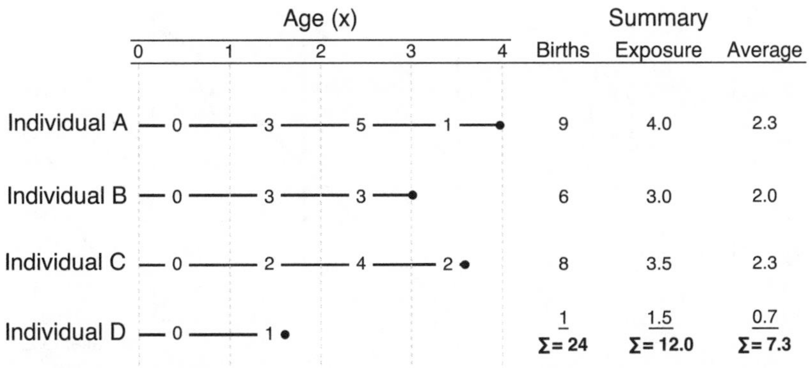

FIGURE 4.2. Schematic illustrating the measurement of individual-level births by age and of life-days lived as the exposure metric. Numbers superimposed on life-lines indicate the number of offspring produced within the age interval.

Lexis diagram: Births by age and period

The Lexis diagram can be used to illustrate the measurement of three different categories of birth rates (fig. 4.3).

Lexis square. This is the number of births within a single period for two cohorts in the same age class. In the figure this total equals the number of births from the cohort prior to cohort A (25 births), plus the total from cohort A (32 births) that occurred in the year 2000, when individuals were between 31 and 32 years old. The total is 57 births.

Lexis horizontal parallelogram. This is the number of births through the time that a cohort is a particular age and that age covers two periods. In the figure cohort A produced 32 births in 2000 and 19 births in 2001, when individuals in this cohort were between 31 and 32 years of age. The total is 51 births.

Lexis vertical parallelogram. This is the number of births for a cohort during a single period. In the figure for the year 2000, some individuals from cohort A were between 30 and 31 years and they produced 27 births, and other individuals from cohort A were between 31 and 32 years and produced 32 births. The total is 59 births.

Per capita reproductive rates

Daily rates

The most basic reproductive measure is gross maternity, defined as the average number of offspring produced by a female in the interval x to x + 1. This is designated M_x, where

$$M_x = \frac{\text{total offspring produced from x to x} + 1}{\text{number of individuals alive at midpoint of x and x} + 1}$$

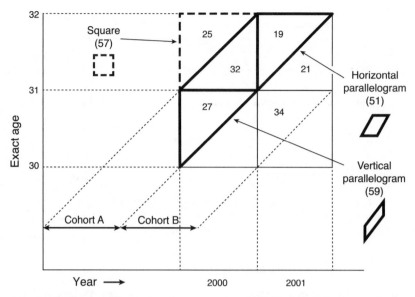

FIGURE 4.3. Lexis diagram illustrating three ways in which births can be counted when age and period are considered—Lexis square, Lexis horizontal parallelogram, and Lexis vertical parallelogram.

The number of females alive at the midpoint of the age interval is

$$\text{midpoint number} = \frac{(\text{number alive at x}) + (\text{number alive at x} + 1)}{2}$$

This relationship implies that the individuals that died in the interval did so midway through the interval. This assumption is identical with that for determining the number of life-days lived in the age interval to compute L_x in the life table (see chapter 2). The parameter L_x is frequently used as a multiplier of M_x and other reproductive parameters to yield *net maternity*, which is the number of offspring produced by females aged x, weighted by the probability of their attaining age x from age 0. An illustration of these measures is given in table 4.1 with a Medfly example.

The number of offspring counted at each exact age represents the total number produced in the age class interval x to x + 1. For example, the 5,616 eggs counted at x = 40 corresponds to the eggs produced by the 352 females (midpoint estimate) living in the age interval 40 to 41 (indexed at x = 40). It is thus necessary to distinguish between *exact age* and designated *age class x*, that is, the index for the interval between two exact ages.

Lifetime rates and mean ages

Maternity schedules may be summed over the lifetime of a cohort to yield a number of different reproductive rates. Maternity schedules may also be combined with age

Table 4.1. Age-specific egg-laying rates and metrics for the Medfly

	Exact age (days) at count			
	40	41	42	43
Number females alive at exact age x	360	344	322	297
Number alive at midpoint of interval		352	333	310
Number offspring counted at exact age	5,616	4,713	4,347	4,158
Age interval	40–41	41–42	42–43	43–44
Age index (x)	40	41	42	43
L_x (midpoint survival)		0.352	0.333	0.310
M_x (offspring at age x)		15.6	14.2	14.0
$L_x M_x$ (net maternity)		5.5	4.7	4.3

Source: Carey et al. (1998a).

weightings and the survival schedule to determine the mean reproductive ages in a cohort.

A hypothetical yet important lifetime reproductive rate is the *gross reproductive rate* (GRR), given by the formula

$$\text{Gross reproductive rate} = \text{GRR} = \sum_{x=\alpha}^{\beta} M_x \qquad (4.1)$$

where α and β denote age of first and last reproduction, respectively. This sum gives the lifetime reproduction by an average female. The *net reproductive rate*, which gives the average lifetime reproduction for a newborn female, can be computed as

$$\text{Net reproductive rate} = \text{NRR} = \sum_{x=\alpha}^{\beta} L_x M_x \qquad (4.2)$$

These two rates can be defined alternatively as the per-generation reproductive contribution of newborn (or newly eclosed) females to the next generation. Net rates give the per capita lifetime contribution. Therefore, when these rates are divided by the number of days lived by the average female, this yields the *average daily reproduction*:

$$\text{Eggs/female/day} = \frac{\sum\limits_{x=\alpha}^{\beta} L_x M_x}{\sum\limits_{x=\varepsilon}^{x} L_x} \qquad (4.3)$$

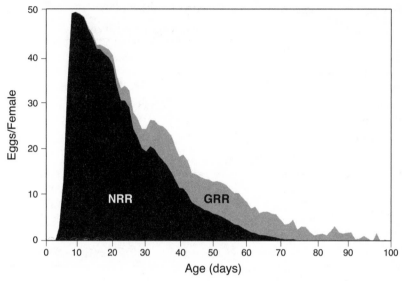

FIGURE 4.4. Gross and net reproduction in the fruit fly *D. melanogaster*.

$$\text{Fertile eggs/female/day} = \frac{\sum\limits_{x=\alpha}^{\beta} L_x M_x}{\sum\limits_{x=\epsilon}^{x} L_x} \qquad (4.4)$$

Note that the denominator for each of these equations equals the expectation of life at birth (or eclosion). The mean age of reproduction for each of these schedules is computed by dividing the sum of each of these schedules, weighted by age class x, by the sum of the unweighted schedules. Data for computing gross and net reproduction for *D. melanogaster* are given in table 4.2 and depicted in fig. 4.4.

Previous and remaining reproduction

The net and gross rates of fecundity given in the previous sections shed light on reproduction at two different points on the age scale of the average individual from age 0 through age ω (oldest age). That is, the net rate represents the expected future reproduction of the average newborn (newly eclosed) individual, and the gross rate represents the past reproduction of an individual that lived to the last possible age (age ω). These basic relations are not restricted to the endpoints and can be extended to every age. The reproductive levels attained by the average female age x are

$$\text{Per capita fecundity to age x} = \sum\limits_{y=\alpha}^{x} M_y \qquad (4.5)$$

Table 4.2. Gross and net reproductive schedule of *D. melanogaster*

Age (days)	Number living	Total eggs laid by cohort	Fraction of cohort surviving to age s	Per capita egg laying	Net maternity
x	N_x	M_x	l_x	m_x	$l_x m_x$
(1)	(2)	(3)	(4)	(5)	(6)
0	666	0	1.0000	0.0	0.0
1	666	0	1.0000	0.0	0.0
2	666	2,133	1.0000	3.2	3.2
3	666	8,960	1.0000	13.5	13.5
4	666	23,239	1.0000	34.9	34.9
5	666	34,114	1.0000	51.2	51.2
6	666	34,304	1.0000	51.5	51.5
7	665	33,978	0.9985	51.1	51.0
8	663	33,498	0.9955	50.5	50.3
9	658	31,682	0.9880	48.1	47.6
10	654	30,542	0.9820	46.7	45.9
11	653	28,868	0.9805	44.2	43.3
12	650	28,710	0.9760	44.2	43.1
13	642	28,048	0.9640	43.7	42.1
14	639	27,564	0.9595	43.1	41.4
15	633	26,467	0.9505	41.8	39.7
16	626	23,317	0.9399	37.2	35.0
17	613	21,070	0.9204	34.4	31.6
18	603	21,108	0.9054	35.0	31.7
19	591	20,053	0.8874	33.9	30.1
20	579	16,858	0.8694	29.1	25.3
21	564	15,539	0.8468	27.6	23.3
22	551	13,944	0.8273	25.3	20.9
23	533	13,514	0.8003	25.4	20.3
24	521	14,246	0.7823	27.3	21.4
25	507	13,809	0.7613	27.2	20.7
26	491	12,864	0.7372	26.2	19.3
27	469	12,148	0.7042	25.9	18.2
28	455	11,402	0.6832	25.1	17.1
29	437	10,416	0.6562	23.8	15.6
30	434	9,355	0.6517	21.6	14.0
31	424	8,087	0.6366	19.1	12.1
32	410	7,999	0.6156	19.5	12.0
33	393	7,129	0.5901	18.1	10.7
34	383	5,897	0.5751	15.4	8.9
35	362	5,482	0.5435	15.1	8.2
36	354	5,150	0.5315	14.5	7.7
37	338	4,774	0.5075	14.1	7.2
38	328	4,590	0.4925	14.0	6.9
39	311	4,186	0.4670	13.5	6.3
40	296	4,036	0.4444	13.6	6.1
41	286	3,793	0.4294	13.3	5.7
42	268	3,434	0.4024	12.8	5.2
43	256	3,028	0.3844	11.8	4.5
44	241	2,644	0.3619	11.0	4.0

(*continued*)

Table 4.2. (*continued*)

Age (days)	Number living	Total eggs laid by cohort	Fraction of cohort surviving to age s	Per capita egg laying	Net maternity
x	N_x	M_x	l_x	m_x	$l_x m_x$
(1)	(2)	(3)	(4)	(5)	(6)
45	227	2,480	0.3408	10.9	3.7
46	213	2,113	0.3198	9.9	3.2
47	197	1,748	0.2958	8.9	2.6
48	188	1,407	0.2823	7.5	2.1
49	168	1,330	0.2523	7.9	2.0
50	161	1,036	0.2417	6.4	1.6
51	149	1,003	0.2237	6.7	1.5
52	138	926	0.2072	6.7	1.4
53	122	767	0.1832	6.3	1.2
54	114	699	0.1712	6.1	1.0
55	105	514	0.1577	4.9	0.8
56	96	372	0.1441	3.9	0.6
57	89	361	0.1336	4.1	0.5
58	81	398	0.1216	4.9	0.6
59	76	209	0.1141	2.8	0.3
60	67	232	0.1006	3.5	0.3
61	65	219	0.0976	3.4	0.3
62	60	130	0.0901	2.2	0.2
63	53	80	0.0796	1.5	0.1
64	51	87	0.0766	1.7	0.1
65	45	84	0.0676	1.9	0.1
66	38	56	0.0571	1.5	0.1
67	34	79	0.0511	2.3	0.1
68	30	105	0.0450	3.5	0.2
69	24	51	0.0360	2.1	0.1
70	19	35	0.0285	1.8	0.1
71	16	33	0.0240	2.1	0.0
72	13	28	0.0195	2.2	0.0
73	11	6	0.0165	0.5	0.0
74	9	7	0.0135	0.8	0.0
75	6	8	0.0090	1.3	0.0
76	5	4	0.0075	0.8	0.0
77	4	0	0.0060	0.0	0.0
78	2	4	0.0030	2.0	0.0
79	2	0	0.0030	0.0	0.0
80	2	1	0.0030	0.5	0.0
81	2	0	0.0030	0.0	0.0
82	1	0	0.0015	0.0	0.0
83	0	0	0.0000		

Note: Data from Robert Arking, with permission.

Table 4.3. Summary of parameters and formulas for any age-specific event at age x, g_x

Description	Formula
Event intensity	
Gross rate	Σg_x
Net rate	$\Sigma l_x g_x$
Daily (yearly) rate	$\Sigma l_x g_x / \Sigma l_x$
Event timing	
Mean age gross schedule	$\Sigma x g_x / \Sigma g_x$
Mean age net schedule	$\Sigma x l_x g_x / \Sigma l_x g_x$

and the expected remaining fecundity is given by

$$\text{Expected remaining fecundity at age } x = \frac{1}{L_x} \sum_{y=x+1}^{x} L_y M_y \qquad (4.6)$$

Therefore, the expected total fecundity for an average female age x is simply the sum of the previous reproduction and expected future reproduction. That is,

$$\text{Expected total fecundity at age } x = \sum_{y=\alpha}^{x} M_y + \frac{1}{L_x} \sum_{y=x+1}^{x} L_y M_y \qquad (4.7)$$

Rate generalizations

Age-specific reproductive rates and reproductive timing can be computed for a cohort if we let g_x denote any age-specific event or property of a cohort (table 4.3). Specifically, g_x might refer to age-specific activity level, lactation rate, mating rate, or migration risk. It can also refer to number of clutches (egg batches) produced by a female age x, as was done for the Mexican fruit fly (*Anastrepha ludens*) where the gross, net, and daily rates of clutch production in this species were 162.1, 112.2, and 1.5 clutches, respectively, with an average of approximately 6.5 eggs per clutch. The mean ages of the gross and net schedules were 54.2 and 45.8 days, respectively (Carey 1993; analysis of data from Berrigan et al. 1988).

Reproductive Heterogeneity

Birth intervals

Daily reproduction by individuals may be thought of as two responses to age: *quantal* (i.e., whether offspring are produced) and *graded* (i.e., the number of offspring produced). As demonstrated in the hypothetical data in table 4.4, an analysis of per capita

Table 4.4. Example of reproductive heterogeneity in three hypothetical cohorts of insects showing reproductive heterogeneity

Day	Cohort A			Cohort B			Cohort C		
	X_1	X_2	Avg	X_1	X_2	Avg	X_1	X_2	Avg
1	10	0	5	2	8	5	5	5	5
2	0	10	5	8	2	5	5	5	5
	10	10	10	10	10	10	10	10	10

daily reproduction, m_x, obscures daily information about variation in individual performance among individuals X_1 and X_2 within different cohorts. Note that the total reproduction for the 2-day period is identical for all individuals and the averages are the same for both days in all three cohorts, and neither sheds light on the true reproductive differences among the three cohorts. Individuals in cohort A produced offspring every other day while those in cohorts B and C produced offspring every day. In terms of similarities in birth interval, cohorts B and C are more alike; but in terms of daily production of offspring cohorts, A and B are most similar because their daily variation is the highest.

In the analysis of birth intervals, it is important to distinguish the period between the age of adulthood and the age of reproductive maturity in which no eggs are laid, and the intermittent days past reproductive maturity during which an individual produces no offspring. The period prior to reproductive maturity is usually thought of as a separate demographic parameter; thus lumping the two types of zero production days confounds the analysis. The complete analysis of birth intervals in a cohort requires the following steps: (1) determine the age of first reproduction for each individual; (2) count the number of days each individual was alive during the observation period; (3) compute the number of mature days by subtracting the age of first reproduction from the total days lived; (4) count the number of days in which at least one offspring was produced; (5) compute the fraction (or percentage) of mature days that at least one offspring was produced, where the inverse of this value yields the birth interval for each individual; and (6) compute the inverse of the dividend of the sum totals of the number of days in which one or more offspring were produced, and the number of mature days lived yields the birth interval.

Data from a Medfly experiment are given in table 4.5, and tallies for each fly in this data set are given in table 4.6 and include the age of first egg (age of reproductive maturity), total days lived out of the 50–60 possible days (of the subset selected), mature days lived, and the number of days when one or more eggs were laid. For example, female 1 produced her first and last eggs at ages 7 and 46 days, respectively (i.e., 39-day reproductive window), she laid eggs on 36 days and did not lay eggs on 15 days, and she had a 4-day post-reproductive period. Female 1 laid a total of 1,103 eggs in her lifetime, which can be expressed as three different daily rates including eggs/day (22.1), eggs/mature day (25.7), and eggs/laying day (30.6).

The reproductive metrics for these 25 female Medflies (table 4.6) show that the average female matured in slightly over one week, but this maturation ranged from 3 to 12 days. As a group, the average female laid eggs for slightly over 43 days, with a range of 36 to 52. Daily egg-laying averages ranged from a low of around 10 eggs/day to over 42 eggs/day. The least fecund female laid 600 eggs and the most fecund produced nearly fourfold more with a lifetime production of 2,200 eggs.

Table 4.5. Individual age-specific reproduction in 25 Medfly females that lived between 50 and 60 days selected from database of 1,000

x	Individual female (eggs/day)																									Σ
	1	2	3	4	5	6	7	8	9	10	11	12	13	14	15	16	17	18	19	20	21	22	23	24	25	
0	0	0	0	0	0	0	0	0	0	0	0	0	0	0	0	0	0	0	0	0	0	0	0	0	0	0
1	0	0	0	0	0	0	0	0	0	0	0	0	0	0	0	0	0	0	0	0	0	0	0	0	0	0
2	0	0	0	0	0	0	0	0	0	0	0	0	0	0	0	0	0	0	0	0	0	0	0	0	0	0
3	0	0	0	0	0	0	0	0	0	0	0	0	0	8	0	0	0	0	0	0	0	0	0	0	0	8
4	0	0	39	0	0	32	0	0	65	0	0	0	0	38	0	0	0	0	0	15	18	34	0	0	0	241
5	0	0	34	0	0	48	0	0	50	0	57	0	0	0	0	0	82	0	19	40	23	33	0	0	0	457
6	0	0	11	0	0	80	0	0	55	0	42	0	0	50	0	0	25	0	22	27	84	27	107	0	0	692
7	38	0	17	55	27	67	55	0	75	0	62	0	0	61	0	0	77	0	24	23	56	32	49	69	0	910
8	45	0	22	32	20	81	50	0	66	0	29	0	0	50	0	0	77	0	40	18	75	40	48	33	11	836
9	37	0	2	19	19	76	65	0	76	0	49	38	0	67	0	79	60	0	40	8	63	47	52	24	0	1,008
10	68	0	0	21	37	80	58	40	78	0	42	9	0	75	0	51	63	0	24	35	68	30	52	47	0	980
11	52	0	66	35	0	71	65	29	76	43	29	36	11	66	0	53	56	41	32	22	50	30	51	36	0	1,006
12	56	22	70	42	28	68	76	19	66	22	45	11	0	65	27	76	58	34	17	6	74	50	71	50	2	1,228
13	46	24	42	25	23	73	69	17	64	13	5	14	11	68	0	45	55	53	16	2	56	27	58	35	0	1,015
14	48	58	16	45	29	63	63	46	57	10	34	9	11	57	0	60	52	51	24	26	36	23	56	40	0	1,031
15	58	38	48	47	20	75	55	29	61	35	11	29	11	75	0	56	52	51	16	24	59	32	52	40	0	1,127
16	48	38	30	26	16	64	56	31	52	28	36	7	18	54	0	81	38	55	24	12	50	27	63	33	0	1,053
17	38	38	15	31	38	64	59	25	46	4	36	7	33	51	0	35	45	52	25	3	53	37	52	30	0	969
18	46	50	2	37	20	60	48	13	58	40	10	28	21	64	140	74	63	53	10	2	38	22	58	20	22	1,193
19	50	36	50	23	27	67	54	17	45	28	38	7	32	56	54	52	58	31	11	3	39	19	42	34	12	1,006
20	40	45	19	7	17	77	43	5	54	22	10	13	30	67	48	62	43	35	39	0	40	22	34	9	15	1,014
21	37	35	9	17	19	53	48	0	41	1	6	14	30	69	40	51	30	46	32	0	34	17	35	4	16	846
22	36	30	36	7	22	52	43	0	49	10	9	39	20	60	63	48	40	49	13	0	21	6	35	4	21	855
23	28	34	48	23	7	48	44	0	51	9	34	34	32	54	75	53	45	43	26	4	36	10	24	7	18	951
24	14	34	9	17	21	53	41	37	43	16	12	24	21	52	50	23	47	42	23	0	23	11	37	44	19	829
25	34	26	0	21	12	43	48	7	43	22	24	13	21	58	48	42	44	38	40	1	15	0	33	18	18	823
26	26	0	0	37	15	30	30	4	43	26	26	39	24	38	53	41	39	16	26	2	15	12	23	20	18	683
27	32	79	0	10	17	39	38	1	57	22	37	38	7	48	44	33	39	35	20	15	23	3	26	23	42	856
28	14	37	74	46	16	44	38	12	56	14	26	21	41	56	44	40	40	59	23	63	11	0	10	32	50	1023
29	15	43	22	13	12	43	34	39	61	14	29	28	14	55	56	31	36	41	23	36	0	0	13	41	65	958

(continued)

Table 4.5. (*continued*)

Individual female (eggs/day)

x	1	2	3	4	5	6	7	8	9	10	11	12	13	14	15	16	17	18	19	20	21	22	23	24	25	Σ
30	27	46	20	15	12	48	24	38	44	31	32	10	20	53	30	37	38	45	32	30	0	0	62	36	17	853
31	15	49	32	40	9	37	26	28	45	23	34	19	20	37	34	35	30	46	21	22	0	4	46	58	47	942
32	20	68	0	0	2	38	28	0	54	36	28	5	31	52	46	30	34	41	38	0	7	3	52	0	37	841
33	16	67	9	34	16	32	25	45	42	28	31	23	33	37	21	36	26	53	30	19	3	26	44	23	40	965
34	14	47	26	22	7	35	26	7	38	35	36	29	24	30	23	29	35	56	31	2	0	0	51	21	17	766
35	18	64	24	43	9	38	21	14	38	40	47	5	44	34	42	25	34	53	23	9	0	22	32	44	22	876
36	11	51	24	35	19	36	24	29	42	5	10	22	15	24	47	17	39	46	17	9	21	2	45	19	5	763
37	13	52	1	17	0	36	23	4	37	30	31	28	19	26	47	24	37	37	27	26	14	4	31	34	11	747
38	32	62	0	40	7	31	14	6	29	12	33	50	22	37	40	18	38	52	10	0	17	27	25	31	6	799
39	0	53	6	21	3	31	23	16	40	37	16	24	10	33	29	25	24	44	18	9	14	7	36	6	27	670
40	0	38	34	14	6	40	25	17	34	3	15	12	0	30	27	24	41	44	17	15	12	0	26	2	48	675
41	0	47	2	3	0	32	18	17	33	32	0	25	12	34	31	26	22	39	14	22	4	21	26	43	16	644
42	19	54	14	9	4	40	27	17	30	15	4	21	12	26	30	20	26	38	15	20	10	17	35	25	33	705
43	2	31	1	3	23	26	18	16	29	7	11	17	12	31	25	18	27	36	14	0	19	18	20	35	36	606
44	0	37	1	0	22	30	15	10	0	11	6	23	17	30	27	22	27	28	14	2	2	5	22	24	33	529
45	7	68	1	5	19	16	23	15	0	15	26	4	14	34	7	19	32	34	14	2	19	14	23	24	39	573
46	3	43	0	7	13	25	9	13	0	25	13	3	31	33	22	14	23	28	16	19	9	7	15	4	27	513
47	0	42	0	4	36	29	24	27	0	29	5	5	0	32	22	12	27	28	16	3	8	14	26	21	44	578
48	0	43	0	0	24	20	4	10	0	17	0	17	12	37	21	11	34	27	18	9	15	15	25	28	33	568
49	0	22	0	0	32	6	6	16	0	17	0	16	12	38	31	9	21	30	12	7	4	14	17	32	23	485
50	0	0	0	0	9	0	6	6	0	24	0	19	8	26	23	14	13	27	22	1	5	10	24	6	16	337
51	—	—	—	—	0	5	0	5	0	11	0	11	9	19	33	13	35	15	8	0	0	12	29	14	23	291
52	—	—	—	—	—	0	0	3	0	2	0	7	14	29	30	19	12	20	8	0	9	19	28	10	25	250
53	—	—	—	—	—	—	—	—	—	0	0	0	5	22	24	9	25	20	0	2	0	7	25	6	22	188
54	—	—	—	—	—	—	—	—	—	0	0	0	6	9	24	9	18	42	0	0	7	5	11	6	13	158
55	—	—	—	—	—	—	—	—	—	—	—	—	0	0	10	13	4	24	0	0	0	5	12	17	19	94
56	—	—	—	—	—	—	—	—	—	—	—	—	—	—	0	0	0	18	0	1	0	0	18	0	12	55
57	—	—	—	—	—	—	—	—	—	—	—	—	—	—	—	—	—	0	0	0	0	0	21	1	16	41
58	—	—	—	—	—	—	—	—	—	—	—	—	—	—	—	—	—	—	—	—	—	—	0	4	0	4
59	—	—	—	—	—	—	—	—	—	—	—	—	—	—	—	—	—	—	—	—	—	—	0	0	0	0
60	—	—	—	—	—	—	—	—	—	—	—	—	—	—	—	—	—	—	—	—	—	—	—	0	0	0

Source: Carey et al. (1998a).

Table 4.6. Summary of individual-level reproduction in Medfly data from table 4.5

Female	Life span (days)	Reproductive window		Egg-laying frequency			Eggs/day	Egg-laying rate		Total eggs
		First egg	Last egg	Laying days	Postlaying days	Post-laying period		Eggs/mature day	Eggs/laying day	
1	50	7	46	36	15	4	22.1	25.7	30.6	1,103
2	50	12	49	37	14	1	33.0	43.4	44.6	1,651
3	50	4	45	36	15	5	17.5	19.0	24.3	876
4	50	7	47	39	12	3	19.0	22.0	24.3	948
5	51	7	50	41	11	1	14.4	16.7	17.9	734
6	52	4	51	47	6	1	42.0	45.5	46.4	2,182
7	52	7	49	43	10	3	30.4	35.2	36.8	1,583
8	53	11	52	39	15	1	13.8	17.4	18.7	730
9	53	4	43	40	14	10	38.2	41.3	50.6	2,023
10	54	12	52	41	14	2	15.7	20.2	20.7	850
11	54	5	47	42	13	7	20.6	22.7	26.4	1,110
12	54	9	52	44	11	2	16.2	19.4	19.9	874
13	55	11	54	41	15	1	14.4	18.0	19.3	790
14	55	3	54	50	6	1	40.6	43.0	44.7	2,235
15	56	12	55	39	18	1	26.5	33.7	38.1	1,484
16	56	9	55	47	10	1	28.8	34.3	34.3	1,614
17	56	5	55	51	6	1	35.5	38.9	38.9	1,986
18	57	11	56	46	12	1	31.5	39.0	39.0	1,796
19	57	5	53	47	11	4	18.0	19.7	21.8	1,025
20	58	4	56	41	18	2	10.4	11.1	14.7	601
21	58	4	55	44	15	3	21.9	23.5	28.9	1,270
22	59	4	56	47	13	3	15.1	16.2	19.0	893
23	59	6	57	52	8	2	32.3	36.0	36.7	1,908
24	60	7	57	49	12	3	21.0	23.8	25.7	1,261
25	60	8	58	43	18	2	17.3	20.0	24.2	1,040
Mean	54.8	7.1	52.2	43.3	12.5	2.6	23.8	27.4	29.9	1,302.7
SD	3.26	2.99	4.17	4.61	3.51	2.16	9.28	10.38	10.39	502.68
Max	60.0	12.0	58.0	52.0	18.0	10.0	42.0	45.5	50.6	2,235.0
Min	50.0	3.0	43.0	36.0	6.0	1.0	10.4	11.1	14.7	601.0

Reproductive parity

Reproductive parity is illustrated in the hypothetical cohort of four individuals over a 4-day period in table 4.7. All four individuals in this cohort produce offspring every day, thus no differences exist among their birth intervals; and average daily reproduction, m_x, is the same over all age classes; and yet, there are differences in reproduction within the cohort. Individual 1 consistently produced only one offspring/day and individual 4 consistently produced two offspring/day. Additionally, daily egg laying for individuals 2 and 3 was different, but their average reproduction was the same over the entire 4-day period.

Daily parity

Differences in daily offspring production of individuals within a cohort can be expressed in terms of daily parity classes, which describe the fraction of the cohort whose offspring production falls into one of several preselected reproductive classes (Carey et al. 1988). For example, the fraction of the hypothetical cohort producing either one or two offspring in any given age class is given in table 4.8. Thus at all ages exactly 50% of the cohort produced one offspring and 50% two offspring, which shows why the average daily production, m_x, equals 1.5 for all age classes.

Cumulative parity

A second parity measure involves the cumulative daily reproduction of individuals and is thus termed cumulative parity. These measures express the fraction of the cohort whose previous and current cumulative offspring production falls into one of several preselected reproductive classes. Daily parity provides insight into the consistency of egg-laying levels in the cohort at a specific age, while cumulative parity shows the long-term consistency. For example, the daily parity of a cohort in which each individual produces one offspring on each of 10 days is strikingly different than a cohort in which the individuals produce 10 offspring once every tenth day. In both cases the mean eggs/female/day in the cohorts are identical (= 1) as is their cumulative parity. However, the true differences in their laying strategies are obscured with these simple descriptive metrics.

Choice of the number of daily parity classes should include a zero class and at least two or three other divisions. We use four daily parity classes, for the Drosophila data summarized in table 4.9. These are 0, 1 to 30, 31 to 60, and > 60 eggs/day.

Table 4.7. Example of reproductive parity in four hypothetical females

| Age | Individual female | | | | |
	1	2	3	4	Average
0	1	1	2	2	1.5
1	1	2	1	2	1.5
2	1	1	2	2	1.5
3	1	2	1	2	1.5
TOTAL	4	6	6	8	6.0

Table 4.8. Summary of cumulative reproductive parity for hypothetical example in table 4.7

	Cumulative production by individual				Percentage in cumulative class			
Age	1	2	3	4	1–2	3–4	5–6	>6
0	1	1	2	2	100	0	0	0
1	2	3	3	4	25	75	0	0
2	3	4	5	6	0	50	50	0
3	4	6	6	8	0	25	50	25

Table 4.9. Percent daily and cumulative parity in *D. melanogaster* in 10-day age intervals

Age (days)	Daily parity (eggs)				Cumulative parity (eggs)			
	0	1 to 30	31 to 60	>60	0	1 to 750	751–1,500	>1,500
0	100.0	0.0	0.0	0.0	100.0	0.0	0.0	0.0
10	7.6	18.0	44.2	30.1	1.2	97.6	1.2	0.0
20	10.4	44.4	37.3	7.9	0.3	55.4	43.4	0.9
30	7.4	66.1	26.5	0.0	0.5	21.7	67.1	10.8
40	14.2	76.4	9.5	0.0	0.3	10.1	67.9	21.6
50	28.0	72.0	0.0	0.0	0.0	4.3	61.5	34.2
60	41.8	58.2	0.0	0.0	0.0	0.0	61.2	38.8

Source: Unpublished data from Robert Arking, by permission.

This table shows that nearly a third (30%) of the females laid over 60 eggs/day. Over 44% produced between 31 and 60 eggs per day at 10 days of age. It also shows that old flies ($\geq$ 40 days) rarely laid over 30 eggs/day. Cumulative parity patterns reveal that a small percentage of females that consistently laid a high number of eggs had laid over 750 in the first 10 days (1.2%) and over 1,500 by 20 days (0.9%). Nearly 40% of all females produced over 1,500 eggs by 60 days. The overall lifetime patterns of both daily and cumulative parity for the *D. melanogaster* reproductive data are shown in fig. 4.5.

Individual-Level Reproduction

Event history graphics

Data sets on the longevity and age-specific reproduction of individuals are often both large and detailed. For example, Partridge and Fowler (1992) monitored the 2-day reproductive rates in 430 individual *D. melanogaster* females throughout their average lifetime of 20–25 days. Their effort produced records of 430 individual life spans and $\approx$ 5,000 fertility records classified by individual and age. An event history chart is a graphical method for visualizing both cohort survival and individual-level reproduction (Carey et al. 1998b) that provides insights into age patterns of reproduction, particularly the inter- and intra-individual variation, that are lost in summary metrics such as mean and variance.

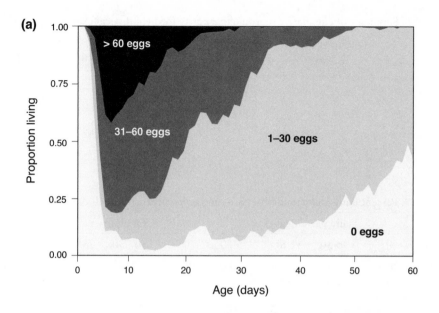

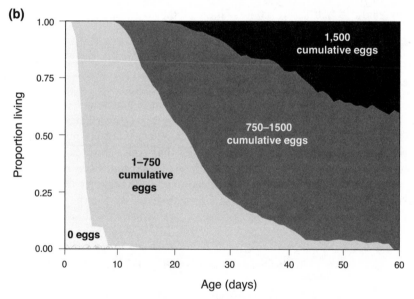

FIGURE 4.5. (a) Daily parity and (b) cumulative parity for *D. melanogaster*. (*Source*: R. Arking, unpublished)

An example graphic for reproduction in 1,000 individual Medfly females is presented in fig. 4.6. Since the average individual in the 1,000-female cohort lived around 35 days and laid approximately 760 eggs, each figure portrays ≈ 35,600 numbers representing the distribution of 760,000 eggs. Because the graphs are constructed from original rather than smoothed or curve-fitted numbers, they allow the data to "speak for themselves." This is important because abrupt changes in the level of reproduc-

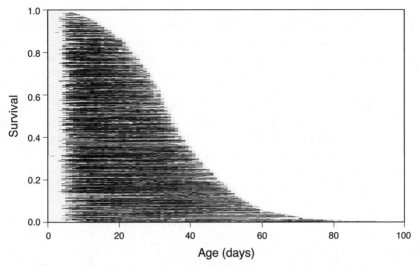

FIGURE 4.6. Age-specific cohort survival and lifetime reproduction for 1,000 individual Mediterranean fruit fly females (Carey et al. 1998b). Each horizontal line portrays the longevity of a single individual, and the color codes designate the level of reproduction for each age (light gray = zero eggs; medium gray = 1–35 eggs; black = > 35 eggs). See appendix I for methods for constructing this graphic.

tion between adjacent age classes (e.g., zero eggs laid at one age followed by 50 eggs laid at the next) appear in these graphs that would not appear in figures designed to show, for example, zones of high and low reproduction. Such subtle patterns and nuances can highlight both cohort and individual patterns.

Cohort versus individual patterns

A broad pattern that is immediately evident in fig. 4.6 is the sigmoidal shape of the cohort survival schedule, showing a gentle decrease from 0 to around 20 days during which the first 100 flies died, followed by a more rapid drop from days 20 through 50 during which around 800 flies died, and ending in a long tail at the oldest ages when the remaining 100 flies died. This pattern is a manifestation of an underlying mortality schedule in female Medflies that accelerates at young and middle ages and decelerates at older ages (see fig. 3.10). A number of other aspects of fig. 4.6 merit comment. First, there is a weak correlation between the age of first reproduction and life span, which is apparent from the absence of any distinct trends between the left-most band depicting pre-reproductive ages versus the life spans depicted by the survival schedule. For these data, the correlation coefficient between the age of first reproduction and life span is only $r^2 = 0.259$. Second, although the figure shows that there are high egg production days throughout the cohort and across all ages, most reproduction is concentrated in a "reproductive window" spanning ages 5 to 25. This 20-day band of high egg production appears at nearly all longevity levels and thus implies that the correlation between early reproduction and longevity is also weak. Note also that high egg production days are noticeably absent from flies living beyond 60 days. Conversely, low levels of egg production are evident at several times during the life cycle, including during the first few days of oviposition and the last several days of life for each fly, during the oldest ages (see survival tail), and also throughout the cohort as shown in the more or less even scatter of low egg production days at all ages across individuals. This reflects the variation in daily egg laying by individual flies. Fig. 4.6 also highlights the distribution of days in which no eggs were laid. Of the 35,600 fly days in the cohort, a total of around 14,500, or 40%, were days in which no eggs were laid. These zero production days can be classified into three types: (1) the immature periods experienced by all flies prior to laying their first eggs (these accounted for roughly 50% of all zero egg days); (2) zero production days resulting from lack of egg laying in 64 sterile flies (these "infertile fly days" accounted for about 10% of the zero egg days in the cohort); and (3) those zero production days due to day-to-day variation in egg laying by fertile flies (these accounted for about 40% of the zero egg days and include post-reproductive periods experienced by many flies, occasional periods when individual flies did not lay for several days, and a clustering of zero egg days for the longest-lived flies in the tail). Finally, the figure shows that egg-laying patterns in the oldest flies (> 60 days) consisted of a mixture of many days of no egg laying and days of low egg production. This lack of reproductive activity may shed light on the period when mortality levels off at older ages and suggests that a reduction in mortality at these oldest ages may be partly due to a reduction in the mortality costs associated with reproductive activity rather than exclusively due to demographic selection (see chapter 3).

Importance of individual-level data

This analysis of the patterns of reproduction across individuals, and the graphical techniques that help visualize individual-level data, highlights several reasons why longitudinal data on individuals are preferred over data that are grouped or cross-sectional. As a study population ages, it becomes more selected owing to attrition. By 40 days the fly population used in this illustration (Carey et al. 1998a, 1998b) had less than half of the original cohort. The loss of individuals over time reveals that cohort means from measurements on young females are based on observations of flies that did not all survive to age 40, while data from older flies obviously represent a select subgroup of the population that did survive. Another reason why individual-level data are important is that they provide insight into the between-fly variation in egg laying and thus reveal compositional influences on the cohort average. For example, individual-level data show whether a decrease in cohort reproduction with age is due to an increase in the fraction of females that lay zero eggs or to an overall decrease in the level of egg laying by each individual. A third reason why individual-level data are important is that if periods of more intensive egg laying vary from fly to fly, this intra-individual variation can be wiped out in the process of averaging across individuals (see peak-alignment averaging in chapter 11, S39). For instance, the shape of a peak of egg laying in the averaged or cross-sectional egg-laying display may not resemble any of the peaks observed in an individual's egg-laying behavior. Finally, reproductive data on individuals are important because they allow between-fly comparisons to be made on lifetime levels of reproduction and, in turn, on the long-term trajectories of reproduction in each individual over a specified period. In particular, they provide important insights into the reproductive age patterns of flies by comparing high versus low lifetime reproductive rates, early versus late ages of first reproduction, or short versus long lifetimes.

Models for individual-level reproduction

Three models have been developed to characterize and describe the stages and schedules of reproduction.

Model 1: Three adult stages

The female adult phase of the life cycle can be divided into three stages: maturation, maturity, and senescence. The first stage is a transient period of achieving a steady state at maturity that can then be maintained until the senescent stage. Novoseltsev and his colleagues (2004) developed a model that incorporated these stages into an example of female Medflies, and they noted that model fitting at the individual fly level allowed for the efficient characterization of individual fecundity with only five numbers for each fly. These numbers include age of first reproduction (X_{onset}), age of senescence (X_{sen}), life span (LS), the rate of daily egg laying in the reproductive window (RC), and the exponential rate of senescence (α_{sen}). These numbers can be used to compute three additional individual-level metrics, including the length of the reproductive window (T, where $T = X_{onset} - X_{sen}$), the egg-laying rate during the senescence phase (F, where $F = RC*\exp(x\alpha_{sen})$), and the senescence phase duration (S, where $S = (LS - T - X_{onset})$). A schematic of this model is given in fig. 4.7, and a comparison of individual-level metrics generated using this model for comparing egg laying and survival in two strains of Medflies is given in table 4.10.

Model 2: Reproductive clock

The trajectory of age-specific change in the rate of reproduction, and its association with longevity, can be compared across individuals. In an analysis of Medfly data,

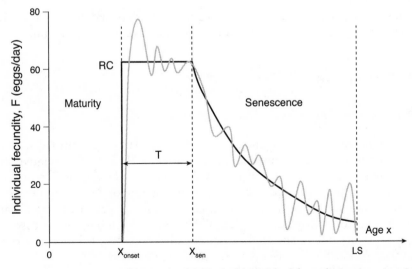

FIGURE 4.7. Three-stage reproduction model. Typical individual fecundity pattern in a Medfly female. The daily fecundity is shown in the gray line and the general pattern according to the three-stage adult model by the thicker, black line. (*Source*: Novoseltsev et al. 2004)

Table 4.10. Parameter values of mean population fecundity patterns for Greek and Israeli Medfly strains

Medfly strain	Eggs/day RC	Days T	Days X_{onset}	Life span LS	Total eggs RS
Greek	9.7	25.5	10	134	334.2
Israeli	40.0	7.3	3	60	737.7

Source: Table 1 in Novoseltsev et al. (2004).

Müller and his colleagues (2001) developed a model of individual egg laying based on the rate of decrease in egg laying between ages 10 and 25 days. They discovered that individual egg-laying trajectories rose sharply after egg laying began 5–17 days after emergence, reached a peak, and then slowly declined. The rate of decline varied between individuals, but this rate was approximately constant for each individual. They modeled this age trajectory of reproductive decline for each according to the exponential function

$$f(x) = \beta_0 \exp\left(-\beta_1 (x - \Theta)\right) \tag{4.8}$$

where $f(x)$ is the fecundity (measured by daily egg count) of the fly at age x days and Θ is the age at peak egg laying. The two parameters β_0, the peak height of the trajectory, and β_1, the rate of decline, varied considerably from fly to fly. A modest but significant negative correlation between β_0 and β_1 indicated that fecundity tends to decline more slowly for flies with higher peak fecundity. They found that the protracted decline in egg laying after the initial sharp rise was reasonably well predicted by this exponential model.

A consequence of these simple egg-laying dynamics was that, for any age x, it was possible to predict the fraction of remaining eggs relative to the total number of eggs from

$$\pi(x) = \int_x^\infty f(s)\,ds \Big/ \int_\Theta^\infty f(s)\,ds = \beta_0 \exp\left(-\beta_1(x - \Theta)\right) \tag{4.9}$$

This function, with values declining from 1 to 0 as the fly ages, provided a simple measure of reproductive exhaustion at age x in terms of remaining (relative) reproductive potential and could thus be loosely described as an individual's reproductive clock, which advances at a speed determined by the rate of decline. In their experiment the likelihood that a fly died increased as the fly's reproductive potential was exhausted and the reproductive clock advanced. This finding, of an association between mortality and exhaustion of reproductive potential, led to a new perspective on the relationship between reproduction and longevity.

The association between remaining reproductive potential and longevity was further demonstrated through their analysis of the fecundity trajectories. Fecundity and mortality are strongly correlated with age—as fecundity decreases with older age, mortality increases; however, in order to prevent confounding effects from this association, they fit the trajectories by only using data prior to day 25, whereas longevity was measured as remaining lifetime after day 25. Thus the fitted trajectories of fecundity were predicted after age 25 days based on the above model. This guaranteed that the fitted trajectories were not influenced by a fly's life span and allowed bona fide predictions of subsequent mortality.

The schematic in fig. 4.8a shows the relationship between the slope of egg-laying decline after age 10 days for individual female flies and their predicted longevity. On average, the flies whose reproductive output decreased slowly after day 10 lived longest and flies whose egg-laying rate decreased rapidly after day 10 lived the shortest. Examples of the relationship of egg laying and longevity is shown in fig. 4.8b for short-, medium-, and long-lived flies.

Model 3: Working and retired flies

In a third model that connects reproduction and future life expectancy, Curtsinger (2015, 2016) divided the life of adult *Drosophila melanogaster* into two functional stages: *working* and *retired*. The working stage was characterized by relatively high levels of oviposition and survival, and the retired stage by low levels of oviposition and reduced survival. This model showed that the retired stage in the Drosophila typically lasts one-quarter of the total adult life span and the age of transition varies between flies. As a result, cohorts of same-aged flies contain mixtures of working and retired flies. Examples of the survival differences between working and retired flies are shown in fig. 4.9, where stage-dependent conditional survivorship is shown for 15-day-old flies. Retired flies had significantly lower subsequent survival than working flies of the same age. For example, flies that entered the "retired" stage at age 15 were 19 times more likely to die by day 20 than working flies of the same age.

Parity Progression

Parity progression ratios (PPR), known more generally in statistics as continuation ratios, are the proportion of females with a certain number of offspring who go on to have additional offspring. That is,

$$PPR_i = \frac{\text{females with at least } i+1 \text{ offspring}}{\text{females with at least } i \text{ offspring}} \tag{4.10}$$

where PPR_i denotes the ratio at parity i. For example, if 100 females have produced 10 offspring and 85 females have produced more than 10 offspring, then

$$PPR_{10} = \frac{85}{100}$$

$$= 0.85$$

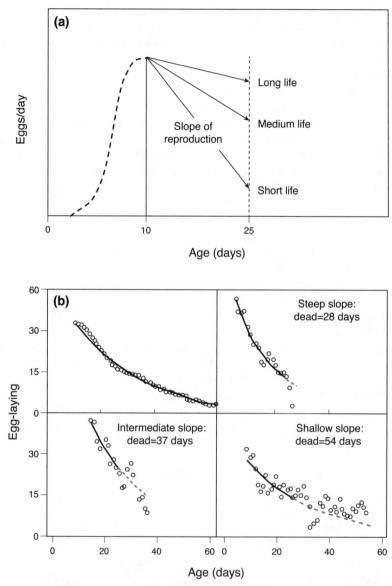

FIGURE 4.8. (a) Schematic to visualize the concept that the trajectory (slope) of egg laying for individual flies between the ages of 10 and 26 days predicts the life span of the fly. (b) The trajectories of fecundity are fitted to data from the peak to day 25 by nonlinear least squares and predicted thereafter (dashed line). (*Source*: Müller et al. 2001)

In other words, the probability that a female who has produced 10 offspring will go on to produce more offspring is 0.85. This probability is the conceptual equivalence of period survival in the life table, p_x (see chapter 2). This means that the number of offspring (or offspring classes) can be used as the vertical dimension of the life table (i.e., replacing age x), and analogs of all life table parameters can be computed from the schedule of PPR_i over all values of i. These include

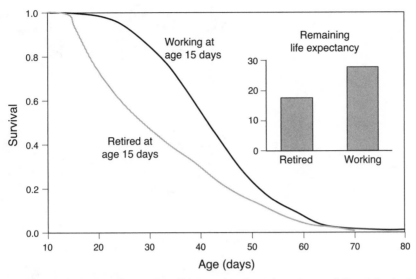

FIGURE 4.9. Survival curves for 15-day-old retired and working Drosophila adults (redrawn from Fig. 4a in Curtsinger 2015).

$_nd_i$ = proportion of individuals that die in the egg-class interval i to i+n

l_i = proportion of individuals that survive to exact egg class i

$_np_i$ = probability of advancing to egg class i+n given that in egg class i

$_nL_i$ = number of female days lived in the egg class interval i to i+n

T_i = number of female days lived beyond egg class i

e_i = expectation of egg class interval attained before death

An example application of parity progression methods is given in table 4.11 for the *D. melanogaster.*

Fertility Models

Age-specific fecundity curves require three general properties in order to be useful in demography: the potential to assume an asymmetrical pattern, the ease with which fecundity level is adjusted, and the insertion of a pre-reproductive period or zero fecundity days. Here we introduce three models that possess these three properties and thus are commonly used to fit reproductive data in humans. However, they are equally useful for modeling reproductive patterns in nonhuman species. (For reproductive models applied to mammalian species, see Gage 2001.)

Table 4.11. Parity progression ratios for *D. melanogaster*

Parity i	Number dying at parity i D_i	Number surviving to parity i N_i	Fraction surviving from parity 0 to parity i+1 l_i	Fraction that survived to parity i that survive to parity i+1 p_i	Fraction that survived to parity i that die before parity i+1 q_i	Fraction of cohort dying in parity class interval i to i+1 d_i	Expectation of future offspring production at parity i e_i
(1)	(2)	(3)	(4)	(5)	(6)	(7)	(8)
0	4	666	1.0000	0.9940	0.0060	0.0060	1132.6
100	8	662	0.9940	0.9879	0.0121	0.0120	1039.1
200	15	654	0.9820	0.9771	0.0229	0.0225	951.2
300	19	639	0.9595	0.9703	0.0297	0.0285	872.4
400	28	620	0.9309	0.9548	0.0452	0.0420	797.6
500	40	592	0.8889	0.9324	0.0676	0.0601	732.9
600	34	552	0.8288	0.9384	0.0616	0.0511	682.4
700	30	518	0.7778	0.9421	0.0579	0.0450	623.9
800	47	488	0.7327	0.9037	0.0963	0.0706	559.2
900	45	441	0.6622	0.8980	0.1020	0.0676	513.5
1000	42	396	0.5946	0.8939	0.1061	0.0631	466.2

1100	58	354	0.5315	0.8362	0.1638	0.0871	415.5
1200	49	296	0.4444	0.8345	0.1655	0.0736	387.2
1300	52	247	0.3709	0.7895	0.2105	0.0781	354.0
1400	37	195	0.2928	0.8103	0.1897	0.0556	335.1
1500	34	158	0.2372	0.7848	0.2152	0.0511	301.9
1600	30	124	0.1862	0.7581	0.2419	0.0450	271.0
1700	27	94	0.1411	0.7128	0.2872	0.0405	241.5
1800	24	67	0.1006	0.6418	0.3582	0.0360	218.7
1900	15	43	0.0646	0.6512	0.3488	0.0225	212.8
2000	9	28	0.0420	0.6786	0.3214	0.0135	200.0
2100	7	19	0.0285	0.6316	0.3684	0.0105	171.1
2200	6	12	0.0180	0.5000	0.5000	0.0090	141.7
2300	4	6	0.0090	0.3333	0.6667	0.0060	133.3
2400	0	2	0.0030	1.0000	0.0000	0.0000	200.0
2500	1	2	0.0030	0.5000	0.5000	0.0015	100.0
2600	1	1	0.0015	0.0000	1.0000	0.0015	50.0
	0	0	0.0000	0.0000		0.0000	

Source: Unpublished data from Robert Arking, by permission.

The first model is the Pearson type I curve, with origin at the mode given as

$$f(x) = g\left(1 + \frac{x}{a_1}\right)^{m_1}\left(1 - \frac{x}{a_2}\right)^{m_2} \tag{4.11}$$

The fitting procedure consists of approximating the parameters a_1 and a_2 by visual inspection of the fecundity curve to which the equation is to be fit and then using trial and error to obtain the best visual fit.

The second model is the Brass fertility polynomial, given as

$$f(x) = c(x - s)(s + g - x)^2 \tag{4.12}$$

where x denotes age, s denotes the age when fertility begins, $s + g$ is the age when fertility ends, and c is a scalar.

The third model is the Coale-Trussell model (Coale and Trussell 1974), which is essentially a relational method where the level and pattern of a species-specific "standard" age schedule of natural fertility, n_x, is adjusted using a scalar, M, for the former and an exponential, s, for the latter. Fertility at age x, F_x, is given as

$$F_x = n_x M e^{sx} \tag{4.13}$$

Here the parameter s in the exponential term is a single value used to adjust the pattern. However, in their original paper the exponential component was the product of two values, one that measured age-specific fertility control of a population and the other that adjusted the deviation of realized fertility from natural fertility, both of which were specific to particular (human) populations. These three age-specific fecundity models are given in fig. 4.10 for data on *Drosophila melanogaster*.

Consider Further

Content on or related to reproduction in other sections of this book include material in the sections on both human reproduction and family demography in chapter 8 (Human Demography), the section on constructing event history diagrams in appendix I, and entries in both family demography (S49–S51) and contraception (S54) in chapter 11 (Biodemography Shorts). Because the age pattern of reproduction involves peaks, the entry in chapter 11 on peak-aligned averaging (S39) is also relevant to reproduction even though the example involves peaks in mortality curves.

A recommended starting point for the literature on more advanced concepts and methods involving the biodemography of reproduction is the classic book *Fertility, Biology and Behavior* (Bongaarts and Potter 1983). Chapters 4 and 6 in the book *Essential Demographic Methods* (Wachter 2014) contains exceptionally clear explanations of cohort fertility and period fertility, respectively. Additionally, a novel

(a)

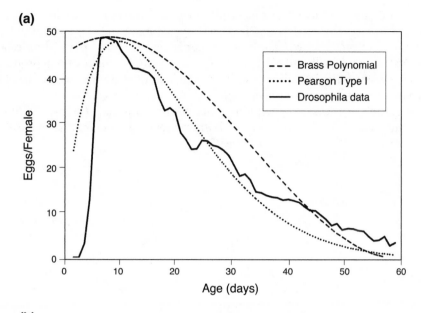

(b)

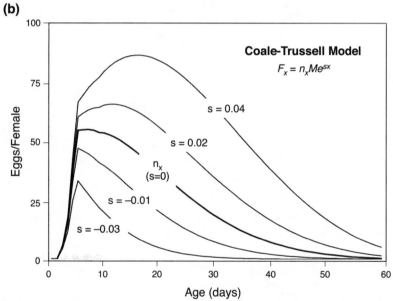

FIGURE 4.10. Models fitted to Drosophila reproductive data: (a) Brass polynomial and Pearson type I; (b) Coale-Trussell model. (*Source*: Data from Robert Arking, unpublished)

method (i.e., implied total fertility rate) for estimating total fertility rate using age pyramids is described by Hauer et al. (2013). Reproductive rates in ecology and population biology are typically contained in the life history literature (e.g., Roff 1992, 2002; Stearns 2002), such as in the overview by Sibley (2002) and the entry on humans by Kaplan (2002) in the *Encyclopedia of Evolution*. The deeper literature on the demography of reproduction in humans includes the book by Wood (1994) on the biology and demography of reproduction as well as articles or encyclopedia entries on reproductivity (Dharmalingam 2004), fertility (Morgan and Hagewen 2005), indirect estimates from child-woman ratio (Hauer et al. 2013), and population replacement (Vallin and Caselli 2006). The most comprehensive sources for reproduction across the Tree of Life include entries in the original four-volume series of the *Encyclopedia of Reproduction* (Knobil and Neil 1998) and more recently in the six-volume second edition of this series, including volumes on both male (Jegou and Skinner 2018) and female (Spencer and Flaws 2018) reproduction in humans as well as non-human species.

Population I: Basic Models

> It seems therefore that there must be a limiting "stable" type about
> which the actual distribution varies, and toward which it tends to
> return if through any agency disturbed therefrom.
>
> F. R. Sharpe and A. J. Lotka (1911, 435)

Foundational Concepts

Population rates

Population models are concerned with rate of change in population numbers. If the
number in a population is 100 at the beginning of a week and 120 at the end of the
week, then its rate of change would be the arithmetic rate of 20 per week (Keyfitz
1985). To convert this to a demographic (geometric) rate, this population is growing
at the geometric rate of 20/100, which means that every individual is increasing its
number by 20%, or 1.2-fold per week.

The arithmetic rate, denoted r^*, is given by the model

$$r^* = N_{t+1} - N_t \tag{5.1}$$

or

$$N_{t+1} = N_t + r^* \tag{5.2}$$

The demographic (geometric) rate, denoted r, is given by the model

$$r = \frac{N_{t+1} - N_t}{N_t} \tag{5.3}$$

or

$$N_{t+1} = N_t(1 + r) \tag{5.4}$$

Projections of these rates yield different results. Arithmetically, a population of 1,000 individuals growing at 250 individuals per year will have grown by 500 at the end of 2 years and by 750 at the end of 3 years. Geometrically, a population growing at 25% per year will have grown by 562 in 2 years and 953 in 3 years. Note that the arithmetic series has a common difference between numbers; for example, 1, 3, 5, 7, and 9 is an arithmetic series with a common *difference* of 2. And the geometric series has a common ratio; for example, the series 1, 3, 9, 27, and 81 is a geometric series with a common *ratio* of 3. Population growth rate involves geometric growth in discrete time and exponential growth in continuous time.

The balancing equation

The simplest population model is the *crude rate model* that relates the total population this year to the total population last year. This is based on the balancing equation, where if we let last year represent the initial population at time 0, denoted Pop_0, then the balancing equation is given as

$$Pop_1 = Pop_0 + births - deaths + in\text{-}migrants - out\text{-}migrants. \qquad (5.5)$$

Assuming that the population is closed to migration and therefore excluding in- and out-migration, we can define the following:

$$births = Pop_0(b) \qquad deaths = Pop_0(d)$$

where b and d denote per capita birth and death rates, respectively. Substituting these terms into eqn. (5.5) (excluding migration terms) yields

$$Pop_1 = Pop_0 + Pop_0 b - Pop_0 d \qquad (5.6)$$

$$= Pop_0(1 + b - d) \qquad (5.7)$$

Note that if $b - d = 0$ the population at time t equals the population at time $t + 1$, if $b > d$ the population is increasing, and if $b < d$ the population is decreasing. The population at time $t = 2$ (i.e., Pop_2) is determined as follows:

$$Pop_2 = Pop_1(1 + b - d) \qquad (5.8)$$

$$Pop_2 = Pop_0(1 + b - d)(1 + b - d) \qquad (5.9)$$

$$Pop_2 = Pop_0(1 + b - d)^2 \qquad (5.10)$$

The relationship for t number of time units is

$$Pop_t = Pop_0 (1 + b - d)^t \qquad (5.11)$$

Substituting N_t for Pop_t and λ for $(1 + b - d)$ yields

$$N_t = N_0 \lambda^t \qquad (5.12)$$

This expression gives the discrete (geometric) version of population growth. This model can be rephrased in terms of exponentials by taking logs of both sides and defining $r = \ln (e^\lambda)$:

$$\ln N_t = t \ln \lambda + t \ln N_0 \qquad (5.13)$$

$$= rt + \ln N_0 \qquad (5.14)$$

Thus

$$N_t = e^{rt} N_0 \qquad (5.15)$$

In continuous notation this becomes

$$N(t) = e^{rt} N(0) \qquad (5.16)$$

Note that the numerical relationship between geometric and exponential growth has two parts. First, at low r-values, $e^r = 1 + r$. Second, geometric and exponential growth are interrelated as

$$N(t) = e^{rt} \text{ (exponential growth)}$$

$$\ln N(t) = rt$$

$$N_t = \lambda^t \text{ (geometric growth)}$$

$$\ln N_t = t \ln \lambda$$

Thus if we set $\ln N(t) = \ln N_t$, then $rt = t \ln \lambda$ and therefore $r = \ln \lambda$ and $\lambda = e^r$. In other words, the geometric rate of population increase equals the exponent e raised to the rth power.

There are a number of assumptions of the crude rate model, including that (1) the population is homogeneous and has no age structure, (2) birth and death rates are fixed, and (3) the population is closed (no migration). Whereas some of these assumptions may be limiting, this model has provided a foundation for the development of more complicated models, as we describe below.

Doubling time and half-life

The expression for the geometric increase of population size to time t is

$$N_t = N_0(1+r)^t \qquad (5.17)$$

where r is the fraction of increase per unit time and N_0 is the initial number in the population. The time it will take the population to double (DT) in size [i.e., when $2 = N_t/N_0$] is the solution to the equation

$$(1+r)^t = 2 \qquad (5.18)$$

$$t = DT = \frac{\ln 2}{\ln(1+r)} \qquad (5.19)$$

Since $\ln(1+r)$ is approximately equal to r except at high values of r, this equation further reduces to

$$DT = \frac{\ln 2}{r} \qquad (5.20)$$

More generally, a population will increase by n-fold according to the equation

$$\text{Time to increase by n-fold} = \frac{\ln n}{r} \qquad (5.21)$$

For example, if $r = 0.1$, then n-values of 3, 10, or 100 will yield increases by 3-fold in 11 days, by 10-fold in 23 days, and by 100-fold in 46 days.

Population growth: Subpopulations and sequence

Subpopulations

There are two ways to compute the growth of the total population at time t (denoted N): one sums the growth of independent subpopulations and the other computes growth rate as the average rate across subpopulations. Consider two examples where population A is increasing by $\lambda_A = 1.1$ and population B is increasing by $\lambda_B = 1.4$.

CASE I: INDEPENDENT POPULATIONS

In this case each subpopulation grows independently of the other, and the total population is simply the sum total of the numbers in the two different populations.

$$N_t = N_{0,A}(\lambda_A)^t + N_{0,B}(\lambda_B)^t \tag{5.22}$$

$N_{0,A}$ and $N_{0,B}$ here denote the numbers in the respective populations at time 0.

Suppose each population begins with a single individual. Projection to time $t = 30$ for this situation is

$$
\begin{aligned}
N_t &= N_{0,A}(\lambda_A)^{30} + N_{0,B}(\lambda_B)^{30} \\[4pt]
&= (1.1)^{30} + (1.4)^{30} \\[4pt]
&= 17.5 + 24{,}201.4 \\[4pt]
&= 24{,}218.9
\end{aligned}
\tag{5.23}
$$

CASE II: AVERAGE GROWTH OF TWO POPULATIONS

The second way to compute growth rate of the total population at time t is to compute the average of the respective growth rates $\lambda_A = 1.1$ and $\lambda_B = 1.4$. That is,

$$
\begin{aligned}
\lambda &= \sqrt{(\lambda_A \lambda_B)} \\[4pt]
\lambda &= \sqrt{1.1 \times 1.4} \\[4pt]
\lambda &= 1.24
\end{aligned}
\tag{5.24}
$$

The projection for this second case is

$$
\begin{aligned}
N_t &= N_0 \lambda \\[4pt]
&= 1.24^{30} \\[4pt]
&= 634.8
\end{aligned}
\tag{5.25}
$$

The nearly 40-fold difference between the results of the two projections in these two cases shows that separate projections of subpopulations always yield a greater total population growth than projecting the average of the growth rates from each subpopulation (Keyfitz 1985).

Sequence of Growth Rates

Population growth can vary through time, and the temporal sequence of growth can be computed. For example, if the finite rates of increase of a population are 1.1, 1.8, 1.2, and 1.5 at times 0, 1, 2, and 3, respectively, and denoted λ_0, λ_1, λ_2, and λ_3, then starting at $t = 0$ with $N_0 = 25$, the number of individuals at $t = 4$ will be

$$N_4 = N_0 \lambda_0 \lambda_1 \lambda_2 \lambda_3$$

$$= 25 \times 1.1 \times 1.8 \times 1.2 \times 1.5 = 89.1$$

(5.26)

Note that the geometric mean of the four finite rates is

$$\sqrt[4]{\lambda_0 \lambda_1 \lambda_2 \lambda_3}$$

$$= 1.374$$

(5.27)

and that

$$N_4 = 25 \times 1.374^4$$

$$= 89.1$$

(5.28)

This is equivalent to the outcome for the projection of N_4 using the separate finite growth rates. This numerical example illustrates a general principle about population growth that varies through time, and that is that the resulting rate of population growth for any temporal sequence is the geometric mean. This relationship shows that the numerical effect of a varying rate of growth on a population total is the same as the arithmetic average rate when it is applied at each moment over the time in question (Keyfitz 1985).

The Stable Population Model

Background

One of the largest singular achievements in demography was the linkage of stable theory to real populations by Lotka (1907, 1928). Lotka originated the terms and the concepts of a *stable age distribution* (SAD) and the *intrinsic rate of population growth* (r) and noted that a population growing with fixed birth and death rates will eventually evolve to the state where growth rate is constant and the fraction of the total population in each age class is fixed. The equation Lotka derived was actually a convolution of the integral equation that the French mathematician Euler had derived 150 years earlier, which had not previously been applied in a population context. This integral equation is now known in demography as the Euler equation, the Lotka equation, or the characteristic equation, and it serves as the foundation of virtually all renewal theory in biology.

Biologists recognized the importance of birth and death rates for understanding the broad properties of biological populations long before demography was introduced to biology and ecology, but the fundamental properties of populations were not understood because there was no model with which to examine the consequences of changes in birth and death rates. The stable population model derived by Lotka (1907) thus provided both a conceptual and an analytical tool for demographic analysis of populations. His model provides the foundation upon which virtually all population models are built.

Uses and assumptions of the model

The objective of the stable model is to trace the dynamic characteristics of a population that starts off with an arbitrary age structure and is submitted from that moment on to a specified demographic regime (Lopez 1961). The most important conclusions of the theory are twofold: first, that the age distribution of a population is completely determined by the history of fertility and mortality rates; and second, that the particular schedule of birth and death rates sets forces in motion that make the age structure and the rate of population growth point toward an inherent steady state independent of initial conditions.

The assumptions of the stable model are basically the same as those of the crude rate model, except that age structure is incorporated into a stable model. Like the crude rate model, the stable model assumes fixed birth and death rates, it is a closed population (no migration), and it has only one sex. These assumptions highlight some of the limitations of the applicability of the theory to natural populations. Specifically, animals often migrate, sex ratios other than unity are commonplace, and the birth and death rates are virtually never fixed. Coale (1972) saw stable theory as providing a population gauging tool. When used properly, it provides a simple starting point for addressing fundamental questions about the process of population renewal, and it identifies the two most important population parameters, age distribution and growth rate, their interdependence, and their relationship with the cohort parameters of birth and death. Additionally, the stable model suggests a direction for population growth, the sign of growth rate, and the general skew of the age distribution. Finally, it can be used to project actual numbers that may be useful as frames of reference and for asking "what if" questions.

Derivation

Consider the simplest age-structured population model, consisting of three age classes:

$$N_{0,t+1} = m_1 N_{1,t} + m_2 N_{2,t} \tag{5.29}$$

$$N_{1,t+1} = p_0 N_{0,t} \tag{5.30}$$

$$N_{2,t+1} = p_1 N_{1,t} \tag{5.31}$$

where $N_{x,t}$ and $N_{x,t+1}$ denote the number of individuals in the population who are age x at times t and t + 1, m_x denotes the number of offspring produced by a female age x ($m_0 = 0$), and p_x is the probability of surviving from age x to x + 1.

The objective is to find the rate at which this population will increase each time step, that is, the finite rate of increase, λ. Since the rate at which the population increases also applies to the rate at which the number in each age class increases, we set the ratio of the number in each age class at time t + 1 to the number in each age class at time t. That is,

$$\frac{N_{0,t+1}}{N_{0,t}} = \frac{N_{1,t+1}}{N_{1,t}} = \frac{N_{2,t+1}}{N_{2,t}} = \lambda \qquad (5.32)$$

This expression can be given as

$$N_{0,t+1} = \lambda N_{0,t} \qquad (5.33)$$

$$N_{1,t+1} = \lambda N_{1,t} \qquad (5.34)$$

$$N_{2,t+1} = \lambda N_{2,t} \qquad (5.35)$$

These equations state that the three age classes 0, 1, and 2 increase each time step by a factor of λ.

Substituting the right-hand sides of equations (5.32), (5.33), and (5.34) for $N_{0,t+1}$ and $N_{1,t+1}$ yields

$$\lambda N_{0,t} = m_1 N_{1,t} + m_2 N_{2,t} \qquad (5.36)$$

$$\lambda N_{1,t} = p_0 N_{0,t} \qquad (5.37)$$

$$\lambda N_{2,t} = p_1 N_{1,t} \qquad (5.38)$$

Since $N_{0,t}$ can be defined in terms of $N_{1,t}$, λ, and p_0 in eqn. (5.39) as

$$N_{0,t} = \frac{\lambda N_{1,t}}{p_0} \qquad (5.39)$$

the right-hand side can be substituted into eqn. (5.38) to obtain an expression containing only $N_{1,t}$:

$$\lambda \left(\frac{\lambda N_{1,t}}{p_0} \right) = m_1 p_0 \left(\frac{\lambda N_{1,t}}{p_0} \right) + p_1 m_2 N_{2,t} \qquad (5.40)$$

Rearranging and dividing by λ yields

$$0 = N_{1,t} \left(-1 + l_1 m_1 \lambda^{-1} + l_2 m_2 \lambda^{-2}\right) \tag{5.41}$$

or

$$1 = l_1 m_1 \lambda^{-1} + l_2 m_2 \lambda^{-2} \tag{5.42}$$

The terms inside the parentheses represent the Lotka equation for three age classes:

$$1 = \Sigma_0^2 \lambda^{-x} l_x m_x \tag{5.43}$$

By setting $\lambda = e^r$, where r denotes the intrinsic rate of increase, then

$$1 = \Sigma_\alpha^\beta e^{-rx} l_x m_x \tag{5.44}$$

where α and β are the first and last ages of reproduction, respectively. This is the discrete version of the Lotka equation.

Population parameters

Intrinsic rates: Birth (b), death (d), and growth (r)

First introduced by Dublin and Lotka (1925), the intrinsic rate of increase is the rate of natural increase in a closed population that has been subject to constant age-specific schedules of fertility and mortality for a long period and that has converged to be a stable population. The intrinsic rate of increase is a special case of the crude growth rate; that is, the ratio of the total number of individuals in a population at two different times yields its growth rate. The population is stable if this ratio does not change over time.

The exact value of r can be determined from data on survival and reproduction using the Newton method, a numerical method for finding successively better approximations of a function, based on the formula $r_1 = r_0 - \dfrac{f(r)}{f'(r)}$, where r_0 is the original estimate of r, and r_1 is the corrected estimate. The expression $f(r)$ is the original Lotka equation

$$f(r) = \left[\left(\Sigma e^{-rx} l_x m_x\right) - 1\right] \tag{5.45}$$

and $f'(r)$ is the derivative of this function as

$$f'(r) = \left[\left(\sum xe^{-rx}l_xm_x\right)\right] \tag{5.46}$$

The steps for computing r include the following (see table 5.1):

Step 1. Enter the basic data, including ages (x), in col. 1, survival in col. 2, and age-specific number of offspring per female in col. 3.

Step 2. Compute net maternity, l_xm_x (col. 4).

Step 3. Estimate an r-value as a starting point for progressively closer approximations. For the example presented in table 5.1, $r_0=0.30$ is the approximation.

Step 4. Compute for all ages $e^{-rx}l_xm_x$ (col. 5) and $xe^{-rx}l_xm_x$ (col. 6).

Step 5a. Determine the first analytical approximations for r, denoted r_1, using the equation

$$r_1 = r_0 - \left[\frac{(\text{sum col. 6}) - 1.0}{(\text{sum col. 7})}\right] \tag{5.47}$$

$$r_1 = 0.30 + \left[\frac{2.1574 - 1.0}{-35.2050}\right] \tag{5.48}$$

$$r_1 = 0.3329 \tag{5.49}$$

Step 5b. Determine the second analytical approximations for r, denoted r_2, using the equation

$$r_2 = r_1 - \left[\frac{(\text{sum col. 8}) - 1.0}{(\text{sum col. 9})}\right] \tag{5.50}$$

$$r_2 = 0.3329 + \left[\frac{1.2677 - 1.0}{-20.32366}\right] \tag{5.51}$$

$$r_2 = 0.3460 \tag{5.52}$$

Step 5c. Determine the third analytical approximations for r, denoted r_3, using the equation

$$r_3 = r_2 - \left[\frac{(\text{sum col. 10}) - 1.0}{(\text{sum col. 11})}\right] \tag{5.53}$$

$$r_3 = 0.3460 - \left[\frac{1.0271 - 1.0}{-16.3629}\right] \tag{5.54}$$

$$r_3 = 0.3476 \tag{5.55}$$

Table 5.1. Computation of intrinsic rate of increase, r, for *D. melanogaster* ($r = 0.3477$), where $f(r) = [(\sum e^{-rx} l_x m_x) - 1]$ and $f'(r) = [(\sum x e^{-rx} l_x m_x)]$

Age	Stage	Survival	Fecundity	Net fecundity	First iteration, r = 0.3000		Second iteration, r = 0.3329		Third iteration, r = 0.3460	
x		l_x	m_x	$l_x m_x$	f_x	f'_x	f_x	f'_x	f_x	f'_x
(1)	(2)	(3)	(4)	(5)	(6)	(7)	(8)	(9)	(10)	(11)
0	Egg	1.0000	0.0	0.0	0.0000	0.0000	0.0000	0.0000	0.0000	0.0000
1	Larva	0.9900	0.0	0.0	0.0000	0.0000	0.0000	0.0000	0.0000	0.0000
2		0.9405	0.0	0.0	0.0000	0.0000	0.0000	0.0000	0.0000	0.0000
3		0.8935	0.0	0.0	0.0000	0.0000	0.0000	0.0000	0.0000	0.0000
4		0.8488	0.0	0.0	0.0000	0.0000	0.0000	0.0000	0.0000	0.0000
5	Pupa	0.8064	0.0	0.0	0.0000	0.0000	0.0000	0.0000	0.0000	0.0000
6		0.7902	0.0	0.0	0.0000	0.0000	0.0000	0.0000	0.0000	0.0000
7		0.7744	0.0	0.0	0.0000	0.0000	0.0000	0.0000	0.0000	0.0000
8		0.7589	0.0	0.0	0.0000	0.0000	0.0000	0.0000	0.0000	0.0000
9		0.7438	0.0	0.0	0.0000	0.0000	0.0000	0.0000	0.0000	0.0000
10	Adult	0.7289	0.0	0.0	0.0000	0.0000	0.0000	0.0000	0.0000	0.0000
11		0.7289	0.0	0.0	0.0000	0.0000	0.0000	0.0000	0.0000	0.0000
12		0.7289	3.2	2.3	0.0638	-0.7654	0.0430	-0.5159	0.0367	-0.4405
13		0.7289	13.5	9.8	0.1985	-2.5804	0.1295	-1.6830	0.1091	-1.4182
14		0.7289	34.9	25.4	0.3814	-5.3394	0.2407	-3.3699	0.2002	-2.8024
15		0.7289	51.2	37.3	0.4148	-6.2213	0.2533	-3.7995	0.2079	-3.1184
50		0.3130	13.6	4.3	0.0000	-0.0001	0.0000	0.0000	0.0000	0.0000
51		0.2933	13.3	3.9	0.0000	0.0000	0.0000	0.0000	0.0000	0.0000
52		0.2802	12.8	3.6	0.0000	0.0000	0.0000	0.0000	0.0000	0.0000
53		0.2638	11.8	3.1	0.0000	0.0000	0.0000	0.0000	0.0000	0.0000
54		0.2484	11.0	2.7	0.0000	0.0000	0.0000	0.0000	0.0000	0.0000
55		0.2331	10.9	2.5	0.0000	0.0000	0.0000	0.0000	0.0000	0.0000
56		0.2156	9.9	2.1	0.0000	0.0000	0.0000	0.0000	0.0000	0.0000
57		0.2058	8.9	1.8	0.0000	0.0000	0.0000	0.0000	0.0000	0.0000
58		0.1839	7.5	1.4	0.0000	0.0000	0.0000	0.0000	0.0000	0.0000

(*continued*)

Table 5.1. (*continued*)

Age	Stage	Survival	Fecundity	Net fecundity	First iteration, r = 0.3000		Second iteration, r = 0.3329		Third iteration, r = 0.3460	
x		l_x	m_x	$l_x m_x$	f_x	f'_x	f_x	f'_x	f_x	f'_x
(1)	(2)	(3)	(4)	(5)	(6)	(7)	(8)	(9)	(10)	(11)
59		0.1762	7.9	1.4	0.0000	0.0000	0.0000	0.0000	0.0000	0.0000
60		0.1631	6.4	1.0	0.0000	0.0000	0.0000	0.0000	0.0000	0.0000
80		0.0175	1.8	0.0	0.0000	0.0000	0.0000	0.0000	0.0000	0.0000
81		0.0142	2.1	0.0	0.0000	0.0000	0.0000	0.0000	0.0000	0.0000
82		0.0120	2.2	0.0	0.0000	0.0000	0.0000	0.0000	0.0000	0.0000
83		0.0098	0.5	0.0	0.0000	0.0000	0.0000	0.0000	0.0000	0.0000
84		0.0066	0.8	0.0	0.0000	0.0000	0.0000	0.0000	0.0000	0.0000
85		0.0055	1.3	0.0	0.0000	0.0000	0.0000	0.0000	0.0000	0.0000
86		0.0044	0.8	0.0	0.0000	0.0000	0.0000	0.0000	0.0000	0.0000
87		0.0039	0.0	0.0	0.0000	0.0000	0.0000	0.0000	0.0000	0.0000
88		0.0020	2.0	0.0	0.0000	0.0000	0.0000	0.0000	0.0000	0.0000
89		0.0010	0.0	0.0	0.0000	0.0000	0.0000	0.0000	0.0000	0.0000
90		0.0000	0.0	0.0	0.0000	0.0000	0.0000	0.0000	0.0000	0.0000
			TOTAL		2.1574	−35.2050	1.2677	−20.3266	1.0271	−16.3629

The value of $r_3 = 0.3476$ is within three decimal places of the exact value of $r = 0.3477$ since the adjustment at the third iteration was small.

Intrinsic rate of increase: Analytical approximations

The value of r in the Lotka equation can be approximated analytically in several ways, three of which are covered here. The first method sets the variable equal to the mean age of net fecundity in the cohort, denoted T:

$$1 = \sum e^{-rT} l_x m_x \tag{5.56}$$

where

$$T = \frac{\sum x l_x m_x}{\sum l_x m_x} \tag{5.57}$$

The denominator in this equation is the net reproductive rate (NRR), or R_0. The exponential in the first equation is a constant and therefore can be brought outside the summation as

$$e^{-rT} = R_0 \tag{5.58}$$

Therefore,

$$r = \frac{\ln R_0}{T} \tag{5.59}$$

A second method for approximating r in the Lotka equation involves two additional components, survivorship to age T (l_T) and the gross reproductive rate (GRR). In this derivation both sides of the Lotka equation are multiplied by e^{rT}:

$$e^{rT} = e^{rT} \sum e^{-rx} l_x m_x \tag{5.60}$$

$$e^{rT} = \sum e^{-r(x-T)} l_x m_x \tag{5.61}$$

Dividing both sides by l_T yields

$$e^{rT} = l_T \sum e^{-r(x-T)} \frac{l_x}{l_T} m_x \tag{5.62}$$

When x equals T,

$$\frac{l_x}{l_T} = 1.00 \qquad (5.63)$$

and therefore

$$e^{-r(x-T)} = e^{-r(0)} = 1.00 \qquad (5.64)$$

This equation reduces to

$$e^{rT} = l_T \sum m_x \qquad (5.65)$$

Given that the sum of m_x over all ages is the gross reproductive rate, GRR, then

$$e^{rT} = l_T GRR \qquad (5.66)$$

The new analytical estimation of r is then given as

$$r = \frac{\ln(l_T) + \ln(GRR)}{T} \qquad (5.67)$$

A third way of estimating r was shown by Preston and Guillot (1997), who noted that if survival, l_x, is linear in the childbearing ages, then NRR reduces to

$$R_0 = GRR \times p(A_M) \qquad (5.68)$$

where

$$GRR = \sum_0^\infty m_x \qquad (5.69)$$

$p(A_M)$ is the probability of surviving from birth to the mean age at childbearing, denoted A_M, and is computed as

$$A_M = \frac{\sum_0^\infty x m_x}{\sum_0^\infty m_x} \qquad (5.70)$$

If the proportion of births that are female is a constant, S, across all mothers' ages, then

$$R_0 = TFR \times S \times p(A_M) \qquad (5.71)$$

where TFR denotes total fertility rate. Substituting the right-hand side of eqn (5.71) into the expression for R_0 in eqn (5.58), taking logs of both sides, and rearranging yields

$$r = \frac{\ln(TFR) + \ln(S) + \ln[p(A_M)]}{T} \qquad (5.72)$$

This equation indicates that mortality and fertility levels influence r in ways that can be separated because their effects are additive rather than related in a more complex fashion. Here r is an additive function of the log of the total fertility rates rather than of the rates itself. In other words, the effect on r of fertility deepens only on the proportionate decline in the TFR and not on the absolute decline. Additionally, in this equation the changes in the mean generation time, T, have a direct and inversely proportional effect on r, where increasing T decreases r and vice versa (Preston and Guillot 1997).

Intrinsic birth and death rates

The intrinsic birth rate, b, is the per capita birth rate of a population that would be reached in a closed female population subject to fixed age-specific birth and death rates. This is also the per capita birth rate in a stable population. Its counterpart, the intrinsic death rate, d, is the per capita death rate of a population subject to the same conditions. The formulas for b and d are

$$b = \frac{1}{\sum e^{-rx} l_x} \qquad (5.73)$$

and

$$d = b - r \qquad (5.74)$$

Putting this together, a stable population will grow at the rate, r, with b births and d deaths for each individual in the population. The intrinsic birth and death rates can be used to express the population growth rate, given as $(b - d)$, the relative probability that a birth will occur relative to a death, given as the ratio (b/d), and the total per capita vital events, given as $(b + d)$, also known as *population metabolism* (after Ryder 1973).

Net reproductive rate

The net reproductive rate, denoted either R_0 or NRR, is the average number of female offspring that would be born to a birth cohort of females during their lifetime if the cohort experienced a fixed pattern of age-specific birth and death rates, given as

$$R_0 = \sum_\alpha^\beta l_x m_x \qquad (5.75)$$

This parameter expresses the per-generation growth rate of the population and is related to the discrete daily growth rate, λ, as given in the following example. If $R_0 = 100$ offspring/female and mean generation time is $T = 25$ days, then the daily growth rate is the twenty-fifth root of 100, or

$$\lambda = \sqrt[25]{100} \qquad (5.76)$$

$$\lambda = 1.2023 \qquad (5.77)$$

This value of λ can be verified by noting that if the initial number in a population is $N_0 = 1$, then 25 days later (i.e., T) the population has increased by 100-fold. That is,

$$N_T = N_0 \lambda^{25}$$

$$= 1.2023^{25} \qquad (5.78)$$

$$= 100 \text{ females/newborn female}$$

Stable age distribution

The stable age distribution in a stable population emerges when the birth and death rates are fixed and the distribution of ages in a population are stable. Consider a hypothetical stable population with two age classes increasing twofold each day starting at time 0. Suppose further than no mortality occurs from age class 0 to age class 1. Then the number in age class 0 and age class 1 will always differ by twofold (i.e., 4/2, 8/4, and 16/8), as will the total number between two time steps (i.e., 6/3, 12/6, and 24/12) (table 5.2). Here the fraction in age class 0 will always be two-thirds and

Table 5.2. Hypothetical examples of a stable population with two age classes increasing by twofold each time step

Age class	Time step			
	0	1	2	3
0	2	4	8	16
1	1	2	4	8
TOTAL	3	6	12	24

Note: A constant fraction of the population is contained in each age class, and thus the population is considered at the stable age distribution.

the fraction in age class 1 will always be one-third of the total population number; in other words, the distribution of ages will be stable. Note that the fraction of the total population in age class 1 is always smaller than the fraction in age class 0 due to the growth rate and not, in this case, as a result of mortality. For any other age classes, a combination of both growth rate and mortality determines the exact fraction of the total population for that age class in a stable population.

The stable age distribution (SAD) is defined as the schedule of the fractions each age class represented in the stable population. The formula for this fraction, denoted c_x, is given as

$$c_x = \frac{e^{-rx} l_x}{\sum e^{-rx} l_x} \qquad (5.79)$$

This formula can also be used to show the relationship between age structure and the finite rate of growth, λ. Consider the relationship between the fraction in age class x, c_x, and the fraction in age class $x + 1$, c_{x+1}:

$$\frac{c_{x+1}}{c_x} = \frac{\lambda^{-(x+1)} l_{x+1}}{\lambda^{-x} l_x} \qquad (5.80)$$

Given that

$$\frac{l_{x+1}}{l_x} = p_x$$

then

$$\frac{c_{x+1}}{c_x} = \frac{p_x}{\lambda} \qquad (5.81)$$

Thus

$$\lambda = \frac{p_x c_x}{c_{x+1}} \tag{5.82}$$

In words, the growth rate of a stable population is the product of survival from age x to age $x+1$ and the ratio of the fractions in the population at age x and age $x+1$. Thus, if $\lambda = 1.2$ and survival is 100% (i.e., $p_x = 1.00$), then the proportion in each adjacent age class will differ by 1.2-fold.

Mean generation time

The mean generation time, T, is defined in two ways. The first definition is the mean age of reproduction, which characterizes T as the mean interval separating the births of one generation from those of the next (Pressat 1985). The formula for this definition is given as

$$T = \frac{\sum x l_x m_x}{\sum l_x m_x} \tag{5.83}$$

A second definition of mean generation time is the time required for a population to increase by a factor equal to the net reproduction rate, R_0, in other words, the time for a newborn female to replace herself by R_0-fold. The formula for this definition of T is

$$T = \frac{\ln(R_0)}{r} \tag{5.84}$$

For example, using $r = 0.3302$ and a net reproductive rate $R_0 = 40.5$ gives a T value of

$$T = \frac{\ln(40.5)}{0.3302}$$

$$= 11.2 \text{ days}$$

This population will increase by a factor of 40.5 individuals every 11.2 days.

Population projection

Leslie matrix

Leslie (1945) reframed the continuous time Lotka model as a discrete time model using matrix algebra. Known as the Leslie matrix, this model provides a numerical tool for determining growth rate in age-structured populations that can be used to study the transient properties of populations as they converge to the stable state.

The Leslie matrix is of the form

$$\begin{pmatrix} F_0 & F_1 & F_2 \\ P_0 & 0 & 0 \\ 0 & P_1 & 0 \end{pmatrix} \begin{pmatrix} N_{0,t} \\ N_{1,t} \\ N_{2,t} \end{pmatrix} = \begin{pmatrix} N_{0,t+1} \\ N_{1,t+1} \\ N_{2,t+1} \end{pmatrix}$$

where the top row of the matrix contains the birth elements, F_x, and subdiagonals, P_x, are the survival elements for the different age classes. The number of individuals age x at time $t+1$, denoted $N_{x,t+1}$, is the product of the matrix and the age vector containing the numbers at age x and time t, $N_{x,t}$. The birth elements are computed as follows (Caswell 2001):

$$F_x = \frac{m_x + P_x m_{x+1}}{2} \tag{5.85}$$

where m_x and m_{x+1} denote the number of offspring produced by females age x and $x+1$, respectively. Note that the formula for fertility depends on the distribution of births and deaths within an age class relative to the timing of the census. Additional fertility formulas are presented in table 6.4.

Example iteration

A population is projected through time by first entering an initial number of individuals into one or more age classes and multiplying the Leslie matrix by the age vector through a process of one-step iteration. For example, consider a population with three age classes starting with

$$\begin{pmatrix} F_0 & F_1 & F_2 \\ P_0 & 0 & 0 \\ 0 & P_1 & 0 \end{pmatrix} \begin{pmatrix} N_{0,t} \\ N_{1,t} \\ N_{2,t} \end{pmatrix} = \begin{pmatrix} N_{0,t+1} \\ N_{1,t+1} \\ N_{2,t+1} \end{pmatrix}$$

Time 0 (initial):

$$\begin{pmatrix} 0.0 & 5.0 & 3.0 \\ 0.8 & 0 & 0 \\ 0 & 0.5 & 0 \end{pmatrix} \begin{pmatrix} 1.0 \\ 1.0 \\ 1.0 \end{pmatrix} = \begin{pmatrix} 8.0 \\ 0.8 \\ 0.5 \end{pmatrix}$$

Time 1:

$$
\begin{pmatrix} 0.0 & 5.0 & 3.0 \\ 0.8 & 0 & 0 \\ 0 & 0.5 & 0 \end{pmatrix}
\begin{pmatrix} 8.0 \\ 0.8 \\ 0.5 \end{pmatrix} =
\begin{pmatrix} 5.5 \\ 6.4 \\ 0.4 \end{pmatrix}
$$

Time 2:

$$
\begin{pmatrix} 0.0 & 5.0 & 3.0 \\ 0.8 & 0 & 0 \\ 0 & 0.5 & 0 \end{pmatrix}
\begin{pmatrix} 5.5 \\ 6.4 \\ 0.4 \end{pmatrix} =
\begin{pmatrix} 33.2 \\ 4.4 \\ 3.3 \end{pmatrix}
$$

Table 5.3 shows the results of this hypothetical population projected over 20 time periods. Note that initially there are an equal number of individuals in each age class, but after a few time periods the largest percentage of individuals shifts to age classes 0 and 1. There is a numerical pattern of convergence to a stable (fixed) rate of increase in table 5.3, and the variation in λ from $t = 0$ until λ reaches a constant value is shown in fig. 5.1.

Note that from this relatively simple projection model several fundamental properties of populations can be identified, including oscillations, convergence, ergodicity, and stability. First, the model projections show oscillations where both the growth rate and age structure proportions wax and wane due to the lag in reproductive output. These oscillations are caused by the time required for newborn individuals to begin reproducing such that when the proportions of the population in the reproductive age classes are maximal, births surge and growth rate is at its highest, and when individuals are primarily in the non-reproducing age classes the growth rate is at its lowest. Second, convergence of both age structure and growth rates occurs when births are spread out over at least two age classes. This means that the peaks and troughs in birth sequences are thrown together and thus smoothed out in the replacement process (Arthur 1981). A third property of populations that is demonstrated in this example is ergodicity; in other words, the property that the present state of a population is independent of its makeup in the remote past, and is determined only by the birth and death rates recently experienced (Cohen 1979; Arthur 1982; Wilson 1985). Finally, stability, the state at which the age structure is unchanging because the fixed fraction of reproducing adults will generate a fixed number of offspring relative to the whole population, is also evident. In this example, as the population approaches its stable growth rate, λ, at $\lambda = 2.1$, its stable age distribution (SAD) is approximately 67%, 27%, and 6% in age classes 0, 1, and 2, respectively.

Projection for *Drosophila melanogaster*

The results of a projection with empirically determined vital rates (contained in table 5.4) and observed numbers of age classes for the various stages are given in table 5.5. This projection is purely theoretical because in the wild many constraints

Table 5.3. Results of Leslie matrix projection of a hypothetical three-age-class population

Time	Number in age class			Total number	Growth rate	Percentage in age class		
	0	1	2			0	1	2
t	N(0)	N(1)	N(2)	N_{Tot}	λ	%N(0)	%N(1)	%N(2)
0	1.0	1.0	1.0	3.0	3.1	33.3	33.3	33.3
1	8.0	0.8	0.5	9.3	1.3	86.0	8.6	5.4
2	5.5	6.4	0.4	12.3	3.3	44.7	52.0	3.3
3	33	4	3	41	1.5	81.4	10.8	7.8
4	32	27	2	60	2.9	52.4	44.0	3.6
5	139	25	13	178	1.6	78.3	14.2	7.5
6	166	112	13	290	2.7	57.2	38.4	4.4
7	596	133	56	784	1.8	75.9	17.0	7.1
8	832	476	66	1,375	2.5	60.5	34.6	4.8
9	2,582	666	238	3,486	1.8	74.1	19.1	6.8
10	4,044	2,065	333	6,442	2.4	62.8	32.1	5.2
11	11,325	3,235	1,033	15,592	1.9	72.6	20.7	6.6
12	19,272	9,060	1,617	29,950	2.3	64.3	30.3	5.4
13	50,152	15,418	4,530	70,100	2.0	71.5	22.0	6.5
14	90,679	40,122	7,709	138,509	2.3	65.5	29.0	5.6
15	223,735	72,543	20,061	316,339	2.0	70.7	22.9	6.3
16	422,898	178,988	36,272	638,158	2.2	66.3	28.0	5.7
17	1,003,755	338,318	89,494	1,431,568	2.0	70.1	23.6	6.3
18	1,960,074	803,004	169,159	2,932,237	2.2	66.8	27.4	5.8
19	4,522,499	1,568,059	401,502	6,492,061	2.1	69.7	24.2	6.2
20	9,044,801	3,617,999	784,029	13,446,830		67.3	26.9	5.8

Note: Growth rate (λ) is computed as $\lambda = N_{Tot}(t+1)/N_{Tot}(t)$.

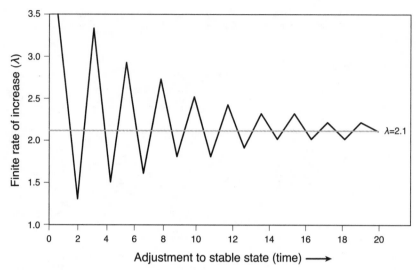

FIGURE 5.1. Illustration of the convergence of the finite rate of increase to a fixed rate (from projections of the data in table 5.3).

limit growth rate, including physical conditions (temperature; humidity), nutrition (food shortages; quality), and natural enemies (predators; parasites; pathogens). However, the numerical projection captures characteristics of growing populations that can be generalized. Moreover, at certain times even natural populations of *D. melanogaster* may experience optimal conditions of resources for short periods and thus experience near exponential growth rates. In the laboratory colonies of flies that are maintained under optimal rearing conditions (diet; mating; climate control) may also experience rates of population increase that approximate the rates described below. Cross-sectional perspectives on this hypothetical *D. melanogaster* population growing at maximal rates are given in table 5.5 and visualized in fig. 5.2.

> *Population at time t = 10 days.* The Drosophila population increased from an initial total number of $N(0) = 4$ to a total at 10 days of $N(10) = 332$, which is an increase of 83-fold/10 days or an average of 1.56-fold per day. Of this total number of 332 individuals there were 52 eggs, 165 larvae, 111 pupae, and 4 adults, respectively. In other words, slightly over 1% were in the adult stage and 99% were in the preadult stages. Since the preadult stages are found either in fruit (eggs; larvae) or in the soil (pupae), only a tiny fraction of the population (adults) would be visible flying around.
>
> *Population at time t = 20 days.* In only 20 days the population expanded by over 3,250-fold from 4 individuals to over 13,000. The 10-day growth rate at this time was an increase of 40-fold, or a 1.45-fold daily increase. Across life stages there were 5,304 eggs, 7,526 larvae, 180 pupae, and 213 adults in the population, which translates to 40% eggs, 57% larvae, and less than 2% for both the pupal and adult stages.
>
> *Population at time t = 40 days.* In around 6 weeks, the hypothetical fly population exploded to over 3.5 million or around 40-fold over the last

Table 5.4. Elements of a 90 by 90 Leslie matrix for population projection of *Drosophila melanogaster*

Age	Stage	Subdiagonal	Fertility	Top row	Age	Stage	Subdiagonal	Fertility	Top row
x		P_x	m_x	F_x	x		P_x	m_x	F_x
0	Egg	0.9000	0.0	0.0	46	Adult	0.9779	15.1	14.7
1	Larva	0.9036	0.0	0.0	47		0.9548	14.5	14.0
2		0.9036	0.0	0.0	48		0.9704	14.1	13.9
3		0.9036	0.0	0.0	49		0.9482	14.0	13.4
4		0.9036	0.0	0.0	50		0.9518	13.5	13.2
5		0.9036	0.0	0.0	51		0.9662	13.6	13.2
6	Pupa	0.9791	0.0	0.0	52		0.9371	13.3	12.6
7		0.9791	0.0	0.0	53		0.9552	12.8	12.1
8		0.9791	0.0	0.0	54		0.9414	11.8	11.1
9		0.9791	0.0	0.0	55		0.9419	11.0	10.6
10		0.9791	0.0	0.0	56		0.9383	10.9	10.1
11	Adult	0.9985	0.0	0.0	57		0.9249	9.9	9.1
12		0.9985	0.0	1.6	58		0.9543	8.9	8.0
13		0.9985	3.2	8.3	59		0.8936	7.5	7.3
14		0.9985	13.5	24.1	60		0.9583	7.9	7.0
15		0.9985	34.9	43.0	61		0.9255	6.4	6.3
16		0.9985	51.2	51.3	62		0.9262	6.7	6.5
17		0.9985	51.5	51.3	63		0.8841	6.7	6.1
18		0.9970	51.1	50.7	64		0.9344	6.3	6.0
19		0.9925	50.5	49.2	65		0.9211	6.1	5.3
20		0.9939	48.1	47.3	66		0.9143	4.9	4.2
21		0.9985	46.7	45.4	67		0.9271	3.9	3.8
22		0.9954	44.2	44.1	68		0.9101	4.1	4.3
23		0.9877	44.2	43.7	69		0.9383	4.9	3.7
24		0.9953	43.7	43.3	70		0.8816	2.8	2.9
25		0.9906	43.1	42.3	71		0.9701	3.5	3.4
26		0.9889	41.8	39.3	72		0.9231	3.4	2.7
27		0.9792	37.2	35.5	73		0.8833	2.2	1.8
28		0.9837	34.4	34.4	74		0.9623	1.5	1.6
29		0.9801	35.0	34.1	75		0.8824	1.7	1.7
30		0.9797	33.9	31.2	76		0.8444	1.9	1.6
31		0.9741	29.1	28.0	77		0.8947	1.5	1.8
32		0.9770	27.6	26.1	78		0.8824	2.3	2.7
33		0.9673	25.3	24.9	79		0.8000	3.5	2.6
34		0.9775	25.4	26.0	80		0.7917	2.1	1.8
35		0.9731	27.3	26.9	81		0.8421	1.8	1.8
36		0.9684	27.2	26.3	82		0.8125	2.1	1.9
37		0.9552	26.2	25.5	83		0.8462	2.2	1.3
38		0.9701	25.9	25.1	84		0.8182	0.5	0.6
39		0.9604	25.1	24.0	85		0.6667	0.8	0.8
40		0.9931	23.8	22.6	86		0.8333	1.3	1.0
41		0.9770	21.6	20.1	87		0.8000	0.8	0.4
42		0.9670	19.1	19.0	88		0.5000	0.0	0.3
43		0.9585	19.5	18.4	89		0.5000	1.0	0.5
44		0.9746	18.1	16.6	90		0.0000	0.0	0.0
45		0.9452	15.4	7.7					

Table 5.5. Results of Leslie matrix projection of *D. melanogaster* at times t=0, 10, 20, 30, and 40 and (for reference) the stable state (at t=∞)

Age (days) x	Number of individuals age x at times t, N(x,t)					Stable fraction age x at t=∞
	t=0	t=10	t=20	t=30	t=40	
0	1	52	5304	18470	1139766	0.31802
1	0	40	3674	14320	828487	0.21257
2	0	35	2292	11641	585412	0.14301
3	0	32	1098	9390	396299	0.09657
4	0	30	361	7575	253806	0.06551
5	1	28	101	6169	151394	0.04465
6	0	27	40	5050	82438	0.03056
7	0	27	39	4476	44136	0.02274
8	0	27	40	3891	22680	0.01695
9	0	30	35	3287	12912	0.01264
10	0	0	26	2643	9204	0.00940
30	0	0	0	0	22	0.00003
31	0	0	0	0	19	0.00002
32	0	0	0	0	18	0.00001
33	0	0	0	0	17	0.00001
34	0	0	0	0	18	0.00001
35	0	0	0	1	18	0.00001
36	0	0	1	0	18	0.00000
37	0	0	0	0	18	0.00000
38	0	0	0	0	18	0.00000
39	0	0	0	0	19	0.00000
40	0	0	0	0	0	0.00000
50	0	0	0	0	0	0.00000
51	0	0	0	0	0	0.00000
52	0	0	0	0	0	0.00000
53	0	0	0	0	0	0.00000
54	0	0	0	0	0	0.00000
55	0	0	0	0	0	0.00000
56	0	0	0	0	0	0.00000
57	0	0	0	0	0	0.00000
58	0	0	0	0	0	0.00000
59	1	0	0	0	0	0.00000
60	0	0	0	0	0	0.00000
TOTALS	4	332	13,223	91,719	3,580,493	1.00000

10 days and nearly a millionfold over the initial population size. This is equivalent to 20 population doublings in a 40-day period. At 40 days the population is approaching its stable state with a daily finite growth rate of 1.3-fold and a stage age (stage) distribution of 31.8%, 61.9%, 4.8%, and 1.5% of the total population in the egg, larval, pupal, and adult stages, respectively. These values are all quite close to the stable state described next.

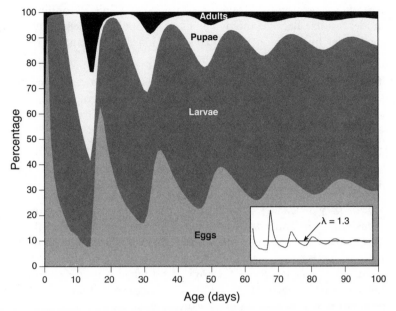

FIGURE 5.2. Stage structure of *D. melanogaster* population converging to the stage age (stage) distribution. Inset: Population growth rates, λ, converging to a constant rate of growth, the intrinsic growth rate r (= ln λ).

> *Population at* $t = long$ *(i.e., stable) state.* The characteristics of this population when it has converged to the stable state include a daily finite growth rate of 1.41-fold and a stable stage distribution of 31.8%, 61.9%, 4.8%, and 1.5% for eggs, larvae, pupae, and adults, respectively.

Two aspects of these projection results merit comment. First, the fraction of adults in the population is extremely small. Inasmuch as this stage is typically the most easily sampled in the wild (e.g., banana baits), observations of changes in the numbers of adults are not unlike the iceberg metaphor: most of the population is out of sight. Second, with a doubling time of only 2 to 3 days, a population doubling from 10 to 20 or from 50 to 100 may go unnoticed; however, several doublings from a few thousand flies can become millions of flies very quickly. Thus what appears to be a sudden population explosion of flies may have nothing to do with changes in population growth rates but is rather an issue of detectability.

Questions to consider when evaluating matrix models

When evaluating populations with matrix models, there is a set of broad questions that can be addressed for many different types of analysis (Caswell 2001).

(1) *Asymptotic properties.* What happens if the set of processes contained in a model operate for a very long time? Does the population grow, decline, persist, go extinct, converge to an equilibrium, oscillate, or become chaotic?

(2) *Ergodicity properties*. How important are the initial conditions to the model results? Because the dynamics of a population depend not only on the model but also on the initial conditions, understanding the ergodic properties of a model is important.

(3) *Transience properties*. What is the importance of short-term dynamics? Are the transient properties of a model more relevant than asymptotic analysis in characterizing the response of populations to perturbations?

(4) *Perturbation properties*. What is the sensitivity of the conclusions derived from a model if there are changes in the model parameters? For example, how do growth rate and asymptotic dynamics change when birth and death rates are modified?

(5) *Stationarity properties*. We add this property to Caswell's list of four previous questions and ask, What does the analysis reveal about the population properties (e.g., age structure) and behavior when its growth rate is zero; which is to say, when birth and death rates are equal? For example, what is the age structure of a population when it is stationary? What is the sensitivity of growth rate and age structure at stationarity when the population is subject to minor perturbations?

These last two questions about perturbation properties and sensitivity to minor perturbations are addressed with more detail in chapter 6.

Fundamental Properties of Populations

There are a number of noteworthy properties of population dynamics and structure that emerge from these basic matrix models, several of which were originally described in Carey (1993).

Age structure transience

Short-term, transient variation in age structure and growth rates will exist in populations with either fixed or variable vital rates. If two populations, one with fixed birth and death rates and the other with variable rates, begin with a narrow age distribution, say, a single female, each would undergo a sequence of demographic changes. The oscillations of both the growth rates and the age structure will occur as the current newborns reach maturity and a new surge of births occurs. In the population with fixed schedules, these echo patterns of new births would initially be periodic and distinct and would eventually become dampened. In the population with variable birth and death rates, the surges of new births may or not may not be distinct, nor periodic, but will depend instead on the specific features of the changing schedules of the vital rates. Additionally, this population with variable rates would, by definition, never become stable. Most importantly, variation in age structure and growth rate will exist in both populations. A population growing with fixed schedules, which has not yet converged to a stable state, may produce patterns that are virtually indistinguishable over a short period from those produced by a population subject to variable schedules.

Convergence to a stable state

Population dynamics decompose into a growth process and a smoothing of the initial birth sequence over the generations that force the age composition toward a limiting form (Arthur 1981, 1982). The process of smoothing averages out the peaks and troughs in the birth sequence, which in turn points both the age structure and the growth rate toward a steady state. The ergodic property of losing information of the past shape of the birth sequence, and thus of the age structure, emerges. Once the birth sequence reaches an exponential increase the age composition must assume its stable shape.

Independence of initial conditions

Over time, the more remote a past age distribution becomes, the less impact the form of that past distribution has on the shape of the current age distribution (Coale 1972). Similarly, the same factors that cause the transient effects of an initial age distribution to disappear from the stable population will also operate for any time path of the fecundity and mortality rates. After a suitably long period, the effect of an initial age distribution is swamped by the cumulative effect of the age pattern of vital rates. This means that the age distribution of any closed population is entirely determined by the most recent survival rates. It is thus impossible to determine either the initial population size, or its initial age structure, from an observed age structure after a population is survived forward and begins reproducing, even over a short time period.

Fertility and mortality

Fertility differences usually have a far greater impact on current age distribution than do mortality differences (Coale 1972). This is because the role of fertility in shaping age distribution is simpler than mortality because fertility operates in a single direction, starting with the newborn age class 0. On the other hand, mortality differences have only second-order effects on age distributions. This is because mortality tends to concurrently change all cohorts and thus has a minimal effect on the immediate age distribution. When mortality changes in a gradual and monotonic fashion, the age distribution tends to change continuously and therefore closely reflects current conditions.

Changing schedules and unchanging age structure

There are several instances where a change in vital rate schedules is not reflected in the age distribution. For example, if two populations are characterized as exhibiting *equivalent differences* in birth and death rates, this means that a change in the birth rate in one has the same effect on age structure as a change in the death rate in the other. This concept was introduced by Coale (1972) to apply to growth rate, but it also applies to age structure. A second example where there is no change in age structure following a change in vital rates is when only survivorship schedules are changed. If survivorship in all ages were to be uniformly reduced by a constant fraction, population growth rate would decrease but there may be no change in age structure. This example helps explain why relatively inconspicuous changes in age profiles may occur despite mortality differences across populations.

Determinants of age structure

The age structure of closed stable populations is an outcome of four interconnected, mutually affecting factors (Coale 1957; Keyfitz et al. 1967). First, *birth rate* determines how many individuals are entering the population as newborns. If birth rate is high, the fraction in the first and other early age classes will be high relative to other age classes. Conversely, if birth rate is low, the fraction in younger age classes will be low relative to older age classes. Second, the *age-specific mortality*, which determines how many individuals survive to each age and, consequently, the relative ratio of one age class to another, is a determinant of age structure. Age-specific mortality determines population death rates, which along with birth rates determine the population growth rate. *Net maternity* is a determinant of stable age structure as it governs the contribution of birth rate to population growth rate in which age-specific reproduction is weighted by age-specific survival. Finally, the *population growth rate*, which is an outcome of the pattern, level, and timing of net maternity as it relates to intrinsic birth rate relative to intrinsic death rate, shapes the age structure by skewing it toward youth when growth rate is high and shifting it toward the older ages when growth rate is low. Examples of how population growth rate affects age structure in stable *D. melanogaster* populations are given in fig. 5.3.

Effect on r of reproductive timing

Lewontin (1965) was the first to ask the question, What is the effect of changes in life history parameters on the intrinsic rate of increase? There are three interrelated effects of r owing to shifts in reproductive timing that become easier to understand if viewed as effects on the two components of the intrinsic rate of increase—intrinsic birth rate and intrinsic death rate. First, the shift in reproductive timing moves the highly fecund females into younger age classes, which, in increasing populations, are more abundant than older age classes. Second, a change in the fraction of the population in the highly fecund age classes will, in turn, alter the age distribution due to the effect of this age shift on the population growth rate. This means that the effect on birth rate of a decrease in the age of first reproduction is self-reinforcing. Third, because of a modified population growth rate due to the shift in age structure, and because individuals in different age classes usually have different probabilities of dying, the change in age of first reproduction changes the frequency distribution of population members in different death risk groups.

In short, a change in development time alters the intrinsic birth and death rates in a population owing to age shifts and age class weightings of the vital schedules. Since peak fecundity usually occurs in younger adults, a decrease in the age of first reproduction will increase the intrinsic birth rate, b, substantially in growing populations. On the other hand, a shortening of developmental time will increase the intrinsic death rate when mortality is greater in young individuals than in old individuals. But when mortality is lower in the young than in the old, the intrinsic death rate will decrease. This perspective also sheds light on why changes in developmental time have little

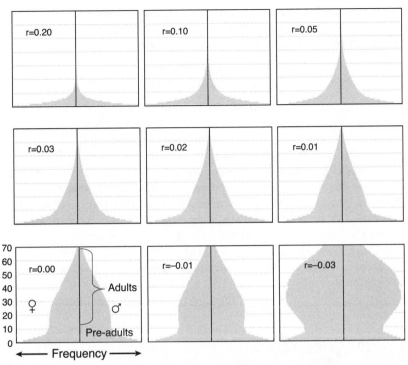

FIGURE 5.3. Age structure of *D. melanogaster* at different population growth rates, r. The value of r is adjusted by scaling the reproductive schedule, leaving mortality rates unchanged. Male and female birth and death rates are assumed to be identical. All distributions sum to unity. Note that there is an extremely high fraction of individuals in the preadult age classes when population growth rate is high, as in the top row of distributions. Conversely, a large proportion of the population is in the older age classes (adults) when population growth rate is equal to or less than zero, as in the bottom row of distributions.

effect on population growth rates in slowly growing or stationary populations. The age distribution for these cases is much flatter, and therefore shifts in reproductive timing do not drastically affect age weightings. This helps explain the findings of Snell (1978) that Lewontin's result, showing that r is most sensitive to changes in developmental time, does not hold for slowly growing populations.

Speed of Convergence

Kim (1986) provided the definitive determinant of the rate of convergence of a population to a stable age distribution and a fixed growth rate. She concluded that (1) for a fixed value of net reproductive rate, the speed of convergence increases as the mean of the net maternity function decreases; (2) for a fixed shape of the net maternity function, the speed of convergence increases as the value of net reproduction rate increases—in other words, the higher the rate of increase, given the same pattern of reproduction, the faster the population converges to stability; and (3) the speed of convergence to stability depends not on the shape of the net maternity function but on the shape of the *stable maternity function*—the smaller the mean of the stable net maternity function, the faster the speed of convergence

Population Momentum

Like physical objects that have a tendency to continue moving once in motion, increasing populations have a tendency to continue growing, which is referred to as *population momentum*—the extent to which a population continues to change in size after it adopts replacement-level rates of mortality and fertility (Kim et al. 1991). As Pressat and Wilson (1987) note, the momentum of a population can be regarded as the opposite of the intrinsic rate of increase, which indicates the growth rate implicit in a set of vital schedules and independent of initial age structure. In contrast, population momentum describes the growth potential due to age structure alone. Keyfitz (1971) published the primary work on momentum in an initially stable population. The population momentum concept is illustrated in fig. 5.4 for a stable population of *D. melanogaster* that is suddenly subject to birth and death rates that confer zero population growth.

Preston and Guillot (1997) generalized the mathematical definition of momentum with the formula

$$M = \sum_0^\omega \frac{c_x}{c_x(s)} w_x \qquad (5.86)$$

where c_x and $c_x(S)$ denote the age structure of the initial population and that of the stationary population, respectively, and where

$$w_x = (\sum_a^\omega l_x m_x)/\bar{x} \qquad (5.87)$$

This gives the ratio of net reproduction above age x to the mean age at childbirth in a stationary population.

Consider Further

Inasmuch as this chapter lays the groundwork for and extensions to stable theory, it follows that the content contained in the following two chapters—chapter 6 (Population II: Stage Models) and chapter 7 (Population III: Extensions of Stable Theory)—is closely related. This includes the age/stage models (Leslie and Lefkovitch) in the former and the two-sex stochastic and hierarchical models in the latter. Perturbation analysis is also discussed in more detail in chapter 6. Population-related content in other parts of the book include coverage in chapter 9 (Applied Demography I: Estimating Parameters) of estimating population growth rates and making use

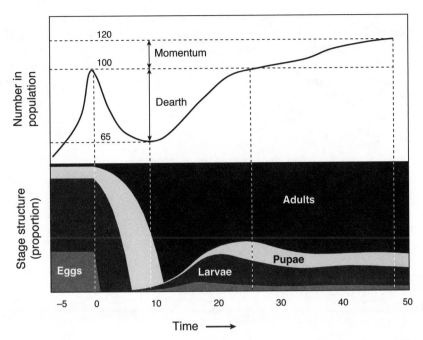

FIGURE 5.4. Momentum dynamics in *D. melanogaster*. Upper panel depicts the number in the population normalized at 100 when, at t = 0, the fecundity schedule is switched from rapid growth of a stable population to replacement-level growth (zero population growth). Because over a third of the population is in the 1-day egg stage in the rapidly growing population, the switch from a high fecundity schedule to a replacement-level fecundity schedule reduces the population by nearly this fraction. This accounts for the population decrease immediately after the switch. The decrease continues until the disproportionately large number of preadults, particularly individuals in the egg stage at the time of the fecundity switch, mature to begin laying eggs. At this time the population begins positive growth due to the disproportionately large number of young individuals from the preswitch cohorts that are now mature. Eventually, the population converges to the replacement-level age structure and growth rate of around 20% greater than the population level at the switch.

of the population life table—a stationary version of stable populations. Population concepts are also covered in chapter 11 (Biodemography Shorts), including the subsection in Group 2 concerned with population examples (S26–S35).

Basic population theory is covered in the seminal books by mathematical demographers Ansley Coale and Nathan Keyfitz and their colleagues, including *The Growth and Structure of Human Populations* (Coale 1972), *Introduction to the Mathematics of Populations* (Keyfitz 1977, 1985), and *Applied Mathematical Demography* (Keyfitz 1985; Keyfitz and Caswell 2010).

Other excellent books on stable theory include Kenneth Land's contribution to mathematical demography (Land et al. 2005) in the *Handbook of Population* (Poston and Micklin 2005) and Stephen Perz's chapter on population change (Perz 2004) in the edited book *The Methods and Materials of Demography* (Siegel and Swanson 2004). A number of chapters in the four-volume *Treatise on Population* are concerned with basic population theory and models, including population models (Caselli et al. 2006b), population replacement (Vallin and Caselli 2006c), population replacement and change (Vallin 2006a), population increase (Wunsch et al. 2006), and population dynamics (Caselli and Vallin 2006). Chapter 10 in Wachter (2014) includes sections on stationary equivalent populations, Lotka's r, the Euler-Lotka equation, and population momentum.

6

Population II: Stage Models

*Mathematics without natural history is sterile, but natural history
without mathematics is muddled.*

John Maynard Smith (1982, 5)

In many situations, an individual's stage, where transitions from stage to stage are possible, may be better than age as an indicator of demographic characteristics, such as the chance of surviving or reproducing or the number of offspring produced. For example, in trees and perennial plants, arthropods, and mollusks, and in fish, amphibians, and reptiles that exhibit indeterminate growth, size is a more important factor than age in determining survival and reproductive rates (Barot et al. 2002). A second group of organisms that require stage-based models are those with *multiple modes of reproduction*. Many organisms have both sexual and vegetative reproduction, including many species of plants as well as invertebrates such as rotifers and daphnia. If same-aged offspring from each reproductive category differ in survival and reproduction, then age is an inadequate demographic category. Matrix stage-structured models are used in these cases to describe the population size, structure, and the population dynamics, at any one point in time (Caswell 2001). The use of matrices is important because their long-term behavior provides a connection to stable population theory (as illustrated in chapter 5) and because matrix elements do not have to be constant but can be equations. Matrices are especially useful in the context of size- and stage-structured models because, as the mathematical component of a model, they integrate seamlessly into the conceptual and visual components, such as the life cycle graphic (Horvitz 2011). This chapter focuses on the basics of constructing and analyzing stage-based models for species with different life cycles and illustrates the connections between age models and stage models. We also introduce recent refinements and advances to matrix modeling approaches that include continuous state variables.

Model Construction and Analysis

Basic stage-structured model

Lefkovitch (1965) generalized the Leslie matrix model (Leslie 1945), discussed in chapter 5, by using stage instead of age groupings. Fig. 6.1 parses out the multiple transitions that may be included in a stage-based model with three stages, including a newborn stage (stage 1), and two reproductive stages (stages 2 and 3) with both life cycle diagrams and matrices. The difference between Leslie and Lefkovitch matrices is structural, that is, the permissible locations for nonzero entries (Horvitz 2016).

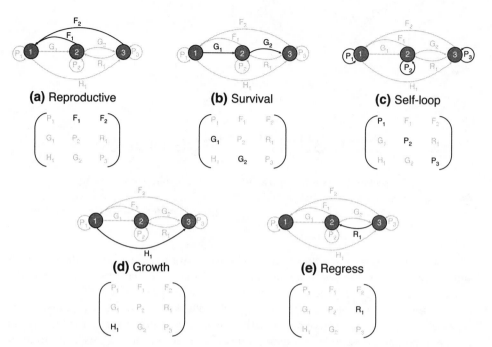

FIGURE 6.1. Classification of elements for transition matrices, including (a) reproductive, (b) survival, (c) self-loop, (d) growth, and (e) regress.

In the Leslie matrix the only nonzero entries are survival probabilities from one age to the next in the subdiagonal and in the top row where newborns are all produced into the first age class. In the Lefkovitch model any entry can be nonzero. For example, survivors can be in any stage at the next time step (same, advance, regress) and newborns may occur as different sizes or types.

Moving from left to right in fig. 6.1, the two reproductive stages contribute to the newborn class with stage-specific fecundities (F_x). All of the nonzero, nonfecundity transition probabilities represent the proportion of individuals moving from one stage to another in one time period. The survival probabilities (G_x) denote the transition probabilities from stage i to stage i + 1, which includes both survival and growth, and self-loop probabilities (P_x) denote the probability that an individual continues to survive in stage i. Individuals may also skip from stage i to stage i + 2 or greater, which is designated as a probability (H_x), and includes both growth and survival. Finally, individuals may regress (R_x) from stage i to stage i − 1 or greater.

A critical assumption of this basic stage-structured model is that there is little variation among individuals in the same stage with respect to their demographic characteristics. This also means that what an organism will do depends on the stage it is in now and not on what stage it was in at a previous time step or how long it has remained in each stage. Other assumptions of this basic model are that the population is closed (i.e., no migration), the vital rates are constant (i.e., no demographic or environmental stochasticity), and the vital rates are not density dependent. As discussed later in this chapter, this basic model can be modified to relax these assumptions and thus be applied to more robust situations. Here we continue with this basic model to discuss, first, how these matrices can be used to determine how different vital rates influence population growth. We then consider how models can be constructed for species with different life cycles.

Perturbation analysis

Perturbation analysis of matrix models can be used to evaluate the impact of changes in vital rates on population growth, which may be important for how different vital rates influence fitness, for evaluating alternative management strategies (e.g., harvesting in conservation) (Caswell 2001), or for predicting the intensity of natural selection (vanTienderen 1995). There are two approaches, which can be classified as either prospective or retrospective, to consider the impact of vital rates on population growth (Caswell 1997; Horvitz et al. 1997; Caswell 2001). Prospective perturbation analysis addresses the effects of potential future changes and can be used to determine which vital rate, if changed, would have the largest impact on population growth rate. Prospective analyses are independent of any actual past or future change in vital rates. Sensitivity and elasticity analyses are prospective analyses where the sensitivity of λ is the effect of a small additive change in one of the vital rates and the elasticity of λ is the effect of a small proportional change in a vital rate. Prospective analyses are often used for management interventions where the goal is to identify an approach to increase λ in the case of a threatened species or to decrease λ for the control of a pest species.

A population matrix A has a corresponding right eigenvector w that represents the stable stage/age distribution of the population such that $Aw = \lambda w$. The left eigenvector of this matrix, v, gives the reproductive values of each stage and is defined by $v'A = \lambda v'$. The sensitivity index (Caswell 1978) is then given by

$$S_{ij} = \frac{v_i \, w_j}{<v, w>} \tag{6.1}$$

where the sensitivity of a matrix element a_{ij} is the product of the ith element of the vector of reproductive values v and the jth element of the vector of the stable age/stage distribution w, divided by the scalar product of the two vectors. The sensitivity, s_{ij}, compares the *absolute* change in λ resulting from an infinitesimal absolute change in the matrix transition, a_{ij}, relative to the impact of equal absolute changes in other elements.

The elasticity, e_{ij}, quantifies the *proportional* change in λ resulting from an infinitesimal proportional change in the matrix transition (deKroon et al. 1986).

$$e_{ij} = \frac{a_{ij}}{\lambda} \times S_{ij} \tag{6.2}$$

The equations for the elasticities of λ presented here apply to cases with density-independent population growth and time-invariant demographic parameters. Elasticities for density-dependent populations with time-invariant or stochastic demographic parameters are also possible (Grant 1997; Grant and Benton 2000).

The other type of perturbation analysis is retrospective, and it requires data on the vital rates under two or more sets of environmental conditions. The goal of this analysis is to determine the contribution of each of the vital rates to the variability in λ. Retrospective analyses look backward at the observed variation in vital rates, across environments, to determine how that variation impacted variation in λ; this method is known as a life table response experiment (LTRE) (Caswell 2001). A consideration of the roles of prospective and retrospective perturbation analysis for conservation biology is discussed in Caswell (2000).

Stage-Based Models for Modular Organisms: Plants

The first step when building a stage-structured model is to determine how to divide the population into stages that reflect the most important factor that the demography of the species depends on. Body size strongly impacts survival and reproduction of individuals, and discrete size classes are often used in stage-based models. If survival and reproduction have nothing to do with size, for example, with a bird species, the stages might be juveniles, nonbreeding adults, and breeding adults; or for an insect it may be developmental stages, such as egg, larva, pupa, and adult. The life cycle of a species, for example, modularity, also influences model structure. Budding hydra, fragmenting corals, and clonal plants are examples of species that can divide into modular, independent units. This modular structure provides opportunities to apply the tools of demography at multiple levels of organization, and the plasticity of this modular structure can lead to individuals shrinking such that age and size may have independent effects on mortality (Harper and White 1974). These effects, as well as the importance of dormancy, are considered in the models here.

Three idealized models are presented with variation in life cycles and critical demographic processes. These models are referred to here as plant models, but many aspects of the structure of these stage models could be applied to some animal species. Stage is size for these models, which for plants may be quantified, depending on the species, as the number of leaves, size of rosettes, or the diameter at breast height. Each of these three models builds on the next and adds traits, including shrinkage, dormancy, and dispersal among populations.

Model I: Growth and renewal

Model I shows growth and renewal and includes four stages: seed (S), juvenile (J), small reproductive (SR), and large reproductive (LR) (fig. 6.2a). The corresponding matrix elements for the transitions between each of these stages are given (fig. 6.2b), and the transitions in and out of each class are diagrammed (fig. 6.2c). At the seed stage, individuals either transition into the juvenile stage, with a probability of 0.8, or they die with a probability of 0.2; in other words, there is no dormancy. The input into this seed class is provided through either of the two reproductive classes (SR and LR). Individuals in the juvenile, non-reproductive class can either transition to the small flowering stage (probability 0.7) or remain as non-reproductive juveniles (probability 0.2). For many plant species this transition to flowering is size dependent, but for other species this transition is environment dependent, in which case no individuals remain in the juvenile class if all individuals experience the same envi-

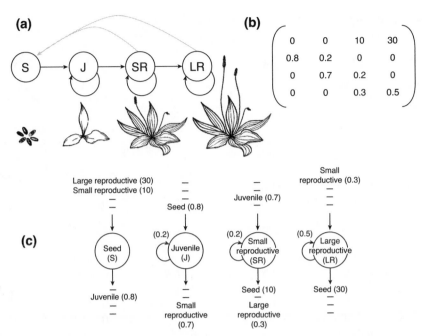

FIGURE 6.2. Model I with growth and renewal showing (a) life cycle graphic, (b) matrix, and (c) transitions between stages. Note that in (c) the stages listed above designate where individuals coming into this stage are coming from, and the stages listed below denote the fate of individuals moving out of this stage. The growth rate of this model is $\lambda = 2.18$, and the fractions in the seed, juvenile, small reproductive, and large reproductive stages are 0.637, 0.257, 0.091, and 0.016, respectively.

ronment. From the juvenile class this model specifies two reproductive classes, where small reproductive (SR) individuals have a 0.3 probability of transitioning into the LR class and the LR individuals have a 0.5 probability of remaining in the LR class.

The life cycle pictured in model I could apply to many animal species that lay eggs and experience a juvenile non-reproductive stage. Most animals, however, show determinate growth and thus have minimal or no growth after the age of first reproduction; thus there will only be one reproductive class, with some probability of remaining in that reproductive class if individuals can reproduce more than once.

With the values specified in this matrix for model I, the growth rate (λ) of this population is 2.18 and the stable stage distribution is 0.637, 0.257, 0.091, and 0.016 for the seed, juvenile, small reproductive, and large reproductive classes, respectively. This matrix model can be considered as a life table (table 6.1) by setting all elements in the top row of the matrix (i.e., reproduction) to zero.

Model II: Growth, shrinkage, and dormancy

Modular organisms, such as plants, can shrink in size, and this may impact their survival or reproduction. Dormancy, a period in an organism's life cycle when growth, development, or physical activity are temporarily stopped, can also influence demography because a population will consist of two parts—the growing individuals that

Table 6.1. Life table for plant model 1—growth and renewal

| Age | Stage | | | | Survival |
	1	2	3	4	
0	1.0000	0.0000	0.0000	0.0000	1.00
1	0.0000	0.8000	0.0000	0.0000	0.80
2	0.0000	0.1600	0.5600	0.0000	0.72
3	0.0000	0.0320	0.2240	0.1680	0.42
4	0.0000	0.0064	0.0672	0.1512	0.22
5	0.0000	0.0013	0.0179	0.0958	0.11
6	0.0000	0.0003	0.0045	0.0533	0.06
7	0.0000	0.0001	0.0011	0.0280	0.03
8	0.0000	0.0000	0.0003	0.0143	0.01
9	0.0000	0.0000	0.0001	0.0072	0.01
10	0.0000	0.0000	0.0000	0.0036	0.00
11	0.0000	0.0000	0.0000	0.0018	0.00
12	0.0000	0.0000	0.0000	0.0009	0.00
13	0.0000	0.0000	0.0000	0.0005	0.00
14	0.0000	0.0000	0.0000	0.0002	0.00
15	0.0000	0.0000	0.0000	0.0001	0.00
16	0.0000	0.0000	0.0000	0.0001	0.00
17	0.0000	0.0000	0.0000	0.0000	0.00
18	0.0000	0.0000	0.0000	0.0000	0.00
19	0.0000	0.0000	0.0000	0.0000	0.00
20	0.0000	0.0000	0.0000	0.0000	0.00

can reproduce and the dormant individuals that can persist for many and, in rare cases, hundreds of years (Baskin and Baskin 1998). Here we consider plant seed dormancy and note that the number of seeds in the seed bank is a function of the seed rain, the fraction that germinate each year, and the fraction of seeds that decay and never germinate. Building on the basic parameters of model I, the juvenile class in model II is now divided into a small non-reproductive (SN) and large non-reproductive (LN) class, where small and large again delineate a measure of plant size (fig. 6.3a). There are a number of new transitions in this model that reflect two plant traits of seed dormancy and shrinkage, in other words, stasis or regression between stages. All stage classes now show some stasis, that is, some probability of individuals staying in the same stage. The probabilities of remaining within a stage are along the diagonal of the matrix (fig. 6.3b), and it should be noted that these stasis transitions represent individuals that stay in the same stage and remain the same size. Regression of individuals into prior stages may occur through transitions from reproductive to non-reproductive stages and also as plants shrink to a smaller size class. The transitions into and out of each stage are given (fig. 6.3c), and a life table has been constructed (table 6.2).

The basic structure of model II could also be applied to a species that, instead of shrinkage, shows fragmentation, where larger individuals break apart to form smaller clones of themselves. Fragmentation may occur not only for some plant species, where modules break off and become physiologically independent, but also for corals or other modular animals.

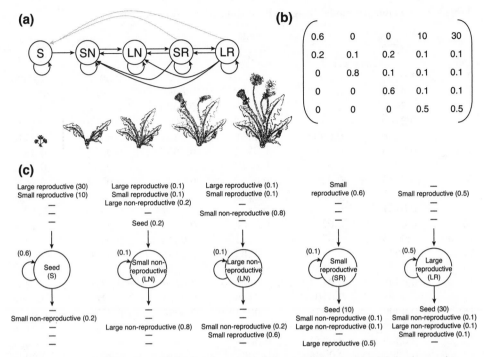

(a)

(b)

$$\begin{pmatrix} 0.6 & 0 & 0 & 10 & 30 \\ 0.2 & 0.1 & 0.2 & 0.1 & 0.1 \\ 0 & 0.8 & 0.1 & 0.1 & 0.1 \\ 0 & 0 & 0.6 & 0.1 & 0.1 \\ 0 & 0 & 0 & 0.5 & 0.5 \end{pmatrix}$$

(c)

FIGURE 6.3. Model II with growth shrinkage and dormancy showing (a) life cycle graphic, (b) matrix, and (c) transitions between stages. The growth rate of this model is $\lambda = 1.543$, and the fractions in the seed, small non-reproductive, large non-reproductive, small reproductive, and large reproductive are 0.768, 0.119, 0.069, 0.030, and 0.014, respectively.

With the values specified in model II, the growth rate (λ) of this population is 1.543 and the stable stage distribution is 0.768, 0.119, 0.069, 0.030, and 0.014 for each of the stage classes, respectively.

Model III: Connecting populations

As an extension to the population projection models for plants considered here, we now look at movement among populations due to seed dispersal, although plant gene flow can also occur through pollen dispersal. In model III each subpopulation has the same transition probabilities as in model I, with the added dispersal of seed between these subpopulations (fig. 6.4). With this structure of connected populations, it is possible to consider either the dynamics of the changes in the number of populations with time or the population growth rate for the whole population matrix (Horvitz and Schemske 1986). Understanding the impact of dispersal dynamics on population growth rates is particularly important for conservation questions in fragmented habitats (Damschen et al. 2014). In the example here, $\lambda = 2.3$.

Dispersal plays a key role in the dynamics of species that are subdivided into discrete subpopulations, and ecologists refer to a population of populations as a *metapopulation* (Levins 1969, 1970; Hanski and Gilpin 1997). A metaphor for metapopulations is that they are a spatially distributed set of asynchronously blinking lights, where a "light on" indicates an occupied patch and a "light off" indicates an empty

Table 6.2. Life table for model II—growth, shrinkage, and dormancy

Age	Stage 1	2	3	4	5	Survival
0	1.0000	0.0000	0.0000	0.0000	0.0000	1.0000
1	0.6000	0.2000	0.0000	0.0000	0.0000	0.8000
2	0.3600	0.1400	0.1600	0.0000	0.0000	0.6600
3	0.2160	0.1180	0.1280	0.0960	0.0000	0.5580
4	0.1296	0.0902	0.1168	0.0864	0.0480	0.4710
5	0.0778	0.0717	0.0973	0.0835	0.0672	0.3975
6	0.0467	0.0573	0.0822	0.0734	0.0754	0.3349
7	0.0280	0.0464	0.0689	0.0642	0.0744	0.2819
8	0.0168	0.0379	0.0578	0.0552	0.0693	0.2370
9	0.0101	0.0312	0.0485	0.0472	0.0622	0.1992
10	0.0060	0.0258	0.0407	0.0401	0.0547	0.1673
11	0.0036	0.0214	0.0342	0.0339	0.0474	0.1405
12	0.0022	0.0178	0.0287	0.0286	0.0406	0.1180
13	0.0013	0.0149	0.0241	0.0241	0.0346	0.0990
14	0.0008	0.0124	0.0202	0.0203	0.0294	0.0831
15	0.0005	0.0104	0.0169	0.0171	0.0249	0.0698
16	0.0000	0.0055	0.0048	0.0096	0.0170	0.0370
17	0.0000	0.0026	0.0020	0.0048	0.0072	0.0166
18	0.0000	0.0012	0.0009	0.0022	0.0034	0.0077
19	0.0000	0.0006	0.0004	0.0010	0.0016	0.0036
20	0.0000	0.0003	0.0002	0.0005	0.0007	0.0017

patch due to local extinction. In the simplest metapopulation model, we assume that occupied patches are the source of colonists for vacant patches and that the spatial arrangement of occupied and empty patches makes no difference to the colonization of empty patches. The dynamics of patch occupancy for an idealized metapopulation can be described by

$$\frac{dP}{dt} = cP(1-P) - eP \tag{6.3}$$

where P is the fraction of patches occupied and c and e are the colonization and extinction rates, respectively (Levins 1969). At equilibrium,

$$P^* = 1 - (e/c) \tag{6.4}$$

P^* is positive as long as $(e/c) < 1$, which gives the conditions for persistence in terms of the extinction and colonization rates, where the metapopulation will persist as long as the colonization rate is greater than the extinction rate. This basic model has been

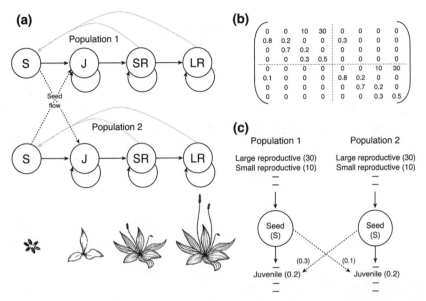

FIGURE 6.4. Plant model with between-population seed flow showing (a) life cycle graphic, (b) matrix, and (c) transitions between stages and across populations. Population growth rate $\lambda = 2.30$. When this data is projected forward to stability, the fractions in each stage for seeds, juveniles, small reproductive, and large reproductive are 0.386, 0.179, 0.059, and 0.010, respectively, for population 1, and 0.223, 0.103, 0.034, and 0.006, respectively, for population 2. The stable stage distribution is 0.609, 0.282, 0.094, and 0.016, respectively, for the fraction in each of the four stages summed over both populations.

expanded, with more relaxed assumptions, and has formed the foundation for understanding the population dynamics consequences of migration, and the persistence of species, in unstable local populations (Hanski and Gilpin 1997; Hanski 1998).

It is important to note that not all patchy populations are truly metapopulations where each subpopulation has a probability of going extinct or being recolonized. For example, with plants that occupy patchy environments, recolonization following genuine extinction does not qualify as a metapopulation scenario if the plant species has a buried seed bank and recolonization is simply the result of the germination of seeds following habitat restoration. Recolonization by dispersal is a prerequisite for a true metapopulation.

Stage-Based Vertebrate Models

The basic stage-based models for vertebrates are similar to model I presented for plants, where individuals can either stay in their current stage class or transition to the next stage. Transition probabilities and fecundities can be estimated by following individuals longitudinally in what is referred to as a dynamic or cohort life table, or as a static life table where at one point in time the number of individuals in each stage category are used to estimate survivorship and reproduction.

Model I: Sea turtles

Model construction of a threatened species

Crouse and her colleagues (1987) used a Lefkovitch matrix model to examine trade-offs in management strategies for loggerhead sea turtles (*Caretta caretta*). They used stage class modeling because accurate age estimation techniques for sea turtles have not been developed and thus preclude the use of age-based population modeling. Demographic data for individual sea turtles are difficult to collect because only the adult nesting females, eggs, and hatchlings, and stranded, dying turtles are ever seen on the beaches. Additionally, long-term monitoring of individual animals is necessary to obtain accurate estimates of fecundity and survival, yet females often create nests over a number of beaches and may nest only once every several years (remigration).

Seven stages were specified for their model, including stage 1—eggs and hatchlings (<1 year), stage 2—small juveniles (1–7 years), stage 3—large juveniles (8–15 years), stage 4—subadults (16–21 years), stage 5—novice breeders (22 years), stage 6—first-year remigrants (23 years), and stage 7—mature breeders (24–54 years). Let F_i denote the number of eggs per year laid by females, P_i denote the probability of survival while remaining in the same stage, and G_i denote the probability of surviving while growing to the next stage. The life cycle graphic for a population model of this species along with the corresponding matrices with either the notation or the values are shown in fig. 6.5. This schematic illustrates the stage-specific biology and specifies the coefficients for the transition matrix. For example, for each time step, all active preadult individuals, except eggs, can either remain in their stage or transition to the next stage. This emphasizes the importance of both within- and between-stage survival for the younger individuals. Note that annual survival in small juveniles is estimated to be higher than survival in both large juveniles and subadults.

Matrix projection

Matrix models can be used to project populations if one assumes that vital rates do not change and that environmental conditions are the same as those that occurred during data collection (Caswell 2001; Coulson et al. 2001; Crone et al. 2011). A 20-year projection of the loggerhead sea turtle population is presented in fig. 6.6, starting with an identical number of individuals in each of seven stages. Note that mature breeders constitute only around 20% of this subgroup and less than 0.2% (2 of 1,000) of the overall population. In other words, the vast majority of the populations consist of pre-reproductive individuals.

Perturbation analysis

Lambda for this population is $\lambda = 0.9450$, thus this is a declining population (Crouse et al. 1987). The management question is then if protection efforts were to focus on one stage over the others, because of available technology or ease of access to a particular life stage, which stage would show the largest impact on population growth in this population? The elasticities of λ to changes in the matrix elements F_i, P_i, and G_i (fig. 6.7) all sum to unity (deKroon et al. 1986), thus their relative contributions to λ can be compared. Crouse et al. (1987) showed that increases in fecundity have

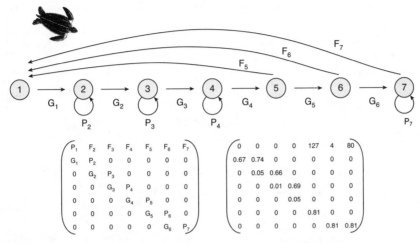

FIGURE 6.5. Loggerhead sea turtle life cycle graphic and corresponding matrices. (*Source*: Matrix notation and values from Crouse et al. 1987)

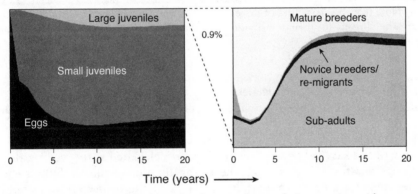

FIGURE 6.6. Stage structure of the loggerhead sea turtle population starting with one individual in each of the seven stages projected through 20 years. (Left) all stages; (right) the 0.9% of the population in the four stages from subadult through mature breeders. (*Source*: Data from Crouse et al. 1987)

a minimal effect on λ, and the probability of survival in the same stage (P_i), particularly the juvenile and subadult stages (2–4), has the greatest impact on population growth rate.

Model II: Killer whales

Model construction of a species with a post-reproductive life stage

One modification to the vertebrate model is to consider a long-lived species with a post-reproductive stage, such as the killer whales, where females exhibit reproductive aging (e.g., menopause). Females stop reproducing when they are 35–45 years old but can survive into their 90s, which is the longest post-reproductive life span of

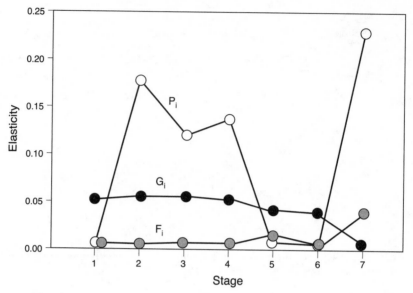

FIGURE 6.7. The elasticities of λ for the sea turtle to changes in the matrix elements F_i, P_i, and G_i. Note that the elasticities of these matrix elements sum to unity, and they can thus be compared directly for their contribution to the population growth rate (redrawn from Crouse et al. 1987).

any nonhuman animal. Killer whales live in social groups, and the post-reproductive females in these family "pods" provide foraging assistance that is critical to reducing the mortality risk of sons, and to a lesser extent daughters (Foster et al. 2012).

The life cycle graph and transition matrix for the killer whale are given in fig. 6.8. The growth and survival probabilities (G_i's) were calculated as the reciprocals of the mean stage durations. For example, the 1-year duration of the yearlings is equal to 1/1. Adjusting for a newborn mortality of 0.0225 yields a survival (transition) rate of $G_1 = 0.9775$. Since the durations of the juvenile and reproductive adult stages are 13.6 and 22.1 years, respectively, their reciprocals are $G_2 = 0.0736$ and $G_3 = 0.0452$. The self-loop (within-stage) yearly survival rates (P_i) for the juvenile, reproductive adult, and post-reproductive adult stages are 0.9110, 0.9534, and 0.9804, respectively. Brault and Caswell (1993) note that this relatively simple model, with only four stages, yields similar estimates of λ and reproductive value as a previous analysis of an age-structured model with 90 age classes for this same population.

Perturbation analysis

An elasticity analysis shows that the population growth rate of this population is most sensitive to changes in adult survival, which is expected in a long-lived species (Brault and Caswell 1993). Juvenile survival and fertility are also critical to population growth. These data also illustrate the value of doing both a sensitivity and an elasticity analysis, as Brault and Caswell (1993) found that when perturbations were done on an incremental rather than a proportional scale, it was demonstrated that even small changes in survival probabilities could have a large impact on population

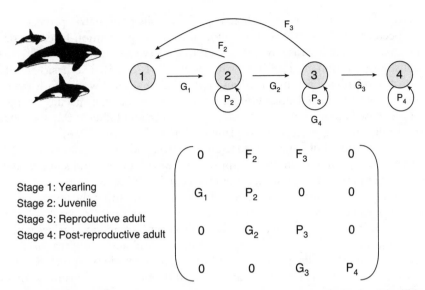

Stage 1: Yearling
Stage 2: Juvenile
Stage 3: Reproductive adult
Stage 4: Post-reproductive adult

FIGURE 6.8. Life cycle graphic and transition matrix for stage-structured killer whale model presented in Brault and Caswell (1993). Values of G_1, G_2, and G_3 are 0.9775, 0.0736, and 0.0452, respectively; for P_2, P_3, and P_4 are 0.9110, 0.9534, and 0.9804, respectively; and for F_2 and F_3 are 0.0043 and 0.1132, respectively.

growth. These data on killer whale pods also demonstrate that perturbations to vital rates may not be independent and a change in one matrix element may be associated with a change in the opposite direction in another element. *Integrated sensitivities* and elasticities measure the total effect of matrix elements on the population growth rate and identify the direct and indirect effects due to the correlations with other elements (vanTienderen 1995). For the killer whales, the probability of staying reproductive, P_3, and becoming post-reproductive, G_3, are negatively correlated, and this negative correlation leads to a negative integrative elasticity for G_3. In other words, a small increase in becoming post-reproductive would lead to a smaller rate of increase of the pod. Additionally, although the post-reproductive females do not directly contribute to population growth rate, an integrated elasticity analysis shows that an increase in P_4 may contribute positively to λ because of the positive covariation among transition parameters P_4 and F_3, adult fecundity (vanTienderen 1995). Correlations among vital rates, which can be either positive or negative, can have significant effects on the sensitivity analyses of population growth to variation in traits (Doak et al. 2005). It is thus critically important that these correlations are accurately estimated when making demographic predictions.

Beyond the Basic Stage Model

Extensions and refinements of the model

Matrix projection models combine multiple vital rates into integrative measures of population dynamics. They thus provide a method to access population status and extinction risk to address basic and applied management questions (Caswell 2001;

Morris and Doak 2002; Crone et al. 2011). The simplicity of matrix projection models has been criticized as not realistic enough to make population forecasts or projections, particularly because in reality vital rates between individuals and years are not constant (Coulson et al. 2001; Crone et al. 2011, 2013). The basic population projection model has been modified and extended in several ways. With respect to connecting age effects and stage effects, a theory to extract age-specific information from stage-based models has been developed (Cochran and Ellner 1992) and extended to temporally varying environments (Tuljapurkar and Horvitz 2006). In some cases, vital rates may show both age and stage dependence, and for this Caswell (2012) has developed a vec-permutation matrix approach. These joint age and stage models are particularly important for questions about the evolution of senescence in size-classified species and for human demographers to address questions about the impact of factors other than age in determining vital rates (Caswell and Salguero-Gómez 2013). Among other realistic extensions, these models have also been extended to species with complex demographic attributes, such as the dormant and active life stages discussed earlier in this chapter (Ellner and Rees 2006), serial correlations among vital rates in fluctuating environments (Tuljapurkar et al. 2009), and nonequilibrium, short-term, transient dynamics (Ezard et al. 2010).

Integral projection model

Matrix projection models characterize individuals into a discrete set of classes, which in some cases may be clear, such as age or developmental stage; but in other cases these classes are continuous variables, such as size, and the division into discrete groupings is artificial. Easterling et al. (2000) introduced the *integral projection model* (IPM) as an alternative that uses continuous individual state variables and thus avoids the need to group individuals into discrete stage classes. An IPM also yields size-specific sensitivities and elasticities that are not biased by stage duration, as they are in matrix projection models (Enright et al. 1995). A basic IPM is deterministic and density independent, which is analogous to a matrix projection model with a constant matrix. Using an example of size-dependent vital rates, the core of an integral projection model consists of regressions that connect the state (size) of an individual to its vital rates—survival, growth, and fertility. Covariates that explain variation in vital rates beyond strictly an individual's state can also be included in these regressions. Additionally, an IPM can incorporate both discrete and continuous state variables. A detailed description of the steps involved in the basic construction of an IPM can be found in Rees et al. (2014) and Merow et al. (2014). This model has been generalized by Ellner and Rees (2006) for species with complex life cycles and density dependence, and species that have both size- and age-dependent demography. Additional extensions of this basic IPM have been made to include, among other situations, time lags, environmental covariates, and environmental and demographic stochasticity (see Rees et al. 2014).

Relationship between Leslie Matrix and Lefkovitch Models

With complete longitudinal demographic data about individuals in a population, either the Leslie or Lefkovitch model can be used to evaluate population structure and projections for population growth. In this section we compare the results from these two approaches using the same hypothetical population with a given life table (table 6.3). We highlight the differences in the parameterization of these two population models and explore the interconnectedness of life tables and population models.

Model construction using the same data

Life cycle graphs and life tables

The life cycle graphs of the equivalent age and stage models are given in fig. 6.9. The braces under the Leslie life cycle graph (top) denote different age classes that have been grouped into stage classes. For example, the preadult stage includes age classes 1–4.

Fertility parameterization: Leslie and Lefkovitch models

The fertility parameters for the age and stage matrices for this population depend on the survival of adults and offspring relative to the timing of the census (table 6.4).

For an age matrix in birth pulse populations with a prebreeding census, the youngest individuals are those produced at the previous birth pulse. Fertility here is the number of newborn offspring an individual produces times the survival of newborns for one census interval. For an age matrix in birth pulse populations with a postbreeding census, the census is done immediately after births and the youngest individuals are those just born. The reproductive adult must survive the entire interval and then reproduce. Reproduction is thus the product of the survival of an adult age class times the number of offspring produced by an individual in that class. When the population is a birth flow—in other words, reproduction occurs continuously over a census interval—then the simplest approach is to assume that all offspring are born at the midpoint of the interval. Reproduction is then the product of the probability that a newborn individual survives to age 0.5 ($l(0.5)$) and the average number of offspring produced over the interval x to x + 1. For the birth flow case the age-specific survival, p_x, is adjusted for the fact that not all adults survive to the next interval.

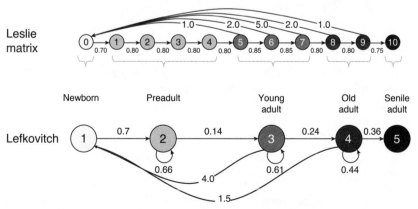

FIGURE 6.9. Life cycle graphics for equivalent Leslie and Lefkovitch models using data from the life table (table 6.3).

Table 6.3. Hypothetical life history data used to compute Lotka's r (=0.14949) and to parameterize Leslie and Lefkovitch matrix models

Age x	Survival l_x	Reproduction m_x	Net reproduction $l_x m_x$	Exponential $e^{-rx} l_x m_x$
0	1.0000	0.0000	0.0000	0.0000
1	0.7000	0.0000	0.0000	0.0000
2	0.5600	0.0000	0.0000	0.0000
3	0.4480	0.0000	0.0000	0.0000
4	0.3584	0.0000	0.0000	0.0000
5	0.2867	1.2500	0.3584	0.1697
6	0.2437	2.3529	0.5734	0.2339
7	0.2072	5.8824	1.2186	0.4279
8	0.1657	2.5000	0.4143	0.1253
9	0.1326	1.2500	0.1657	0.0432
10	0.0994	0.0000	0.0000	0.0000
11	0.0000	0.0000	0.0000	0.0000
		13.2	2.7	1.0000

Table 6.4. Calculation of fertility based on survival of adults and offspring relative to timing of the census

Matrix type	Fertility formula
Age matrix	
Birth pulse, prebreeding census	$F_x = m_x p_0$
Birth pulse, postbreeding census	$F_x = m_x p_x$
Birth flow, midpoint	$F_x = \dfrac{m_x + p_x m_{x+1}}{2}$
Birth pulse, midpoint	$F_x = (l_{0.5}) \dfrac{m_x + p_x m_{x+1}}{2}$
Stage matrix	
Birth pulse, prebreeding	$F_x = l_1 m_x$
Birth pulse, postbreeding	$F_x = P_i m_i + G_i m_{i+1}$

For a stage matrix the fertility parameter for a birth pulse prebreeding census is simply the product of survival to the first age and m_x. For a birth pulse postbreeding census, the reproductive output depends on the survival and possible transitions to other stages.

Constructing the Leslie matrix

Data from the life history table (table 6.3) can be used to compute the survival and reproduction parameters (table 6.5) for the construction of the Leslie matrix (table 6.6).

Table 6.5. Parameter computations for Leslie matrix based on birth and death rates in table 6.6

x	Cohort survival l_x	Subdiagonal survival p_x	Reproduction m_x	Top row reproduction F_x
0	1.000	0.700	0.000	0.000
1	0.700	0.800	0.000	0.000
2	0.560	0.800	0.000	0.000
3	0.448	0.800	0.000	0.000
4	0.358	0.800	0.000	0.500
5	0.287	0.850	1.250	1.625
6	0.244	0.850	2.353	3.676
7	0.207	0.800	5.882	3.941
8	0.166	0.800	2.500	1.750
9	0.133	0.750	1.250	0.625
10	0.099	0.000	0.000	0.000
11	0.000		0.000	0.000

Constructing the Lefkovitch matrix

For the stage matrix, the duration of each stage is estimated assuming an age-related effect such that the fraction of the members of a particular stage who have lived long enough in that stage will move to the next stage if they survive. If we set the proportion of individuals alive in the first cohort of the stage class as 1 and the stage-specific survival rate as p_i, then the survival probability for the entire stage class for individuals surviving d_i years is $p_i^{d_i}$. In the interval t to t + 1, the abundance of groups within a stage class is then 1, p_i, $p_i^2 \cdots p_i^{(d_i-1)}$. This assumes that the age distribution within stages is stable. Note that the sum of those within age class groups, $1+ p_i, p_i^2 \cdots$ is a geometric series that can also be written as

$$\frac{(1-p_i^{d_i})}{(1-p_i)} \tag{6.5}$$

which can then be used to calculate the transitions of individuals. The proportion remaining and surviving, P_i, is

$$P_i = p_i \left[\frac{(1-p_i^{d_i-1})}{(1-p_i^{d_i})} \right] \tag{6.6}$$

To calculate the fraction of individuals in a particular stage that have lived long enough to transition to the next stage, we assume that the oldest individuals in the stage will move to the next stage if they survive and the younger individuals will

Table 6.6. Transition matrix for Leslie model using values from table 6.5

	Age 0	Age 1	Age 2	Age 3	Age 4	Age 5	Age 6	Age 7	Age 8	Age 9	Age 10
Birth rate	0	0	0	0	0.500	1.625	3.676	3.941	1.750	0.625	0
Survival 0 to 1	0.700	0	0	0	0	0	0	0	0	0	0
Survival 1 to 2	0	0.800	0	0	0	0	0	0	0	0	0
Survival 2 to 3	0	0	0.800	0	0	0	0	0	0	0	0
Survival 3 to 4	0	0	0	0.800	0	0	0	0	0	0	0
Survival 4 to 5	0	0	0	0	0.800	0	0	0	0	0	0
Survival 5 to 6	0	0	0	0	0	0.850	0	0	0	0	0
Survival 6 to 7	0	0	0	0	0	0	0.850	0	0	0	0
Survival 7 to 8	0	0	0	0	0	0	0	0.800	0	0	0
Survival 8 to 9	0	0	0	0	0	0	0	0	0.800	0	0
Survival 9 to 10	0	0	0	0	0	0	0	0	0	0.750	0

remain. The proportion that is in the oldest cohort within the stage and is ready to leave is

$$G_i = \frac{p_i^{d_i-1}}{(1 + p_i + p_i^2 + \cdots p_i^{d_i-1})} \tag{6.7}$$

This can be simplified by substituting the expression $(1 - p_i^{d_i})/(1 - p_i)$ for the denominator to yield

$$G_i = \frac{p_i^{d_i-1}(1 - p_i)}{(1 - p_i^{d_i})} \tag{6.8}$$

Moving to the Lefkovitch life cycle graph in fig. 6.9 (bottom), note that the braces under the Leslie life cycle graph denote different age classes that have been grouped into stage classes. For example, the preadult stage includes age classes 1–4. Note that within each of these stage classes the probabilities of surviving are constant over time. The stage duration (d_i) can be counted from the top graph, for example, $d_2 = 4$. The stage-specific survival probabilities (p_i) are the constant survival probabilities within a class, $p_2 = 0.8$. P_i and G_i, the probabilities of surviving and growing from stage i to stage i + 1, are calculated as follows.

$$P_i = \left[\frac{1 - p_i^{d_i-1}}{1 - p_i^{d_i}} \right] p_i \tag{6.9}$$

$$G_i = \frac{p_i^{d_i-1}(1 - p_i)}{(1 - p_i^{d_i})} \tag{6.10}$$

and

$$p_i = (G_i + P_i) \tag{6.11}$$

The parameter values for this Lefkovitch model are then given in table 6.7 for the transition matrix, as shown in table 6.8. This final equation shows that the sum of the probabilities for the within- and between-stage transitions is equal to the overall survival probability in each iteration.

The birth elements F_i in the top row of the matrix in table 6.8 are given by the average of the offspring production in stage i and the offspring production in all stages to which the adult in stage i may move during the interval, weighted by the probabilities of transition. Fertility is given by births times the probability of offspring survival. Table 6.9 gives a summary of the calculations for the parameters of both the Leslie and Lefkovitch matrices.

Table 6.7. Parameter values for Lefkovitch age-stage model based on data in table 6.8

Stage	Within-stage, age-specific survival p_i	Stage-specific development time d_i	Stage-specific transition rate G_i	Within-stage survival P_i	Stage-specific fertility F_i
1	0.70	1	0.7	0	0
2	0.80	4	0.14	0.66	0
3	0.85	3	0.24	0.61	4.0
4	0.80	2	0.36	0.44	1.5
5	0.75	1	0.75	0	0

Table 6.8. Transition matrix for Lefkovitch model using values from table 6.7

	Stage 1	Stage 2	Stage 3	Stage 4	Stage 5
Births	0	0	4.0	1.5	0
Growth 1 to 2/ Self-loop 2 to 2	0.70	0.66	0	0	0
Growth 2 to 3/ Self-loop 3 to 3	0	0.14	0.61	0	0
Growth 3 to 4/ Self-loop 4 to 4	0	0	0.24	0.44	0
Self-loop 5 to 5	0	0	0	0.36	0

Table 6.9. Summary of parameter calculations for Leslie and Lefkovitch matrices

		Model	
Step	Parameter	Leslie matrix	Lefkovitch
1	Reproduction	$F_x = \left(\dfrac{m_x + P_x m_{x+1}}{2} \right)$	$F_i = p_i^{1/2} \left(\dfrac{m_i + \sum_j t_{ji} m_j}{2} \right)$
			$p_i = \sum_i t_{ij}$
2	Survival/growth	$P_x = p_x$	$P_i = \left[\dfrac{1 - p_i^{d_i - 1}}{1 - p_i^{d_i}} \right] p_i$
3	Self-loop	—	$G_i = \dfrac{p_i^{d_i - 1} (1 - p_i)}{(1 - p_i^{d_i})}$

Model synthesis

Lotka equation as the foundation

Population models, including both the Leslie and the Lefkovitch models, and the life tables that are used to derive the elements of the matrices for these models have much in common. Most importantly, the life table is a discrete form of the Lotka equation

$$1 = \sum_{x=0} \exp^{-rx} l_x m_x \qquad (6.12)$$

where r is the intrinsic rate of growth and l_x and m_x denote cohort survival and age-specific reproduction, respectively. This equation shows that inserting a column containing m_x-values beside the column of l_x-values from the life table and computing their age-specific products $(l_x m_x)$ constitutes the first step in transforming a life table into a population model. The sum of the products of these two life table columns $(\Sigma l_x m_x)$ yields the per-generation population growth rate, or net reproductive rate, R_0. By setting the sum of the products of $l_x m_x$ and the exponential term $\exp^{-rx}$ equal to unity, $1 = \sum_{x=0} \exp^{-rx} l_x m_x$, one can solve for the intrinsic rate of increase (r); from this value, computing the stable age distribution (SAD) completes the transformation from a life table to the Lotka stable population model. Setting r to zero then sets the model to replacement-level growth $(1 = \Sigma l_x m_x)$, which can then be reconceived as a stationary population model and, in turn, a life table (Preston et al. 2001).

In the next section we extend and apply this concept of the interconnectedness between the basic life table and the Lotka population model to all types of life tables and all types of population models. We show that all life tables can be conceived of as stationary models, and the reverse—that all stationary population models can be conceived of as life tables.

Life tables as stationary population models: Single-decrement life tables

There are two ways of viewing life tables: either as a tool for tracking the survival of a birth cohort (i.e., the cohort life table from chapter 2) or as a stationary population subject to specified mortality rates and replacement-level reproduction (i.e., $R_0 = 1.0$) (Jordan 1967; Land et al. 2005). A stationary model is a special case of the more general population model (Preston et al. 2001).

Converting a life table into a stationary population draws on the same notation but with new meanings, where l_x is the number of individuals who reach age x in any given year, d_x is the number of deaths between x and $x+1$, and e_0 is the mean age at death for individuals dying in any particular year. The number of individuals at a particular age as well as in the population as a whole becomes constant over time because they are the products of a constant number of births and constant survival probabilities. The stationary population model connects the age of death and expectation of life (e_0), birth (b) and death (d) rates, and the age distribution, which means that estimates can be made for one parameter on the basis of another. Because every plant and animal population is subject to its underlying life table, this means that every population can form the basis for a stationary population model (Preston et al. 2001). It follows that the converse is also true, in other words, that every plant and animal population model can form the basis for a life table.

Life tables as stationary population models: Multiple-decrement and multistate life tables

Life table models that extend the single-decrement concepts can be placed into one of two groups: multiple-decrement life tables or multistate life tables. In *multiple-decrement life tables* (Lamb and Siegel 2004), "state" refers to the state of being alive with individuals subject to different age-specific death rates. Considered another way, multiple-decrement life tables allow for different ways to exit life (dying). *Multistate life tables* (Ledent and Zeng 2010), on the other hand, include all life tables in which individuals exist in different states subject to different age- or state-specific risks of dying. Multistate life tables constitute a subset of the broader area of multistate demography involving the study of (mostly human) populations stratified by age, sex, and other attributes such as region of residence, marital status, number of children, employment status, and health status (Willekens 2003). A stratified population is considered multistate, and the individuals within it who occupy the same state(s) constitute subpopulations. Transitions to different states in some life table models are not possible (e.g., two-sex and frailty life tables). However, transitions to other states are possible in other life table models (e.g., married to divorced to married to widowed).

The process of converting a multistate life table into a stationary population is similar to that for a single-decrement life table, where we draw on the same notation but with new meanings for each parameter. For this case l_x again denotes the number of individuals who reach age x in any given year, d_x the number of deaths between x and x + 1, and e_0 the mean age at death for individuals dying in any particular year. However, each of these single-state life table parameters consists of subcategories according to state i, such that

$$l_x^i = \text{fraction of cohort surviving to age x in state i}$$

$$d_x^{ik} = \text{fraction of cohort transitioning from state i to state k at age x} (k \neq i)$$

$$d_x^{i\delta} = \text{fraction of cohort dying, denoted } \delta, \text{ in state i at age x}$$

$$g_x^{ik} = \frac{d_x^{ik}}{L_x^i} = \text{mobility rate from state i to state j} (k \neq i)$$

where

$$L_x^i = \frac{(l_x^i + l_{x+1}^i)}{2}$$

$$g_x^{i\delta} = \frac{d_x^{i\delta}}{L_x^i} = \text{death rate in state i at age x}$$

Linking the number l_{x+1}^i of survivors in state i at age $x + 1$ with the corresponding number l_x^i at age x involves subtracting the deaths $d_x^{i\delta}$ and the moves d_x^{ik} out of state i to all other states and also adds those transitioning into state i from all other states, d_x^{ki}. Thus

$$l_{x+1}^i = l_x^i - d_x^{i\delta} - \sum_{k \neq i} d_x^{ik} + \sum_{k \neq i} d_x^{ki} \qquad (6.13)$$

A key property of multistate life tables is that within an age interval they permit individuals to exit from the state (stage) they occupied at the beginning of the interval to another state, and then they can return to the original state (Land and Rogers 1982; Land et al. 2005).

A general population model derived from the multistate life table in which age- and stage-specific columns of reproduction are paired with the age- and stage-specific reproduction is of the form

$$1 = \sum_{i=1} \sum_{x=0} e^{(-rx)l_x^i m_x^i} \qquad (6.14)$$

If $r = 0$ (stationary), then the exponential term in this expression drops out and the model is given as

$$1 = \sum_{i=1} \sum_{x=0} l_x^i m_x^i \qquad (6.15)$$

Stationary population models as life tables: Leslie model

Just as life tables can be viewed both as a tool for ordering and computing the actuarial properties of a cohort and as a stationary population, the Leslie model (Leslie 1945) can also be viewed in these ways. The Leslie model can be considered a life table by setting all elements in the top row of the transition (T) matrix equal to zero,

$$T = \begin{pmatrix} 0 & 0 & 0 \\ p_0 & 0 & 0 \\ 0 & p_1 & 0 \end{pmatrix} \qquad (6.16)$$

initializing the vector N at $N_0 = 1.0$,

$$N = \begin{pmatrix} N_0 \\ 0 \\ 0 \end{pmatrix} \qquad (6.17)$$

Table 6.10. Life table analogue of a stationary population based on Leslie matrix with zero growth

Age x	Survival l_x
0	$1.00 \ (= N_0)$
1	p_0
2	$p_0 p_1$
3	0.0

and advancing this initial newborn group forward n iterations (n = number of age classes). This process yields the basic life table in table 6.10.

This life table can then be conceived of as a stationary population, as described in the section above, with the fraction age x in the population denoted c_x $\left(\text{where } c_x = \dfrac{l_x}{\sum_0 l_x} \right)$ The Leslie matrix population model achieves stationarity when its parameters are adjusted to replacement-level growth (i.e., $\lambda = 1.0$). Assuming initial birth and death rates that confer positive population growth, the elements of the transition matrix (T) are modified by either decreasing reproduction, decreasing survival, or a combination of the birth and death elements such that the net reproductive rate (R_0) equals 1.0.

Stationary population models as life tables: Lefkovitch model

Just as all nonzero elements (i.e., survival) in the Leslie transition matrix beneath the top row (i.e., reproduction) can be used to construct a life table survival curve, all the nonzero elements beneath the top row in Lefkovitch models can also be used in this manner and thus be conceived of as a multistate (stationary) population. Consider the Lefkovitch transition matrix

$$T = \begin{pmatrix} 0 & 0 & 0 \\ G_1 & P_2 & R_1 \\ 0 & G_2 & P_3 \end{pmatrix} \tag{6.18}$$

This matrix, T, can be used to iterate the vector N initialized with a birth cohort of $N_0 = 1.00$ through time and therefore through age classes within and between stages.

$$N = \begin{pmatrix} N_0 \\ 0 \\ 0 \end{pmatrix} \tag{6.19}$$

Table 6.11. Life table analogue of Lefkovitch three-stage model

Age (x)	Survival by age x and stage i (l_x^i)		
	Stage i = 1	Stage i = 2	Stage i = 3
0	$l_0^1 = 100{,}000$	—	—
1	—	$l_1^2 = P_1 l_0^1$	—
2	—	$l_2^2 = S_2 l_1^2$	$l_2^3 = P_2 l_1^2$
3	—	$l_3^3 = P_2 l_2^2 + S_3 l_2^3$	$l_3^3 = S_2 l_2^2 + R_1 l_2^3$
4	—	$l_4^3 = P_2 l_3^2 + S_3 l_3^3$	$l_4^2 = S_2 l_3^2 + R_1 l_3^3$
⋮	—	⋮	⋮
y	—	$l_y^3 = P_2 l_{y-1}^2 + S_3 l_{y-1}^3$	$l_y^2 = S_2 l_{y-1}^2 + R_1 l_{y-1}^3$
⋮	—	⋮	⋮
TOTALS	Σ years in stage 1	Σ years in stage 2	Σ years in stage 3

Note that the sum totals for the number of years lived in each of the three stages is equal to the stable stage distribution for the Lefkovitch model at stationarity (table 6.11).

The column sums of this multistate life table can be given as

$$\Sigma_{i=1} \Sigma_{x=0} l_x^i \qquad (6.20)$$

Inserting a term for age- and stage-specific reproduction into this equation (m_x^i) and setting it equal to unity (1.00) yields the stationary multistate population model:

$$1 = \Sigma_{i=1} \Sigma_{x=0} l_x^i m_x^i \qquad (6.21)$$

Unique issues with the conversion of a Lefkovitch model to a life table include both open-ended stages and indeterminate maturation. By virtue of a single stage, details about development and maturation are lost. Specifically, the model does not differentiate between 2-year-old and 35-year-old juveniles, the former of which cannot reproduce and the latter of which can reproduce. Yet all individuals in this stage are capable of contributing offspring in the model. Additionally, there is an assumption of immortality because there is no way to build into the model an end age for individuals within any stage except yearlings due to the self-loops. Furthermore, a fraction of individuals remains in each of the last three stages indefinitely. Consequently, at replacement-level or negative growth rates, there will be juveniles that are older than post-reproductive individuals and reproductive adults that are decades younger than juveniles.

Comparison of model properties

Plant and animal matrix projection models have been used in this chapter to illustrate how the basic structure of stage models can be used for a wide range of species with different life cycles. The Leslie and Lefkovitch models can both be used to project population growth rates and evaluate population structure, but as we have noted, there are differences in the parameterization of these two population models. There are, however, a number of general properties that are common across both age- and stage-structured models, including ergodicity, momentum, sensitivity and perturbation analysis, and convergence dynamics. Convergence is faster in a stage-structured than an age-structured model because the number of stages is usually smaller than the number of ages. More rapid convergence in the stage model also occurs because of the multidirectional flows and the self-loops. In a stage model, the youngest individuals to enter a stage can begin entering the next stage in the next iteration. Therefore, in a model in which the third stage is a reproducing adult (e.g., Orca model), a newborn can be reproducing in 3 years even though the actual maturation period is over a decade. Another difference between these models is that stage models do not have a maximal age or final terminal stage. The theoretical mixing of the age classes within a stage means that some individuals can stay within a stage forever. Finally, when the multistate model is converted to a multistate life table with the identical transition probabilities and self-loops, the number of individuals that survive to a given age is dependent upon the initial cohort size.

Consider Further

This chapter further develops the concepts in chapter 5 (Population I: Basic Models), and these concepts are applied in different contexts in chapter 7 (Population III: Extensions of Stable Theory) and in the later chapters on applied demography. The analysis of connected populations, as discussed in the plant model III, can be extended to models of multiregional demography in chapter 7 and to concepts of mark-recapture that are considered in chapter 9 (Applied Demography: Estimating Parameters).

An important reference for a more advanced theoretical discussion of stage-based models is the book by Caswell (2001) titled *Matrix Population Models*. These models have been increasingly used to predict the future status of populations and species conservation, and a good source to get started in this area is a book by Morris and Doak (2002) titled *Quantitative Conservation Biology*. The recent expansion of integral projection models (IPMs) makes it possible to include both continuous and discrete traits in these analyses. These models provide a link between field demographic data and models that can be analyzed to generate new hypotheses, and to develop predictions, about how populations will respond to future environmental changes (Coulson 2012). The paper by Griffith et al. (2016) titled "Demography beyond the Population" from a recent symposium highlights 20 papers that link demography to other disciplines in ecology and evolution. In so doing, these papers innovatively expand the basic demographic tools that are described in this chapter.

Population III: Extensions
of Stable Theory

The task of a scientific theory is clear: It must explain; it may predict.
The demand that a scientific theory predict has been a stumbling block to
any serious study of invention.

Susan Gill (1986, 21)

Classical stable population theory that emerges from both the continuous Lotka model and the discrete Leslie matrix model serves as the foundation for extending these concepts to a wide range of other models, several of which are described in this chapter. These include two-sex models, stochastic rather than deterministic models, models that divide the population spatially into regions (multiregional models), models that specify both age and stage (age/stage models, discussed in chapter 6), and colony-level models that approach population concepts hierarchically.

Two-Sex Models

Classic stable population theory is essentially a one-sex theory because it posits two age-specific schedules—reproduction and mortality—but only for females. The two-sex problem in stable population theory requires taking into account the male-specific survival—the fraction of all births that are male and the role of males in reproduction (Pollak 1986, 1987). Here we describe simpler, more conventional two-sex models (Goodman 1953, 1967) that are *female dominant*; that is, births of both male and female offspring are attributed to the mothers.

Basic two-sex parameters

The starting point for an analysis of a two-sex system is the tabulation of sex-specific birth and death rates for the female cohort and the death rates for the male cohort. Survival rates from birth to age x are denoted l_x^f and l_x^m for females and males, respectively, and the fraction of offspring that are males is denoted s and assumed to be age independent (i.e., constant).

Sex ratio at age x and intrinsic sex ratio

Consider a model in which male and female life tables differ but the rate of population increase of the sexes is the same. If the ratio of male to female births is s, then the sex ratio at age x can be expressed as

$$e^{-rx}l_x^f = \text{number of females age x per newborn female}$$

$$se^{-rx}l_x^m = \text{number of males age x per newborn female}$$

The ratio of males age x to females age x is then

$$\frac{se^{-rx}l_x^m}{e^{-rx}l_x^f} \tag{7.1}$$

or simply

$$\frac{sl_x^m}{l_x^f} \tag{7.2}$$

Note that the sex ratio at age x depends on the sex-specific life tables and the sex ratio at birth, but not on the common rate of increase (Keyfitz and Beekman 1984). The rate of population increase, r, is determined from the Lotka equation using only the female rates.

The intrinsic sex ratio (ISR) is the ratio of males to females that will eventually emerge in a population with fixed age- and sex-specific survival schedules at which male and female offspring are produced (Goodman 1953). It is computed as the ratio of the sums of the exponentially weighted sex-specific survival schedules (male to female) over all ages:

$$ISR = \frac{s\sum_0^\omega e^{-rx}l_x^m}{\sum_0^\omega e^{-rx}l_x^f} \tag{7.3}$$

Examples of the stable age-by-sex distribution of a hypothetical Drosophila population are shown in fig. 7.1. Note the effect of population growth rate on both the within- and between-sex distribution by age. When growth rate is positive, the sex ratio differences are minimal even though adult female expectation of life is nearly twice that for adult males. The middle graph in fig. 7.1 shows the effects on sex ratio due only to differences in survival, in other words, a stationary population.

Age-specific sex ratio

The sex ratio of the stable age distribution is defined as the ratio of the sum of all males at each age in the stable age distribution to the sum of all females at each age in the stable age distribution. This is expressed as

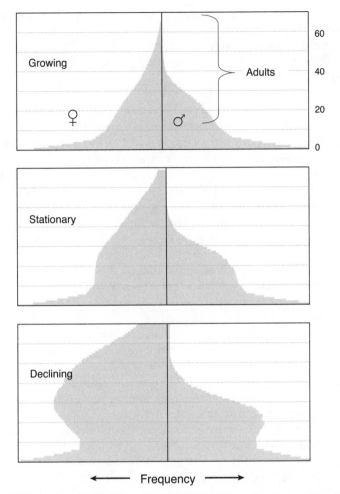

FIGURE 7.1. Age pyramids (0 to 70 days) for hypothetical *D. melanogaster* populations in which preadult survival was assumed to be identical but age-specific mortality in adults differed such that male (right) and female (left) life expectancy, at eclosion, was 20 and 38 days, respectively.

$$SR = \left[\sum_0^\omega e^{-rx} l_x^f m_x^m \right] \frac{\sum_0^\omega e^{-rx} l_x^m}{\sum_0^\omega e^{-rx} l_x^f} \qquad (7.4)$$

where m_x^m denotes the number of male offspring produced by a female age x. Use of this expression assumes that the number of offspring produced depends only on numbers of females in the population and not on the number of males. The expression for sex ratio has an instructive, intuitive interpretation as the product of two terms. One term, the *survival term*, is the ratio of sums over the stable age distribution of males and females, which would be the sex ratio of a population with a primary sex ratio of 1-to-1. The other term, the *reproductive term*, reflects the way in which a primary sex ratio that differs from 1-to-1 distorts the sex ratio. This term is the ratio of the reproductive value of males at age 0 to the reproductive value of females at age 0, the latter of which is unity.

The dependence of the sex ratio on the rate of increase, r, can be understood by examining the effect of r on the survival and reproductive terms. At r = 0, this ratio is determined directly by the difference in survival between males and females. If survival schedules are equal, this ratio would equal 1.0. In a decreasing population, the exponential term emphasizes differences between survival schedules, especially at higher ages, while in an increasing population these differences are diminished.

The reproductive term is an exponentially weighted sum over the product of male fecundity—in other words, the age-specific production of males—and of female survival. This term goes to unity when male fecundity is the same as female fecundity. For unequal fecundities where the growth rate equals zero, this term is the average number of males produced in the lifetime of each female recruited into the population (R_0 for males). For an increasing population this term will be less than R_0 for males, and for a decreasing population the converse will be true.

The dependence of the reproductive term on the rate of increase is demonstrated by the demographic parameters measured in an example with spider mites, where the resulting overall sex ratio depends on both reproductive and survival terms in a way that is not predictable from either term alone (figs. 7.2 and 7.3). In the spider mite, even though males greatly outlive females and males are produced over a longer period, the sex ratio favors females. This results from the fact that the relatively high rate of increase skews each of the three sums in eqn. (7.4) toward the origin. As a result, the disparity in survival schedules in the survival term and the effect of male fecundity in the reproductive term are reduced. This model demonstrates that the sex ratio at age x, and the intrinsic sex ratio, are both conditional on population growth rate if the primary sex ratio is age dependent. The importance of this finding is that it provides at least one possible explanation for the failure of field biologists to find a unique, normal sex ratio in haplo-diploid species such as spider mites in which fertilization, or the lack of fertilization, determines offspring sex.

Stochastic Demography

General background

Stochastic demography is concerned with the theoretical and empirical study of random variation in demographic rates and processes (Tuljapurkar and Orzack 1980; Tuljapurkar 1984; Pressat and Wilson 1987). Understanding the effects of stochasticity on the dynamics of populations is important for the simple reason that real-world populations of both humans and nonhuman species are subject to stochastic conditions. Thus knowledge of the extent to which uncertainty affects birth, death, and/or migration rates is fundamental to demography in general and, more specifically, to the formulation of population policies.

Stochasticity may be introduced into population models in basically two ways. The first is referred to as *demographic stochasticity*. This concept of stochasticity differs

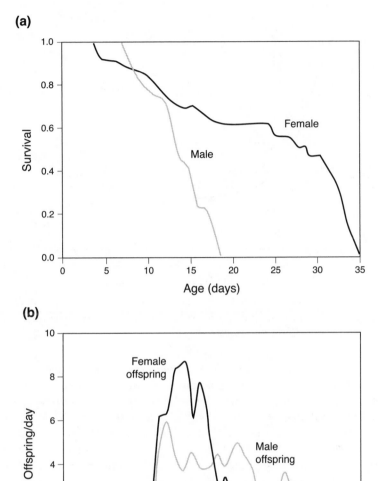

FIGURE 7.2. Survival and reproduction rates for the spider mite, *Tetranychus urticae*: (a) survival, l_x, curves for males and females; (b) age-specific rates of production of male and female offspring. (*Source*: Carey and Bradley 1982)

from the concept of deterministic rates because in a deterministic model each member of the population gives birth to a tiny fraction of an individual in each small interval of time, but in a stochastic model only whole animals are born with specified probabilities. A population goes extinct when its last member dies and this death may be due to chance alone (Goodman 1971). Similarly, the population is reduced to its last member when its second-to-last member dies, and this death may also be due to chance alone.

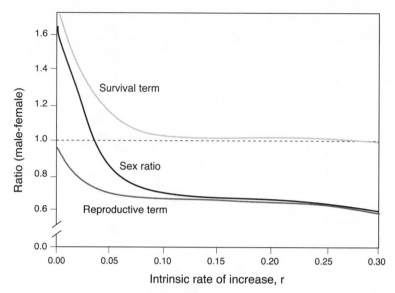

FIGURE 7.3. Relationship of the model's reproductive term, survival term, and sex ratio (male-to-female) to the intrinsic rate of increase, r, using experimental data on the spider mite, *T. urticae*. (*Source*: Carey and Bradley 1982)

The central questions regarding demographic stochasticity have to do with the extent to which chance alone plays a role in population change. Clearly, at low population numbers a single addition or subtraction event may have a substantial impact on the population. A single individual is a small proportion of a population of 1,000 or even of 100. However, this same individual is 10% of a population of 10 and 100% of a population of one. When this one individual dies, the population is extinct, thus demographic stochasticity is especially important at low population numbers and where there is a high likelihood of extinction.

The second way that stochasticity is important to population models is through *environmental stochasticity*. Environmental variation arises when the demographic rates themselves are externally driven by stochastic processes due to environmental randomness, such as bad winters or failures of food supplies. The basic distinction between demographic and environmental stochasticity is that the former arises from fixed vital rates while the latter arises from vital rates that vary over time. Cohen (1982) provides the following example to illustrate the differences between the two and their magnitudes in the case of mortality. Suppose for a population of $N = 1,000,000$ individuals, that the probability of dying within one time period is $q = 0.002$. The variance among populations in the number that die after one time period is $Nq(1-q) = 1,996$. However, if q is a random variable with a mean of 0.002 and a standard deviation of 0.0001, then the variance in the expected number of deaths among populations of a million individuals is $\text{var}(Nq) = N^2\text{var}(q) = 10,000$; thus the environmental variation is over five times the demographic variation.

The difference between demographic and environmental stochasticity for fertility is illustrated by the following example. Suppose that the fertility of a particular age class is 1.5 offspring; however, only whole animals are born, thus this value of fertility might arise because a single individual produces either one or two offspring with equal probability. In this case, if there are N individuals reproducing, then the average number of offspring is 1.5 N. The variance of the total number of offspring is the sum of the variances for each individual, and because individuals reproduce independently, the variance equals

$$N[0.5(0.5)^2 + 0.5(0.5)^2] = 0.25N \qquad (7.5)$$

which is the variance on fertility due to demographic stochasticity. Alternatively, environmental stochasticity may be present if the level of fertility for the entire age class shifts with changes in the environment. For instance, suppose that fertility is either 1.0 or 2.0 with equal probability. The average number of offspring is still 1.5 N, but given that all individuals experience either high or low fertility, the variance is now

$$N^2[0.5(0.5)^2] + [0.5(0.5)^2] = 0.25N^2$$

which is the variance due to environmental stochasticity. Given that these two variances differ by a factor of N^2, it is clear that environmental stochasticity is much more important than demographic stochasticity in this scenario.

Environmental variation in vital rates

The pattern of temporal variation in the environment impacts the correlations in vital rate values within and between years. Here, with a schedule of age-specific fecundities (table 7.1), we illustrate the impact of random or temporally correlated schedules of vital rates on the variance of reproduction.

In the first case, where the schedules of fertility are random across years and thus either two schedules (A or B) are equally likely, there are four possible combinations of rates with a 25% chance for each rate to occur on any given time step. These rates are

$$m_{1A} + m_{2A} = 50 \text{ offspring/day}$$

$$m_{1A} + m_{2B} = 32 \text{ offspring/day}$$

$$m_{1B} + m_{2A} = 23 \text{ offspring/day}$$

$$m_{1B} + m_{2B} = 5 \text{ offspring/day}$$

$$\Sigma = 110 \text{ eggs}$$

The average fecundity of a population experiencing these conditions is $(110/4) = 27.5$ offspring with a variance σ^2 of

Table 7.1. Two hypothetical reproductive schedules used to illustrate the impact of different patterns of environmental variation in vital rates

Age	Schedule A	Schedule B
1	m_{1A} (= 30)	m_{1B} (= 3)
2	m_{2A} (= 20)	m_{2B} (= 2)

$$\sigma^2 = \frac{[(50-27.5)^2 + (32-27.5)^2 + (23-27.5)^2 + (5-27.5)^2]}{4} = 263.3$$

and a standard deviation (SD) of

$$SD = \sqrt{263.3}$$
$$= 16.3$$

In the second case where the schedules of fertility are correlated over time, then both age classes have rates from the same schedule at each time, with the occurrence of schedules A and B equally likely. The two possibilities are

$$Schedule\ A = m_{1A} + m_{2A}$$
$$= 50\ offspring/day$$

$$Schedule\ B = m_{1B} + m_{2B}$$
$$= 5\ offspring/day$$
$$\Sigma = 55\ offspring$$

The average fecundity for this case is $55/2 = 27.5$, which is the same as the random case, but the variance is

$$\sigma^2 = \frac{[(50-27.5)^2 + (5-27.5)^2]}{2}$$
$$= 506.3$$

and the standard deviation (SD) is

$$SD = \sqrt{506.3}$$
$$= 22.5$$

The reason that the variance is lower in the first, random, case is that on average, half the time a high rate in one age class is offset by a low rate in the other; only 25% of the time is fecundity extremely low. In contrast, when the schedules are correlated, the elements within each age are either uniformly high or uniformly low and the daily variance in vital rates is much greater.

Stochastic rate of growth

Two critical demographic problems emerge from these hypothetical examples. First, how do stochastic vital rates impact population growth rate? For example, how does a twofold difference in daily fecundity change growth rate relative to, say, a fivefold difference? Second, how will the average growth rates differ when individual elements are subject to stochastic processes (i.e., demographic stochasticity) versus when individual schedules are subject to stochastic processes (i.e., environmental stochasticity)? Tuljapurkar (1990) developed the framework for examining these questions, including the formula for computing the stochastic rate of population increase, r_s (notation different than in Tuljapurkar 1990). This parameter is not a strict analog of the deterministic "little r" because where r_s is an average of several possible rates, the conventional population growth rate, r, is a singular rate. The formula for the average population growth rate subject to stochastic rates is

$$r_s = r - \frac{c}{2\lambda^2} \tag{7.6}$$

where r denotes the intrinsic rate of increase for a Leslie matrix population whose elements are the average of all rates for each element and c is a term that depends on how the vital rates are correlated as a result of stochasticity. When vital rates vary independently of each other as in the case of environmental stochasticity, then

$$c = \Sigma_0^\infty \sigma^2 \left(\frac{d\lambda}{dm_x} \right)^2 \tag{7.7}$$

Two general points of stochastic growth rate merit comment. First, the value of r_s is never greater than r because the age structure of a population subjected to the changing vital rates is never adjusted to the conditions of the moment (Namboodiri and Suchindran 1987). For example, given a rapid change in vital rates, a population of aphids in which the majority of individuals are in the older adult age classes would not respond as quickly as a population with primarily young individuals. Second, the stochastic growth rate is affected by time filters that modulate variance. For example, long-term growth rates are less affected by large variance in survival in older adults for the same reason that growth rate is affected less by changes in mortality in older adults. This is because these older age groups constitute a small proportion of the total population and, in particular, are in the lower reproductive age classes that have low prospects for continued reproduction.

Strong and weak ergodicity

Population models can incorporate variable vital rates that vary either stochastically or deterministically. The salient result of population models with fixed vital rates is that they "forget" their past and thus exhibit the property referred to as *strong ergodicity*, as age structure and growth rate converge to numerical fixation. Population models with variable rates exhibit *weak ergodicity* because, although population models also "forget" their past, they do not converge to constant growth rates and fixed age structures. Weak ergodicity, with variable vital rates, can be classified in one of two subcategories, either *weak stochastic ergodicity* or *weak deterministic ergodicity*. With weak stochastic ergodicity two populations with identical stochastic rates exhibit similar long-term patterns of growth and age structure. The key concept for this case is "long term"; that is, there are no similarities between the two populations in particular short-term patterns due to the stochastic nature of the vital rates. With weak deterministic ergodicity, two populations subject to the same deterministic sequence of vital rates eventually exhibit identically varying age structures and growth rates independent of differences in their initial conditions. The key concept for this ergodic case is "identically varying"; that is, because changes in growth rate and age structure of populations are subject to the same deterministically varying vital rates, they eventually become identical.

These demographic patterns and properties are illustrated with example projections using vital rates on an aphid species from table 7.2, the results of which are shown in figs. 7.4–7.5. Two important patterns are evident. First, age structures in populations skewed toward adults due to low growth rates may produce major spikes in growth rates when, either by design (deterministically) or by chance (stochastically), the transition to the reproductive schedule containing the highest rates is abrupt. For example, all of the growth spikes in the deterministically varying population shown in fig. 7.4c and fig. 7.5c occur when the reproductive schedule with the highest rates occurs immediately after a sequence of schedules with much lower rates. Likewise, the highest spikes in the population growth rates occur in the populations depicted in fig. 7.4a–c and 7.5a–c immediately after a series of days with low growth rates and thus there are a larger fraction of adults in the population.

The second important aspect from the results of these projections is the potentially profound effect that the length of the deterministic reproductive cycle may have on both growth rate and age structure variability (see figs. 7.4b–c and 7.5b–c), even though both populations are subject to the same set of reproduction schedules. For example, the reproductive schedules that animated the population depicted in fig. 7.4b–c and fig. 7.5c occurred in a 5-day sequence from highest to lowest (i.e., table 7.2 reproductive sequence 3a-3b-3c-3d-3e). In contrast, the population growth rates and age structures depicted in figs. 7.4c and 7.5c were the outcome of a 10-day sequence in which the population was subject to each reproductive schedule for two days in a row (e.g., 3a-3a-3b-3b . . . 3e-3e). Whereas the fraction of the preadults in the population in the shorter cycling population is relatively constant at around 25%, the fraction of preadults in the population with the longer cycle varies between around 10% to over 70%.

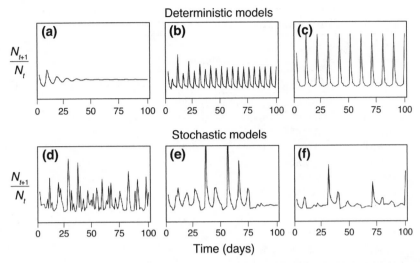

FIGURE 7.4. Population growth rates for matrix projections of aphid populations using constant survival rates and, with the exception of (a), either deterministically or stochastically varying reproductive rates (see table 7.2). All projections are based on the same mean reproductive rate and, with the exception of (a), have equal probability of using one of the same five reproductive schedules—100%, 50%, 25%, 12.5%, and 6.25% of the maximum. Strong and weak ergodicity are properties of the deterministic and stochastic models, respectively.

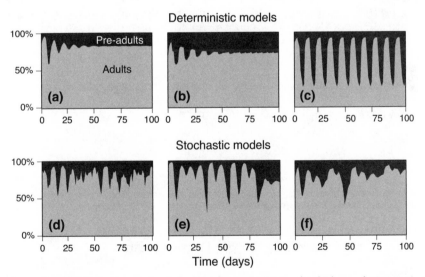

FIGURE 7.5. Population stage structure for matrix projections of aphid populations using constant survival rates and, with the exception of (a), either deterministically or stochastically varying reproductive rates (see table 7.2). The growth rate counterpart of each population is shown in fig. 7.4.

Table 7.2. Survival and reproductive rates for the pea aphid used in population matrix projections that show strong and weak ergodicity in fig. 7.4

Age, x	Survival, l_x	Reproductive levels					Reproductive mean
(1)	(2)	(3a)	(3b)	(3c)	(3d)	(3e)	(4)
0	1.0000	0.000	0.000	0.000	0.000	0.000	0.000
1	0.9700	0.000	0.000	0.000	0.000	0.000	0.000
2	0.9500	0.000	0.000	0.000	0.000	0.000	0.000
3	0.9000	0.000	0.000	0.000	0.000	0.000	0.000
4	0.8800	0.000	0.000	0.000	0.000	0.000	0.000
5	0.8600	0.000	0.000	0.000	0.000	0.000	0.000
6	0.8500	2.200	1.100	0.550	0.275	0.138	0.853
7	0.8200	11.800	5.900	2.950	1.475	0.738	4.573
8	0.8000	12.000	6.000	3.000	1.500	0.750	4.650
9	0.7100	11.800	5.900	2.950	1.475	0.738	4.573
10	0.6500	12.000	6.000	3.000	1.500	0.750	4.650
11	0.5000	11.800	5.900	2.950	1.475	0.738	4.573
12	0.4500	12.200	6.100	3.050	1.525	0.763	4.728
13	0.3400	11.600	5.800	2.900	1.450	0.725	4.495
14	0.3100	10.400	5.200	2.600	1.300	0.650	4.030
15	0.2800	9.600	4.800	2.400	1.200	0.600	3.720
16	0.2500	9.400	4.700	2.350	1.175	0.588	3.643
17	0.2200	8.200	4.100	2.050	1.025	0.513	3.178
18	0.1900	8.000	4.000	2.000	1.000	0.500	3.100
19	0.1400	7.700	3.850	1.925	0.963	0.481	2.984
20	0.1100	6.400	3.200	1.600	0.800	0.400	2.480
22	0.0500	4.000	2.000	1.000	0.500	0.250	1.550
21	0.0700	4.200	2.100	1.050	0.525	0.263	1.628
23	0.0300	3.600	1.800	0.900	0.450	0.225	1.395
24	0.0150	2.200	1.100	0.550	0.275	0.138	0.853
25	0.0060	1.900	0.950	0.475	0.238	0.119	0.736
26	0.0018	1.000	0.500	0.250	0.125	0.063	0.388
27	0.0004	0.840	0.420	0.210	0.105	0.053	0.326
28	0.0000	0.780	0.390	0.195	0.098	0.049	0.302
	12.4	163.62	81.81	40.91	20.45	10.23	63.40

Note: Columns 2 and 3a contain the originally observed demographic data (Frazer 1972). Age-specific reproductive rates in columns 3b–3e are progressively reduced by half of the previous column. Thus, the scalars for reproduction are 1.00, 0.500, 0.25, 0.125, and 0.0625 for columns 3a–3e, respectively.

Multiregional Demography

Rogers (1984, 1995) is recognized as the pioneer of multiregional mathematical demography, an area of demography concerned with the mathematical description of the changes in populations over time and space. This subarea of demography, referred to as multiregional demography, considers persons occupying different regions as being in different states and therefore in different populations, and it "opens" populations to the flow of persons among populations. Rogers demonstrated that generalizing demographic concepts to multiple states was relatively straightforward inasmuch as, assuming constant birth, death, and migration rates, each region (state) would eventually contain a regional share of the overall population and the population itself would eventually converge to fixed age-by-region distribution and a constant rate of growth (Willekens 2003). Many of the approaches to multiregional demography are directly connected to the metapopulation models discussed in chapter 6 and the multistate mark-recapture models discussed in chapter 9. In this section we present three basic two-region models that illustrate the concept of multiregional demography. These include two models not incorporating age structure, one in which location is aggregated by birth origin and another in which location is disaggregated by birth origin. A third model presented here is structured by both age and location.

Location aggregated by birth origin

The simplest population in which migration is considered consists of subpopulations without age structure in each of two regions (Rogers 1984). Let i_A and o_A denote per capita in- and out-migration rates, respectively, for region A, and i_B and o_B denote the corresponding rates for region B. Also let b_A and d_A denote the per capita birth and death rates, respectively, for region A, and b_B and d_B denote the corresponding rates for region B. Suppose that, within one time period, the two regions experienced the changes given in table 7.3.

The rates for region A are interpreted as 0.20 births, 0.05 deaths, and 0.30 out-migrants per individual residing in region A per epoch (table 7.4). The appropriate equations for the numbers in each of the two regions at time $t+1$, denoted $N_{A,t+1}$ and $N_{B,t+1}$, are

$$N_{t+1,A} = N_{t,A}\,(1 + b_A - d_A - o_A) + o_B N_{t,B} \tag{7.8}$$

Table 7.3. Hypothetical data on birth, death, and migration for individuals in a two-region population system

Region	Initial number	Births	Deaths	Migration In	Out
A	1,000	200	50	100	300
B	1,000	100	20	300	100
TOTAL	2,000				

Table 7.4. Per capita birth, death, and out-migration rates

Region A	Region B
$b_A = 200/1000 = 0.20$	$b_B = 100/1000 = 0.10$
$d_A = 50/1000 = 0.05$	$d_B = 20/1000 = 0.02$
$o_A = 300/1000 = 0.30$	$o_B = 100/1000 = 0.10$

Note: Computed from hypothetical data presented in table 7.3.

Table 7.5. Results of the two-region population projection using multiregional population parameters from table 7.4

Time	Number			Growth rate	Percentage	
T	N_A	N_B	Total	λ	% N_A	% N_B
(1)	(2)	(3)	(4)	(5)	(6)	(7)
0	99	1	100	1.15	99.0	1.0
1	84	31	115	1.13	73.3	26.7
2	75	55	130	1.12	57.4	42.6
3	69	77	146	1.11	47.4	52.6
4	66	96	162	1.11	40.9	59.1
5	66	114	180	1.11	36.7	63.3
6	67	131	199	1.10	33.9	66.1
7	70	149	219	1.10	32.1	67.9
8	75	167	242	1.10	30.9	69.1
9	80	186	266	1.10	30.1	69.9
10	87	207	293	1.10	29.6	70.4
11	94	228	323	1.10	29.3	70.7
12	103	252	355	1.10	29.0	71.0

$$N_{t+1,B} = N_{t,B}(1 + b_B - d_B - o_B) + o_A N_{t,A} \qquad (7.9)$$

These equations state that the size of the populations at time $t+1$ result from the increment owing to each region's natural increase, the decrement owing to each region's out-migration, and each region's increment owing to the in-migration from the other region. The results of model iterations through $t=12$ are given in table 7.5.

Note that the percentage of the total population stabilizes at around 29% and 71% for regions A and B, respectively. Also, the overall growth rate approaches a constant rate of $\lambda = 1.1$, which is also the growth rate that each of the two regions will eventually experience. This illustrates that this simple two-region population will eventually possess *fixed regional shares* (29% and 71%) and a *single fixed growth rate* (= 1.1-fold/epoch).

Location disaggregated by birth origin

The preceding model considered only the number in each of the two regions without regard to whether the individuals were born in the region of residence or not. To disaggregate the individuals within a region into those that are native born and those that are immigrants requires modifications. Let $N_{A,t}^B$ denote the number of individuals in region A at time t who were born in region B, where the superscript denotes the region of birth. Analogous terms are $N_{A,t}^A$, $N_{B,t}^A$, and $N_{B,t}^B$. The appropriate accounting relationship for the number of individuals in region A at time $t+1$ who were born in region A, $N_{A,t+1}^A$, consists of three components:

(1) *Net number of individuals.* This component results from the natural increase of individuals in region A who were born in region A and can be expressed as

$$N_{A,t}^{A}\left(1+b_A-d_A-o_A\right) \tag{7.10}$$

(2) *Net number of births by individuals in region A who were born in region B.* This component is expressed as

$$b_A N_{A,t}^{B} \tag{7.11}$$

(3) *Number of emigrants born in region A who, at time t, are located in region B.* This component is expressed as

$$o_B N_{B,t}^{A} \tag{7.12}$$

The number of individuals in region A at time $t+1$ who were born in region A is then

$$N_{A,t+1}^{A}=N_{A,t}^{A}\left(1+b_A-d_A-o_A\right)+b_A N_{A,t}^{B}+o_B N_{B,t}^{A} \tag{7.13}$$

The number of individuals in region A at time $t+1$ who were born in region B is given by

$$N_{A,t+1}^{B}=N_{A,t}^{B}\left(1-d_A-o_A\right)+o_B N_{B,t}^{B} \tag{7.14}$$

Because all births to in-migrants are added to the native population stock, this equation contains no birth rate. The analogous equations for the numbers in region B at time $t+1$ are

$$N_{B,t+1}^{B}=N_{B,t}^{B}\left(1+b_B-d_B-o_B\right)+b_B N_{B,t}^{A}+o_A N_{A,t}^{B} \tag{7.15}$$

$$N_{B,t+1}^{A}=N_{B,t}^{A}\left(1-d_A-o_A\right)+o_A N_{A,t}^{A} \tag{7.16}$$

In matrix form these equations become

$$
\begin{pmatrix}
a_{11} & a_{12} & a_{13} & 0 \\
0 & a_{22} & 0 & a_{24} \\
a_{31} & 0 & a_{33} & 0 \\
0 & a_{42} & a_{43} & a_{44}
\end{pmatrix}
\begin{pmatrix}
N_{A,t}^{A} \\
N_{A,t}^{B} \\
N_{B,t}^{A} \\
N_{B,t}^{B}
\end{pmatrix}
=
\begin{pmatrix}
N_{A,t+1}^{A} \\
N_{A,t+1}^{B} \\
N_{B,t+1}^{A} \\
N_{B,t+1}^{B}
\end{pmatrix}
\tag{7.17}
$$

where

$$a_{11} = 1 + b_A - d_A - o_A \qquad a_{12} = b_A \qquad\qquad a_{13} = o_B \qquad a_{14} = 0$$

$$a_{21} = 0 \qquad\qquad\qquad a_{22} = 1 - d_A - o_A \qquad a_{23} = 0 \qquad a_{24} = o_B$$

$$a_{31} = o_A \qquad\qquad\qquad a_{32} = 0 \qquad\qquad\qquad a_{33} = 0 \qquad a_{34} = 0$$

$$a_{41} = 0 \qquad\qquad\qquad a_{42} = o_A \qquad\qquad\qquad a_{43} = b_B \qquad a_{44} = 1 + b_A - d_B - o_B$$

An example projection is presented in table 7.6, using the parameters from the previous model where the population was not disaggregated by birth type. Similar to the aggregated model (table 7.5), the percentage of the total population in each of the regions stabilizes; in other words, there are *fixed regional shares*. In this second model, the total population in each of the regions stabilizes at 50%. Unique to this two-region population—that is, disaggregated by birth—this population will eventually possess *fixed birth origin shares*. In region A, for example, 21% were native born and 29% were born in region B. There is also a single fixed growth rate. The overall growth rate approaches a constant rate of 1.13, which is also the growth rate that each of the two regions will eventually experience.

Age-by-region projection matrix

The general configuration of a transition matrix for a two-region population with age structure is similar to a Leslie matrix in that it has a top row and a subdiagonal of nonzero elements. However, each is partitioned into groups of 2-by-2 matrices whose elements serve as either the transition from age x in region A to age $x+1$ in region B or a birth in region A to region A, where there is no movement.

Table 7.6. Regional demography disaggregated by birth type

Time t	Number in region A			Fraction in region A		Number in region B			Fraction in region B		Grand total	Growth rate λ
	Born in A	Born in B	Total	Born in A	Born in B	Born in B	Born in A	Total	Born in B	Born in A		
0	10.0	10.0	20.0	0.2500	0.2500	10.0	10.0	20.0	0.2500	0.2500	40.0	0.92
1	11.5	7.5	19.0	0.3026	0.1974	3.0	14.8	17.8	0.0843	0.4157	36.8	1.08
2	11.6	6.4	17.9	0.3228	0.1772	3.5	18.5	22.0	0.0785	0.4215	39.9	1.08
3	11.5	6.0	17.4	0.3284	0.1716	3.5	22.3	25.7	0.0675	0.4325	43.2	1.09
4	11.3	6.1	17.4	0.3242	0.1758	3.4	26.2	29.6	0.0580	0.4420	47.0	1.10
5	11.2	6.6	17.8	0.3142	0.1858	3.4	30.5	33.9	0.0500	0.4500	51.6	1.10
6	11.1	7.3	18.5	0.3015	0.1985	3.3	35.2	38.6	0.0434	0.4566	57.0	1.11
7	11.3	8.3	19.6	0.2881	0.2119	3.3	40.6	43.9	0.0380	0.4620	63.5	1.12
8	11.6	9.4	21.0	0.2753	0.2247	3.4	46.6	50.0	0.0338	0.4662	71.0	1.12
9	12.1	10.8	22.9	0.2638	0.2362	3.5	53.5	57.0	0.0304	0.4696	79.9	1.13
10	12.8	12.4	25.1	0.2538	0.2462	3.6	61.4	65.0	0.0278	0.4722	90.2	—
Stable proportions				0.2135	0.2865				0.0190	0.4810		

The projection matrix for a two-region, four-age-class population is given as

$$
\begin{vmatrix}
0 & b_A^A(1) & b_A^B(1) & b_A^A(2) & b_A^b(2) & b_A^A(3) & b_A^B(3) \\
0 & 0 & b_B^A(1) & b_B^B(1) & b_B^A(2) & b_B^B(2) & b_B^A(3) & b_B^B(3) \\
s_A^A(1) & s_A^B(1) & 0 & 0 & 0 & 0 & 0 & 0 \\
s_B^A(1) & s_B^B(1) & 0 & 0 & 0 & 0 & 0 & 0 \\
0 & 0 & s_A^A(1) & s_A^B(1) & 0 & 0 & 0 & 0 \\
0 & 0 & s_B^A(1) & s_B^B(1) & 0 & 0 & 0 & 0 \\
0 & 0 & 0 & 0 & s_A^A(2) & s_A^B(2) & 0 & 0 \\
0 & 0 & 0 & 0 & s_B^A(2) & s_B^B(2) & 0 & 0
\end{vmatrix}
\begin{vmatrix}
N_{A,t}(0) \\
N_{B,t}(0) \\
N_{A,t}(1) \\
N_{B,t}(1) \\
N_{A,t}(2) \\
N_{B,t}(2) \\
N_{A,t}(3) \\
N_{B,t}(3)
\end{vmatrix}
=
\begin{vmatrix}
N_{A,t+1}(0) \\
N_{B,t+1}(0) \\
N_{A,t+1}(1) \\
N_{B,t+1}(1) \\
N_{A,t+1}(2) \\
N_{B,t+1}(2) \\
N_{A,t+1}(3) \\
N_{B,t+1}(3)
\end{vmatrix}
$$

where the top row of birth elements denote the following:

$b_A^A(x) =$ the number of offspring produced by individuals age x in region A that stay in region A

$b_B^A(x) =$ the number of offspring produced by individuals age x in region A that move to region B

$b_A^B(x) =$ the number of offspring produced by individuals age x in region B that move to region A

$b_B^B(x) =$ the number of offspring produced by individuals age x in region B that stay in region B

Subdiagonals denote

$S_A^A(x) =$ probability of surviving age class x within region A

$S_B^A(x) =$ probability of surviving age class x from region A to region B

$S_A^B(x) =$ probability of surviving age class x from region B to region A

$S_B^B(x) =$ probability of surviving age class x within region B

We used the following values for the matrix elements to project a two-region population from t = 0 through t = 5:

$$M = \begin{vmatrix} 0 & 0 & 2 & 2 & 2 & 2 & 2 & 2 \\ 0 & 0 & 1 & 1 & 1 & 1 & 1 & 1 \\ 0.83 & 0.17 & 0 & 0 & 0 & 0 & 0 & 0 \\ 0.33 & 0.67 & 0 & 0 & 0 & 0 & 0 & 0 \\ 0 & 0 & 0.83 & 0.17 & 0 & 0 & 0 & 0 \\ 0 & 0 & 0.33 & 0.67 & 0 & 0 & 0 & 0 \\ 0 & 0 & 0 & 0.83 & 0.17 & 0 & 0 & 0 \\ 0 & 0 & 0 & 0 & 0.33 & 0.67 & 0 & 0 \end{vmatrix}$$

The results of the first five time steps of the projection, starting with a single individual in each age class for each region, are presented in table 7.7. The results show that the properties of stable theory emerge such that a stable age-by-region distribution is eventually attained with 62% of all residents in region A and 38% in region B. Additionally, the effect of growth within a region supersedes the effect of migration. For example, larger migration rates from region B to region A were swamped by the overall growth rate.

Using values presented in the two-region transitional matrix, M, and initial conditions of single individuals in each of the four age classes within each region, example computations are given below for $N_{A,1}(x)$ and $N_{B,1}(x)$ that are iterated forward from $N_{A,0}(x)$ and $N_{B,0}(x)$:

$$N_{A,1}(0) = (0)(1) + (0)(1) + (2)(1) + (2)(1) + (2)(1) + (2)(1) + (2)(1) + (2)(1) = 12$$

$$N_{B,1}(0) = (0)(1) + (0)(1) + (1)(1) + (1)(1) + (1)(1) + (1)(1) + (1)(1) + (1)(1) = 6$$

$$N_{A,1}(1) = (0.87)(1) + (0.17)(1) + (0)(1) + (0)(1) + (0)(1) + (0)(1) + (0)(1) + (0)(1) = 1$$

$$N_{B,1}(1) = (0.33)(1) + (0.67)(1) + (0)(1) + (0)(1) + (0)(1) + (0)(1) + (0)(1) + (0)(1) = 1$$

$$N_{A,1}(2) = (0)(1) + (0)(1) + (0.87)(1) + (0.17)(1) + (0)(1) + (0)(1) + (0)(1) + (0)(1) = 1$$

$$N_{B,1}(2) = (0)(1) + (0)(1) + (0.33)(1) + (0.67)(1) + (0)(1) + (0)(1) + (0)(1) + (0)(1) = 1$$

$$N_{A,1}(3) = (0)(1) + (0)(1) + (0)(1) + (0)(1) + (0.87)(1) + (0.17)(1) + (0)(1) + (0)(1) = 1$$

$$N_{B,1}(3) = (0)(1) + (0)(1) + (0)(1) + (0)(1) + (0.33)(1) + (0.67)(1) + (0)(1) + (0)(1) = 1$$

Table 7.7. Iterations of the biregional projection model

	Number by age in region A					Number by age in region B					
Time	0	1	2	3	Sum	0	1	2	3	Sum	Grand total
0	1	1	1	1	4	1	1	1	1	4	8
1	12	1	1	1	15	6	1	1	1	9	24
2	12	11	1	1	25	6	8	1	1	16	41
3	46	11	10	1	68	23	8	9	1	41	109
4	81	42	10	10	144	40	31	9	9	90	233
5	224	74	40	10	348	112	54	35	9	210	558

Note: Rounded to whole numbers.

The grand total at time $t = 1$ for the population is 24, which means that the growth rate from time $t = 0$ to time $t = 1$ is 24/8, or threefold, with 15 (61.5%) and 9 (37.5%) individuals in regions A and B, respectively. The stable age-by-region distribution for this two-region population is 37.9%, 15.2%, 6.3%, and 2.7% for ages 0, 1, 2, and 3, respectively, in region A and 18.9%, 11.1%, 5.4%, and 2.5% for ages 0, 1, 2 and 3, respectively, in region B. The stable finite rate of growth for this two-region population is $\lambda = 2.28$.

The addition of age structure to migration models highlights the lag between migration and its manifestations of migrants via their birth and death rates. Keyfitz (1985) commented that the matrix is like a building with a good mixing of air in each room but little circulation between rooms. In other words, we can expect that after any disturbance the within-room variation will settle down to the stable form more quickly than the between-room variation.

Comparison of models

A comparison of the population composition (structure) output of the three multiregional models shows that when there is no age structure (figs. 7.6a and 7.6b) the convergence to fixed fractions in each region is monotonic. Note that fig. 7.6a does not disaggregate the within-region populations by birth origin, whereas in fig. 7.6b there is disaggregation of the population by region into birth origins. When age is added (fig. 7.6c), we see oscillatory behavior due to the age lags as the populations converge to stable age-by-region distributions. These graphs, which represent the simplest of multiregional models, show that dimensionality increases geometrically with each new region, and they illustrate why demographic analysis of multiregional models is generally restricted to closed populations.

Hierarchical Demography

Background

Many biological systems are hierarchically structured, with individuals organized into reproductive units that themselves go through growth cycles, survive or perish, and potentially produce new reproductive units (Al-Khafaji et al. 2008). A good example

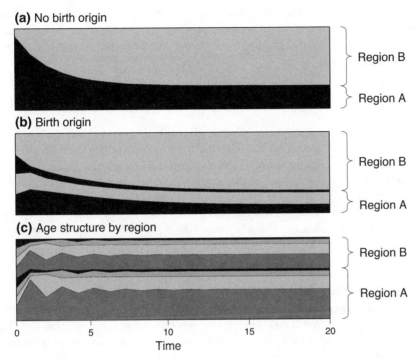

(a) No birth origin

Region B

Region A

(b) Birth origin

Region B

Region A

(c) Age structure by region

Region B

Region A

FIGURE 7.6. Comparison of the structure of the three multiregional models covered in this section.

of this is eusocial insect societies, such as ants and bees; but hierarchical structure is important in many other cases, including modular organisms such as corals and macroalgae, early human groups, some metapopulations, host-parasite dynamics, endosymbionts and their hosts, and assemblages of species. The details of a hierarchically structured population can be quite varied, but the essential commonality is that the dynamics of the population depend on unit-level survival and reproduction. These unit-level dynamics are, in turn, partially determined by the dynamics of individuals within a reproductive unit. For example, differences in growth rates of two populations can result from differences in intra-unit birth and death rates or from differences in unit-level demography, such as time to reproduction. The mapping from individual vital rates and unit-level properties to population growth requires a theory of population dynamics that explicitly incorporates both the hierarchy of individuals within reproductive units and the reproductive units within a population.

Here drawing from the more recent model by Al-Khafaji and his colleagues (2008) and from an earlier version of the model in Carey (1993), we present an analytical framework for the dynamics of hierarchically organized populations in general and for the honeybee (*Apis mellifera*) in particular. Al-Khafaji et al. (2008) developed a renewal equation for reproductive units at the top level in the hierarchy, where individuals comprise reproductive units and the units comprise the population. The growth rate, r, of the number of reproductive units is, then, the growth rate of the population and follows a biological renewal process. In general, the demography (mortality; reproduction) of a reproductive unit may depend on its age and other characteristics, such as multiple queens. For clarity we focus on the case in which age is the only key variable, and we indicate how our analysis can be adapted to a more general setting. Note that when vital rates depend on age, the growth rate r must satisfy an appropriate version of the Lotka equation (see chapter 5).

Honeybee: Individual-to-colony

Concept of superorganism

Wilson (1971, 1975) notes that the term *colony* implies that the members are physically united, differentiated into reproductive and sterile castes, or both. When these two conditions coexist in an advanced stage, the society can be viewed equally well as a superorganism or even as an organism. As Wilson (1971) noted, a dilemma exists that can be stated as follows: At what point does a society become so nearly perfect that it is no longer a society? On what basis do we distinguish the extremely modified zooid of an invertebrate colony from the organ of a metazoan animal?

A superorganism consists of colonies within colonies. An individual within a colony has gonads, somatic tissue, and a circulatory and nervous system. Similarly, the colony of which it is a part may possess features of organization analogous to those physiological properties of the single individual. For example, an insect colony is divided into reproductive castes (analogous to gonads) and worker castes (analogous to somatic tissue). Additionally, it may exchange nutrients by trophallaxis (analogous to the circulatory system), and it may communicate a food source through certain behaviors (analogous to the nervous system). The demographic question in this context is how birth and death rates of the "organ" affect population growth of the "individual."

Assumptions

The model we present makes several assumptions.

(1) A colony contains only one female reproductive queen and nonreproducing female workers. Drone (male) production does not affect worker production rate or death rate and can be excluded from the model, but there are sufficient drones in the population for queen mating.

(2) N_t is the number of workers in a colony at time t, b is the rate at which the queen produces workers, and the worker death rate is μ.

(3) Offspring queens are produced when a swarm is issued and a colony issues a single swarm when it contains N_S workers, the swarming threshold size. After-swarms and absconding are not considered here.

(4) When a swarm is issued, a fraction, g, of the N_S workers present immediately before swarming departs with the resident queen. The fraction $(1-g)$ remains to initiate a growth cycle with the newly emerged queen.

(5) N_0 is the number of workers in a colony at the beginning of a growth cycle. The probability that this colony and its queen survive the growth phase to reach the swarming threshold size is the product of a worker-dependent survival component, $s(N_0)$, and a worker-independent survival component. The worker-dependent component, $s(N_0)$, is the same for all queens

(although N_0 is not). The worker-independent component depends on whether the queen is newly emerged and inheriting her natal colony or is departing in a swarm. These worker-independent components are p_0 for newly emerged queens and q_0 for departing queens. The products of worker-dependent and worker-independent components are called p and q for newly emerged and departing queens, respectively.

(6) There is no regulation of the total number of colonies.

Growth limits

Eusocial insect colonies, such as honeybee colonies, constitute a special kind of population such that virtually all births are directly attributable to a single individual, the queen, while all deaths are attributable to the group (workers). These births and deaths, like any population, are subject to the balancing equation; but unlike populations in which each female has the potential to reproduce, the contribution toward growth through births owing to the individual (i.e., the queen) is offset by the sum of deaths in the group. Consequently, colony size cannot exceed the point where the number of deaths per day in the colony is greater than the maximum daily number of offspring that a queen is capable of producing. Colony growth only occurs after all individuals that die are replaced.

A simple model of this relationship can be derived where, if we let e_0^w denote the worker expectation of life at birth, then $\mu = \dfrac{1}{e_0^w}$ denotes the death rate of workers.

For example, if an individual in a colony of one million workers lives an average of 6 weeks (42 days), then a total of around 25,000 worker deaths occur each day (i.e., $(1/42) \times (10^6)$).

The level of egg production required for queens to maintain the colony is then given as the product of the number of workers in the colony, denoted N, and the per capita worker death rate as

$$\mu N \tag{7.18}$$

Let b denote the maximum number of eggs a queen can produce daily. The product of the per capita death rate of workers and the number of workers yields the number of newborns that the queen must produce each day for replacement, which must equal the maximum number of workers in a colony, denoted N*:

$$N^* \mu = b$$

or

$$N^* = \frac{b}{\mu} \tag{7.19}$$

$$= e_0 b \tag{7.20}$$

Table 7.8. Maximum colony size, given expectation of life for workers and maximum egg production of queens

Expectation of worker life (days)	Daily egg production by queen			
	100	1,000	10,000	100,000
20	2,000	20,000	200,000	2,000,000
50	5,000	50,000	500,000	5,000,000
100	10,000	100,000	1,000,000	10,000,000
200	20,000	200,000	2,000,000	20,000,000

Source: Table 5.12 in Carey (1993).

In words, this expression states that the colony's upper size limit is equal to the product of the expectation of life of workers and the maximum daily egg production rate of the queen. Examples of colony sizes relative to queen productivity and worker life expectancy are given in table 7.8.

There are two implications of the relationship between colony size, queen productivity, and worker life expectancy shown in the table. First, demographic constraints impose an upper limit for colony size. Even when physiological reproductive limits are not considered, a finite amount of time is needed for workers to pick up the eggs for placement in the brood chamber, for example, with ants and termites; or for queens to move between cells and oviposit, for example, with bees and wasps. An example of an upper limit for birth rate in a termite queen is cited by Wilson (1971) for *Odonto-termes obesus*, where the egg-laying capability was reported to be around 86,000 eggs/day (i.e., 1 egg/second). Assuming a 50-day life expectancy for termite workers, this translates to 86,000 deaths daily and thus a maximum colony size of

$$N^* = e_0 b \qquad (7.21)$$

$$= 50 \times 86,000$$

$$= 4.3 \text{ million}$$

A second consequence for limits to colony size is that once the maximum colony size is attained (or the number approached), the only way for growth to continue is by the addition of more reproduction (queens). Colonies with multiple queens are common in some species of termites and ants. Queen addition in honeybees results in, or causes, swarming and is the only way growth can occur once the maximum colony size is attained.

Colony-level dynamics

The rate of change in worker numbers is

$$\frac{dN_t}{dt} = b - \mu N_t \qquad (7.22)$$

where both production and death rates may be functions of worker numbers, age, caste, resource availability, or other environmental variables. The rate at which the queen produces workers (b) is bounded due to physiological limits on the queen, and workers must die at some rate, thus colony growth is intrinsically self-limiting, irrespective of density dependence in b and μ. Without swarming or death, a colony reaches a self-sustaining size N^*, where the number of worker deaths is exactly balanced by the production of new workers:

$$N^* = \frac{b}{\mu} \qquad (7.23)$$

The quantity $(1/\mu)$ is the average life span of a worker, thus the maximum population size (in workers) that a colony is capable of sustaining is the product of the worker production rate and the worker life span. To illustrate the dynamics of this model, we make the simplest assumption—that μ and b are constant—but the demographic structure of our model easily accommodates different functional forms. With this assumption, the number of workers in a colony during a growth cycle is given by

$$N_t = N_0 e^{\mu t} + \frac{b}{\mu}(1 - e^{\mu t}) \qquad (7.24)$$

Here we assume that a colony issues a swarm (see schematic in fig. 7.7) when the worker population reaches a threshold size, N_S, which occurs at some fraction, f, of N^*. Swarming undoubtedly occurs in response to many cues related to within-colony demographic factors (Winston 1987), such as worker population size, worker age distribution, or the number of brood. We assume that swarming is initiated in response to a single cue, N_S. The number of workers at the beginning of a growth cycle, N_0, depends on the parity of the queen (i.e., the number of previous offspring queens). A newly emerged queen (parity zero) inherits her natal colony with $(1-g)N_S$ workers; the departing queen (all higher parities) begins the growth cycle with a swarm of gN_S workers. Using these colony sizes as initial conditions and N_S as the final condition, eqn. (7.23) can be solved to determine that a queen first produces an offspring queen after an interval of time T_1 and then produces offspring queens at intervals of length T_2, where

$$T_1 = \frac{1}{\mu} \ln \left[\frac{1 - (1-g)f}{1-f} \right] \qquad (7.25)$$

$$T_2 = \frac{1}{\mu} \ln \left[\frac{1 - gf}{1-f} \right] \qquad (7.26)$$

The survival probabilities of a queen over the first and subsequent reproductive intervals are given by p and q, respectively. These probabilities have components p_0 and q_0 that are independent of the number of workers (e.g., surviving the nuptial flight) and a component, s, that depends upon the number of attendant workers. Queens are specialized for reproduction and, in honeybees at least, are incapable of surviving without workers to forage and maintain the nest. Colonies with more workers are more successful at defending the nest site, locating and gathering resources, thermoregulating, and performing other functions necessary for the survival of the queen and the colony. The survival of the queen is then an increasing function of initial worker numbers, but is subject to diminishing marginal returns:

$$s(N_0) = \frac{N_0}{\upsilon + N_0} \qquad (7.27)$$

The incremental benefit of each additional worker and the overall importance of workers to queen survival depends on the parameter υ, which reflects environmental conditions. For example, in a relatively benign and constant climate, the thermoregulatory function of workers may be less important and m will be lower than in harsher environments, where more workers would be required to attain a similar survival probability. Worker-independent survival constants p_0 and q_0 may also vary between environments. Thus

$$p = p_0 s[(1-g)N_s] \qquad (7.28)$$

$$q = q_0 s(gN_s) \qquad (7.29)$$

The parameters and equations for this and the next subsection are summarized in tables 7.9–7.10.

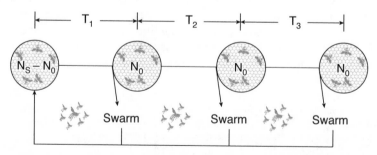

FIGURE 7.7. Honeybee swarming scheme as described by the model. N_S denotes the number at which the colony swarms and N_0 the number that remain at the physical site (parent colony). Therefore, the "newborn" colony depicted in the first circle starts with $(N_S - N_0)$ individuals. (*Source*: Carey 1993; Al-Khafaji et al. 2008)

Table 7.9. Key parameters and notation in hierarchical (honeybee) model

Parameter	Description
Numbers/size	
N_t	Number of workers in colony at time t
N^*	Self-sustaining maximum colony size in absence of fission or death
N_S	Swarming threshold size
Survival rates	
p	Survival of newly emerged queen over T_1 interval
q	Survival of departing queen over T_2 interval
p_0	Worker-independent component of survival for newly emerged queen
q_0	Worker-independent component of survival for departing queen
$s(N_0)$	Worker-dependent component of survival for reigning queen
υ	Strength of worker dependence of queen survival
Worker vital rates	
b	Rate of worker production by queen
μ	Death rate of workers
Swarming proportions	
f	Swarming threshold as a fraction of all workers (N_S/N^*)
g	Proportion of workers departing in swarm
Times	
T_1	Time until a newly emerged queen produces an offspring queen and swarms
T_2	Interswarm interval for departing queens
Rates	
R	Net reproductive rate
r	Population growth rate

Source: Al-Khafaji et al. (2008).

Population-level dynamics

As in the general model, the fertility, m(a), of a queen equals 1.0 when a swarm is issued at the exact ages T_1, $T_1 + T_2$, $T_1 + 2T_2$, $T_1 + 3T_2$, and so forth, and is 0 at all other ages. The survivorship l(a) of a queen to each of these discrete ages is equal to p, pq, pq^2, pq^3, and so forth. Thus the renewal equation reduces to a geometric series whose sum yields

$$1 = qe^{-rT_2} + pe^{-rT_1} \qquad (7.30)$$

Note that this equation for queens and their colonies mirrors what the Euler-Lotka equation is to populations in classical demography. Eqn. (7.30) shows how individual (p and q) and colony (T_1 and T_2) vital rates determine growth in a population of

Table 7.10. Key equations in hierarchical (honeybee) model

Equation	Description
Rates	
$\dfrac{dN_t}{dt} = b - \mu N_t$	Within-colony growth rate
$R = \dfrac{p}{1-q}$	New reproductive rate
$1 = qe^{-rT_2} + pe^{-rT_1}$	Characteristic equation for monogynous, eusocial insects
$r \approx \dfrac{\ln R}{T_1 + \dfrac{qT_2}{(1-q)}}$	Approximation for population growth rate
Number/size	
$N^* = \dfrac{b}{\mu}$	Maximum colony size
$N_t = N_0 e^{\mu t} + \dfrac{b}{\mu}(1 - e^{\mu t})$	Number of workers in colony at time t
Times	
$T_1 = \dfrac{1}{\mu}\ln\left[\dfrac{1-(1-g)f}{1-f}\right]$	Time until a newly emerged queen first swarms
$T_2 = \dfrac{1}{\mu}\ln\left[\dfrac{1-gf}{1-f}\right]$	Interswarm interval for departing queens
Survival	
$s(N_0) = \dfrac{N_0}{\upsilon + N_0}$	Worker-dependent survival
$p = p_0 s\left[(1-g)N_s\right]$	Survival of newly emerged queen
$q = q_0 s(gN_s)$	Survival of departing queen

Source: Al-Khafaji et al. (2008).

colonies. As a fundamental demographic result, eqn. (7.30) applies independently of many details about within-colony and colony-level events. The net reproductive rate of queens, R, can be found by similar methods:

$$R = \frac{p}{1-q} \tag{7.31}$$

The exponential growth rate, r, is greater (or less) than 0 when R is greater (or less) than 1. Expanding the exponential in eqn. (7.29) shows that r can be approximated by

$$r \approx \frac{\ln R}{T_1 + \dfrac{qT_2}{(1-q)}} \qquad (7.32)$$

(Al-Khafaji et al. 2008).

The denominator in eqn. (7.31), by analogy to classical demography, can be regarded as a measure of the cohort generation time appropriate for eusocial queens. Finally, r is not only the growth rate of the number of queens (and colonies) but also the growth rate of the total population. The age composition of queens (and colonies) and workers in our model can be deduced by applying standard demographic methods.

General properties

The hierarchical model described here merits several comments. First, the interswarm intervals (T_1 and T_2) both scale with the swarming threshold size N_S, or equivalently with f, the ratio of swarming threshold to self-limiting colony size. Thus, a large f makes growth cycles long because producing more workers takes a longer time. However, a large f also means that there are more workers when swarms are issued and thus influences queen survival probability. As shown in fig. 7.8, a maximum population growth rate occurs at a swarming threshold well below the self-limiting population size where f = 1.0. Second, the population dynamics are also affected by the trade-off between departing and newly emerged queens. The swarming fraction, g, describes the partitioning of workers between the departing (older) queen and the newly emerged queen. When g is large, many workers will accompany the departing queen and thus T_2 will be short, which allows for a rapid production of subsequent offspring queens. However, the newly emerged queen will be poorly provisioned with workers, which both reduces her survival probability and increases the reproductive maturation times. There is a maximum population growth rate at some intermediate value of g that occurs on a broad fitness plateau rather than at a sharp peak (fig. 7.8). Third, the interplay between levels in the hierarchy is affected by the worker mortality μ and the queen's rate of worker production b. At the individual level, the birth and death rates have straightforward effects; for example, a high μ means higher individual death rates. With respect to population growth rate, however, the impact of μ is more complicated. Both the time until a queen first swarms, T_1, and the time until subsequent swarming, T_2, are inversely proportional to worker mortality such that increasing worker life span—in other words, decreasing worker mortality—for a fixed f increases the time between swarms. However, a longer worker life span also means that N_S is larger, and thus there are more workers available to swarm or to provision the newly emerged queen, which positively impacts their survival. Thus the net effect on population growth rate of reducing worker mortality can be positive or negative due to the balance of impacts on queen survival and interswarm intervals.

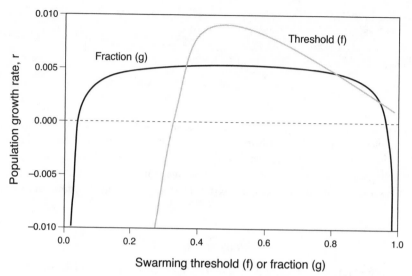

FIGURE 7.8. Effects on population growth rate, r, of the ratio of swarming threshold to self-limiting colony size, f (given g=0.5), and of the proportion of workers departing in the swarm, g (given f=0.9). Worker mortality is $\mu = 1/30$, rate of worker production is b=1,350, worker-independent survival of the newly emerged queen is $p_0 = 0.9$, worker-independent survival of the departing queen is $q_0 = 0.81$, and strength of worker dependence is $\upsilon = 5,000$. (*Source*: Figs. 3–4 in Al-Khafaji et al. 2008)

Consider Further

Content in this chapter involved both analytical extensions of stable theory in the form of two-sex, stochastic, multiregional, and hierarchical models, as well as descriptions of the unifying properties of stable population models—that is, the tendency to forget the past and exhibit ergodic patterns. Because all four of these models extend stable theory they are linked to the content on stable theory in chapter 5 (Population I) and the stage models in chapter 6 (Population II: Stage Models). Many of the narratives in chapter 11 (Biodemography Shorts) are examples of extensions of stable theory. The key literature on the models contained in this chapter as well as on the general properties of stable population models are in Caswell (2001, 2012), Wachter (2003, 2014), Keyfitz and his colleagues (Keyfitz and Beekman 1984; Keyfitz and Caswell 2010), and Tuljapurkar (1989, 1990, 2003).

8

Human Life History and Demography

> Human beings may be unique in their degree of behavioral plasticity and
> in their possession of language and self-awareness, but all of the known
> human systems, biological and social, taken together form only a small
> subset of these displayed by the thousands of living species.
>
> E. O. Wilson (1998, 191)

The overarching goals in this chapter are to provide a synopsis of human traits relevant to biodemography, to describe concepts, methods, and approaches from the
human demography literature that may potentially be useful to scientists studying
nonhuman species, and to introduce demographic content pertaining to humans that
is not often found in conventional demographic literature.

Biodemographic Synopsis

Human demography as evolutionary legacy

The life history of modern humans and, in turn, the demographic components of
human life history are the outcomes of not only our species' evolutionary history,
but also all that preceded it—the four-billion-year history of evolution from the very
first primordial organisms through modern humans (see the 39 levels of evolutionary "rendezvous" in Dawkins 2004).

The demographic manifestations of the primate life history (fig. 8.1) include incipient sociality, sexual dimorphism, low reproductive rates (relative to other mammals),
breast feeding, longer developmental periods and thus later ages of sexual maturation,
and longer life (relative to similar-sized mammals). The great ape evolutionary stage
included demographic innovations, such as increases in sociality and thus in the
concept of community, further reductions in reproduction, extended developmental
and maturation time, and adult longevity, nest building, and recumbent sleeping
posture. The demography of early hominids would likely have been similar to that of
the progenitor of great apes. Based on the life history traits of extant species of apes,
these traits would have included extended maturation times (7 to 13 years), long gestation periods (e.g., 7–9 months), production of singleton offspring at intervals of 3
to 5 years, late weaning age (3–4 years), and life spans from 40 to 60 years (Harvey
et al. 1987; Nowak 1991). Thus it would not have been a great evolutionary stretch
for *Homo sapiens* to have evolved its modern demographic rates in which maturation
and gestation times, as well as life span, are extended further (Jones et al. 1992).

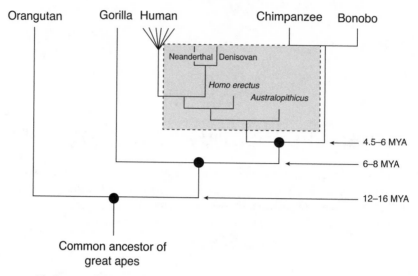

FIGURE 8.1. Phylogeny showing evolutionary relationship of humans within the primate great ape family Hominoidea, a branch of Old World tailless anthropoid primates native to Africa and Southeast Asia—i.e., the chimpanzees (*Pan troglodytes*), bonobos (*Pan paniscus*), gorillas (*Gorilla gorilla*), and orangutans (*Pongo borneo*). Extinct ancestral hominids are shown in shaded box. Timelines for key evolutionary splits are shown in filled circles with times to the right.

Although not demographic per se, the key innovation of the Australopithecine stage of human evolution was the evolution of bipedality—an innovation that allowed for the development of the hand for tool making and more advanced nest building and, in turn, cognitive ability and social complexity. Innovations at the *H. habilis* and *H. erectus* stage included control of fire, hunter-gatherer communities, incipient family concept, reduction in sexual dimorphism, and the beginnings of both childhood and adolescence stages of development. At the *Homo sapiens* (archaic) stage, changes included extended longevity, menopause, pair bonding, child labor, family concept, and secretive (clandestine) mating. Finally, at the *Homo sapiens* (modern) stage, adaptations arose that were adaptive for life in specific climatic and geographic regions ranging from the hot, dry desert to the frigid arctic.

Developmental stages

Anthropologists consider the evolution of four preadult stages of the human life cycle (Bogin 1999; Bogin and Smith 2000), two of which are common in most mammals and all primates (i.e., infant and juvenile) and two of which are unique to humans (i.e., childhood and adolescence). The childhood stage first appeared in *Homo erectus* and the adolescent stage in modern *Homo sapiens*. Based on the growth trajectory shown in fig. 8.2, the duration and age classifications for the evolutionary stages in humans include the following: (1) *infancy* (0–3 years), when maternal lactation provides all or some nourishment to the offspring; (2) *childhood* (3–7 years), a period

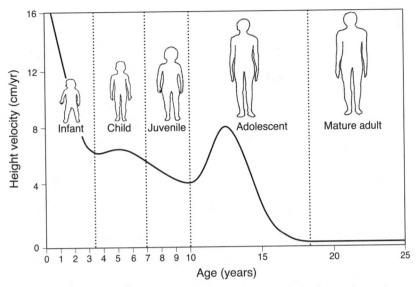

FIGURE 8.2. Human growth trajectory through age 25. Note the high but decreasing growth rate in the early years (0 to 3½ years) followed by an inflection and small "spurt" at 3½ years, which defines the childhood stage. The rate of change in growth rate continues to decrease until age 10 at which time a large spurt occurs (i.e., adolescent spurt). This then decreases to zero at around 18 years at which time adult height is attained. (*Source:* Bogin 1999; Bogin and Smith 2000)

following weaning when offspring still depend on older individuals for feeding assistance and protection; (3) *juvenile* (7–10 years), when the first permanent molars and the brain are maturation milestones but care is still provided by others; (4) *adolescence* (10–19 years), which begins with the onset of puberty and ends with completion of a growth spurt; and (5) *reproductive adulthood* (19–45 years), with the attainment of adult stature, completion of dental maturation, achievement of social maturity and parenthood, and completion of reproduction. Slow early growth followed by a rapid adolescent growth spurt may have facilitated rising human fertility rates and greater investments in neural capital because of the rising demand on parents with multiple, overlapping dependents (Gurven and Walker 2006). Gurven and Walker note that having a greater number of multiple dependents with slow growth and adolescent growth spurts may be a more efficient means of supporting more offspring.

The complexities of modern societies require finer-grained classifications than the framework that emerged from the evolutionary perspective on development. Although in both prehistoric and recent times 18-month-old babies required different levels of parental care than did 30-month-old babies, modern societies are organized differently and thus require different stage classifications that are adaptable to this new social reality. Similar concepts apply to older ages; individuals 50 years old in modern societies are dealing with different sets of work, health, family, and economic issues than are 85-year-olds, thus it is useful to parse the older age groups into more descriptive, functional categories as presented in fig. 8.3.

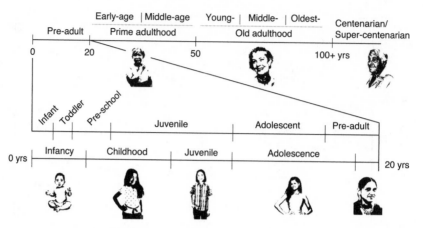

FIGURE 8.3. Stages in the human life course from birth through super-centenarian. Two types of stage classifications are shown from birth through 20 years, including evolutionary (bottom life-line) and sociological/human developmental (top life-line). (*Source*: Bogin 1999; Bogin and Smith 2000)

Reproduction

Basic intervals

Wood (1994) outlines elements describing the general patterns of natural fertility in humans, some of which are part of our primate phylogenetic legacy and some of which are unique to *Homo sapiens*. These elements include late sexual maturation (15–18 years), iteroparity (multiple birth events), low fecundability, long gestation time (9 months), singleton births (usually), prolonged parental care and lactation, long birth intervals (around 3 years), extended reproductive span (from ages 15 to 50 in women), low total reproductive output, and long post-reproductive life for females. All great apes produce only singleton offspring with no overlap in parental care among offspring.

The events determining the reproductive life span and rate of childbearing in modern humans is shown in an idealized schematic (fig. 8.4). The reproductive years start at menarche after which a woman forms a union (e.g., marries) and becomes at risk for childbearing until the onset of menopause (or sterility). Women are reproductive at a rate related to the average duration of the birth interval, the length of which is determined by three components: (1) postpartum infecundable interval, which is primarily a function of breast-feeding behavior; (2) waiting time to conception, which is called the fecundable interval and is a function of natural fecundability and frequency of intercourse; and (3) full-term pregnancy, a segment that is typically 9 months. Time can be added to the birth interval due to intrauterine deaths that adds gestation time from conception to the intrauterine death, the time to return to ovulation, and the conception wait. Thus the timing of menarche and menopause determine the duration of the reproductive period, and the other factors determine the rate of childbearing and the duration of the birth intervals.

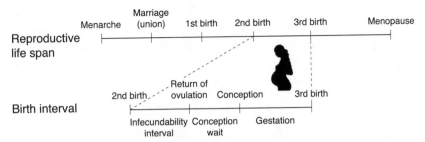

FIGURE 8.4. Events determining the reproductive life span and the rate of childbearing in humans (redrawn from Figure 1.1 in Bongaarts and Potter 1983).

Case study: French-Canadian women

The probability of a woman having a child is strongly influenced by her age (Ivanov and Kandiah 2003), an example of which is shown in fig. 8.5 for French-Canadian women in the seventeenth and eighteenth centuries; but note that the fertility of this specific population of women was extraordinarily high relative to most populations, including indigenous peoples (Howell 1979) and historical populations (Bongaarts and Potter 1983; Ellison and O'Rourke 2000). The onset of reproduction occurs in the early to mid teens, rises rapidly through the early to mid 20s, remains high but relatively constant from the mid 20s through the late 30s, and then decreases to zero by age 50.

The parity compositions by age for these French-Canadian women (fig. 8.6) reveals that by age 30 the majority of women had given birth to from 4 to 6 children, and a small fraction of women had 7 to 9 children. By age 40 nearly three-fourths of the women had from 7 to over 12 children, and by age 50, when their reproductive lives were completed, nearly half had 10 or more children with 17% having over 12 children. With a 9-month gestation period, this meant that for eight of nine children the average woman had both a newborn and a one-year-old (i.e., 9-month gestation + 12-month previous born child). Moreover, for women who had a total of 12 to 23 children this birth interval was even further shortened. Note from fig. 8.6 that over 10% of women age 35 had 10 or more children. By age 50, when child-bearing ended for all women, 17% had 12 or more children whereas only 7% had 1 to 3 children and 5% were childless. Of those who bore children, the average woman had her first child at 22 and her last at 38 with an average birth interval of 21 months (fig. 8.7).

Family evolution

Background

The vast majority of mammalian species produce offspring either singly (ungulates; nonhuman primates) or in litters (canines; felines) and care for their offspring until they are independent before they reproduce again. Some species integrate coopera-tive breeding into their reproductive strategy, where three or more individuals col-lectively raise young in a single brood or litter or assist the breeding pair (Jennions

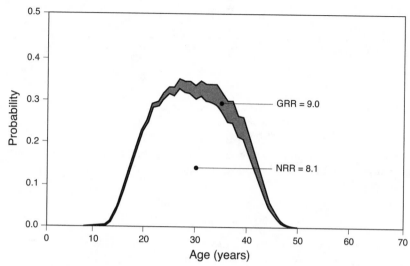

FIGURE 8.5. Age-specific gross reproductive rate (GRR) and net reproductive rate (NRR) in seventeenth- and eighteenth-century French-Canadian women. (*Source*: LeBourg et al. 1993)

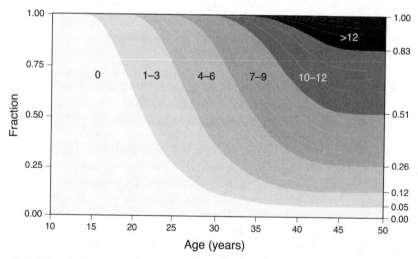

FIGURE 8.6. Historical age-specific parity patterns for French-Canadian women from 10 to 50 years. Each shaded band indicates the fraction of women who have borne the specified number of children by age. For example, at age 40 there were 5%, 8%, 16%, 33%, 22%, and 16% of the women who had 0, 1–3, 4–6, 7–9, 10–12, and >12 children, respectively. (*Source*: LeBourg et al. 1993)

and Macdonald 1994). Evidence for an effect of helpers on breeder fitness has been documented for several nonhuman mammalian species, ranging from coyotes, African lions, and golden jackals to naked mole rats, evening bats, and dwarf mongoose.

Humans evolved a "family model" as a reproductive strategy in which singleton offspring are produced at intervals much shorter than the time needed for them to reach independence. The overlap of the preadult offspring, who stay in the natal home,

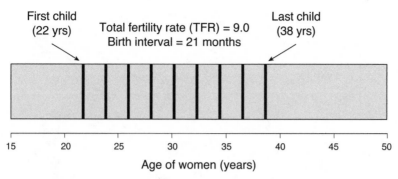

FIGURE 8.7. Summary reproductive metrics of the average French-Canadian woman from the LeBourg et al. (1993) database.

became an intrinsic component of the overall strategy that enabled mothers to reduce their birth intervals and, in turn, increase their reproductive output. Although there is an extensive literature on the contribution of adult female siblings (Croft et al. 2015), grandmothers (Hawkes 2003, 2004; Croft et al. 2015), and males (Hawkes et al. 2001) to the effort of reproducing women, a complementary literature on the critical role of juvenile help has only begun to emerge in the last several decades that provides a more complete explanation for the evolution of the family model (Kramer 2010).

Juvenile help

Because human mothers raise offspring whose ages overlap, they have a built-in set of helpers (Kramer 2011). Although mothers subsidize their children, the children are also important contributors to their mother's fitness. As anthropologist Kathryn A. Kamp notes (Kamp and Theory 2001), children herd, hoe, weed, plant, gather plant foods, trap and hunt animals, collect firewood, bring water, cook, sew, care for other children, clean, and perform much work essential to the economy. Indeed, fitness tasks of hunter-gatherers can be divided into four categories based on strength and skills, including (1) high strength–low skill (e.g., cutting firewood; food processing), (2) high strength–high skill (e.g., tool manufacturing; building shelter), (3) low strength–high skill (e.g., tool manufacturing), and (4) low strength–low skill (e.g., child care; collecting firewood) (Kramer 2011, 536 [box 3]). Adults typically perform tasks in the first three categories. However, juveniles have been observed performing the low strength–low skill category not only for the example tasks above but also for food collection, such as gathering fruits and nuts and hunting shellfish and small game (Bock 2002). This contribution of children to the family group means that the gross requirements of the individual children is partly offset by their contributions, which then reduces their overall net requirements. See the papers by Dorland (2018), Baxter et al. (2017), and Watson (2018) for additional perspectives on the contributions of children to the family and community in prehistoric times.

The *net values* for each of five children and the net balance for a hypothetical mother are shown in fig. 8.8. This figure also shows the *net balance* for this hypothetical mother—that is, the sum of net values for all children alive in each year of a mother's reproductive career. The downward trend from the first to the third birth

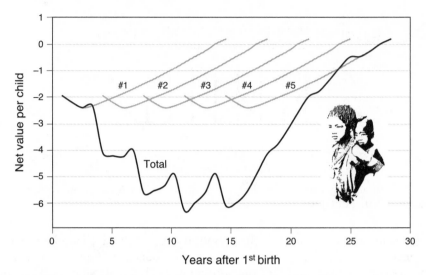

FIGURE 8.8. Net values (gray lines) and total net balance (bold line) for a mother's hypothetical reproductive career, as children (1 to 5) are born into and leave the nest. The unit of measure for these net values is the number of hours mothers have to spend subsidizing children, including time spent producing calories and processing and preparing food, as well as nonfood time, such as shelter construction and clothing manufacturing (plotted from values contained in Table 2 in Kramer 2014).

indicates increasing hourly demands on the mother; and the decade-long "trough" from the third through fifth child shows the offsetting effect of the earlier children from their help (e.g., caregiver inset). Finally, the positive trajectory after the fifth child is born is due to the continued contribution of the older offspring and/or the fact that the older offspring have left the nest.

Human family

The pattern of offspring production and dependency for a hypothetical mother who gives birth to seven children is shown in fig. 8.9a and can be compared to the same information for chimpanzees shown in fig. 8.9b. An average human mother gives birth every 3.1 years from age 19.7 until her last birth at age 39.0, and she survives to age 55.3 ($e_{15} = 40.3$). Each child reaches independence at age 20. The average chimpanzee mother gives birth every 5.9 years from age 14.3, survives until age 29.7 ($e_{15} = 15.4$), and supports one offspring at a time until offspring weaning at 5.3 years (Kramer 2011).

Human life span

Judge and Carey (2000) estimated the evolved life span of both early hominids and modern humans by regressing the average body size (corrected for brain volume) on life span in anthropoid primate subfamilies. They note a major increase in longevity from the 52–56 years of *Homo habilis* to the 60–63 years for *H. erectus* between roughly 1.7 and 2 million years ago. Their analysis indicates that the evolved life span

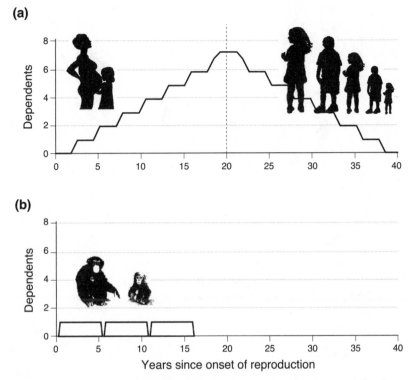

FIGURE 8.9. Comparison of reproductive strategies between (a) humans and (b) chimpanzees. Figure shows the number of dependents in each year across a mother's reproductive career for both species (redrawn from Figure 1 in Kramer 2011).

of modern humans is between 72 and 91 years, noting that it was probably extremely rare for early hominids to live this long (see fig. 8.10). This life span resulted in a primate (*H. sapiens*) with the potential to survive long beyond a mother's ability to birth young. Hammer and Foley (1996) predicted that the life span of *H. habilis* exceeded the age of menopause in extant women by only 7 to 11 years, whereas that of *H. erectus* exceeded menopause by 15–18 years. This literature suggests that post-menopausal survival is not an artifact of modern lifestyle but may have originated between 1 and 2 million years ago, coincident with the radiation of hominids out of Africa. This formulation is important because it takes cessation of reproduction (menopause) as a hominoid given, and it suggests that kin selection was a critical component to prolonged longevity in humans.

Actuarial properties

Human mortality

An example of an age-specific mortality curve for humans is presented in fig. 8.11 using US female mortality in 2006. This is a classic J-shaped mortality curve with several interesting features. First, mortality begins high the first year due to infant mortality, particularly during the first month. Frisbie (2005) thus classifies this very

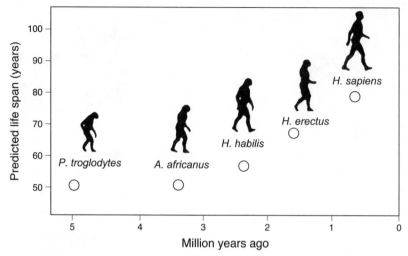

FIGURE 8.10. Predicted life span of selected hominids, including the chimpanzee (*P. troglodytes*), *Australopithecus africanus* ("Lucy"), *Homo habilis*, *H. erectus*, and *H. sapiens* (modern humans) (redrawn from Figure 8 in Judge and Carey 2000).

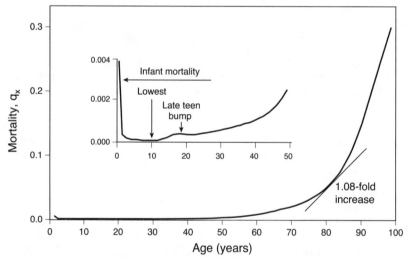

FIGURE 8.11. Age-specific mortality in humans (US females, 2006). (*Source*: Berkeley mortality database)

early infant mortality (i.e., neonatal mortality—deaths under 28 days) separately from the mortality that occurs during the remainder of the first year (i.e., post-neonatal mortality). Mortality drops to a life course low to one-year death probability of around 1 in 8,000 at around 10 or 11 years. Second, there is a teen mortality bump primarily due to an increase in risky behavior (most prominent in males). Finally, after age 30 and continuing until the late 80s or early 90s, there is an exponential increase in mortality at a yearly Gompertzian rate of 8% (Land et al. 2005; Rogers et al. 2005). At the latest years, the mortality probabilities begin to decelerate—in other words, slow their rates of increase over time (Horiuchi 2003; Vaupel 2010). Indeed, Barbi and her colleagues (2018) observed constant hazard curves beyond age 105 in Italian women.

Human life table

An example life table for modern humans in a developed country is presented in table 8.1 for US females in 2006. Several aspects of this table merit comment. First, a remarkable 96% of newborn females survive to age 50. Or expressed conversely, less than 4% of newborns die during the first 5 decades of life. By their early 80s only half of this newborn cohort is dead, as shown in the l_x column, a point that is also reflected in the 80.2-year expectation of life at birth. Additionally, whereas less than 1 in 1,600 5-year-old girls die in their next 5 years, over 25% of 85-year-olds die in their next 5 years and over 90% of centenarians die in their next 5 years. The life table shows that nearly 1 in 5 of all deaths in a newborn cohort occur between 80

Table 8.1. Abridged life table for 2006 US females

Age	Number alive	Fraction alive	Fraction surviving x to x+5	Fraction dying x to x+5	Fraction dying in interval x to x+5	Expectation of remaining life
x	N_x	l_x	$_5p_x$	$_5q_x$	$_5d_x$	e_x
(1)	(2)	(3)	(4)	(5)	(6)	(7)
0	100,000	1.0000	0.9948	0.0052	0.0052	80.2
5	99,482	0.9948	0.9994	0.0006	0.0006	75.7
10	99,426	0.9943	0.9993	0.0007	0.0007	70.7
15	99,352	0.9935	0.9981	0.0019	0.0019	65.7
20	99,162	0.9916	0.9980	0.0020	0.0020	60.9
25	98,966	0.9897	0.9975	0.0025	0.0025	56.0
30	98,714	0.9871	0.9966	0.0034	0.0034	51.1
35	98,374	0.9837	0.9954	0.0046	0.0045	46.3
40	97,925	0.9792	0.9938	0.0062	0.0061	41.5
45	97,319	0.9732	0.9904	0.0096	0.0093	36.7
50	96,389	0.9639	0.9836	0.0164	0.0159	32.1
55	94,804	0.9480	0.9718	0.0282	0.0267	27.5
60	92,130	0.9213	0.9532	0.0468	0.0431	23.3
65	87,818	0.8782	0.9222	0.0778	0.0683	19.3
70	80,989	0.8099	0.8890	0.1110	0.0899	15.7
75	72,001	0.7200	0.8319	0.1681	0.1211	12.3
80	59,895	0.5990	0.7405	0.2595	0.1554	9.3
85	44,355	0.4436	0.5889	0.4111	0.1823	6.6
90	26,122	0.2612	0.3753	0.6247	0.1632	4.5
95	9,803	0.0980	0.1819	0.8181	0.0802	2.9
100	1,783	0.0178	0.0694	0.9306	0.0166	2.0
105	124	0.0012	0.0099	0.9901	0.0012	1.4
110	1	0.0000	0.0000	1.0000	0.0000	0.8
115	0	0.0000				

Source: UC Berkeley mortality database.

and 85 and nearly 1 in 2 of all deaths occur in the 15-year period between 80 and 95 years of age. Finally, less than 2% of newborns survive to 100 years at which time these centenarians have, on average, only 2 more years to live.

Migration patterns

Migration in humans is age-specific (White and Lindstrom 2005). Rogers and Castro (1981) note that young adults in their early twenties show the highest migration rates and young teenagers the lowest. This is because the migration rates of children mirror those of their parents, and therefore the migration rates of infants exceed those of adolescents. The general age pattern of migration (fig. 8.12) is the sum of four mathematical components, including (1) a single negative exponential curve of the pre–labor force ages (from 0 to 30 years); (2) a left-skewed unimodal curve at the labor force ages (from mid teens to late 60s); (3) an almost bell-shaped curve at the post–labor force ages (late 60s to late 70s); and (4) a constant term to improve the overall fit. Both the relative and absolute heights of each of the first three curves, as well as their particular age ranges, depend on a variety of factors ranging from gender and country to era and historical events.

Health Demography I: Active Life Expectancy

According to Pol and Thomas (1992, 1), health demography is "the application of the content and methods of demography to the study of health status and health behavior. Thus, health demography concerns itself with the manner in which such

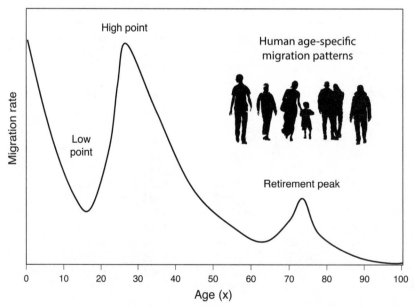

FIGURE 8.12. Model human migration schedule (redrawn from Figure 4 in Rogers and Castro 1981).

factors as age, marital status, and income influence both the health status and health behaviors of populations and, in turn, how health-related phenomena affect health attributes." However, mortality and cause-of-death statistics based on death certificates continue to be the mainstay of health assessment in demography and epidemiology (Kawachi and Subramanian 2005).

Concepts

The study of health biodemography in general and the computation of health expectancies in particular were first developed to address whether or not longer life is being accompanied by an increase in the time lived in good health (i.e., compression of morbidity) or in bad health (i.e., morbidity expansion). The basic concept of health expectancies is to divide life expectancy into life spent in different states of health. *Disability* is a general term for a health condition that limits functioning (WHO 2001). More specific metrics for assessing functional limitations use multiple measures and are termed *activities of daily living* (ADLs), which are concerned with personal care routines, such as eating, dressing, toileting, grooming, transferring in and out of bed, and bathing (Katz et al. 1983; Lamb and Siegel 2004). Disability measures that involve more complex routines associated with independent living are *instrumental activities of daily living* (IADLs), which include using the telephone, shopping, and handling money. Combining data on health or disability status of populations together with mortality data in a life table can be used to generate estimates of expected years of life in various health states. *Disability-free life expectancy* (DFLE) can then be calculated for comparing the health status of different populations, such as across countries during the 1970s and 1980s (WHO 2014).

A general model of mortality, morbidity, and disability is presented in fig. 8.13. The areas under curve A represent the healthy life expectancy, in other words, the number of person-years lived both morbidity- and disability-free (Hayward and Warner 2005). The areas between curves A and B and curves B and C represent the number of person-years lived with a chronic condition (morbidity) and the number of person-years lived with a disability, respectively. The combined area between curves

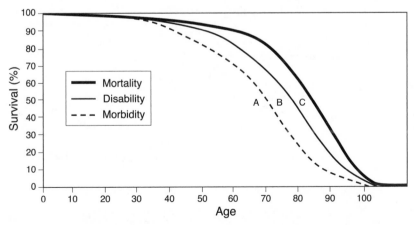

FIGURE 8.13. The general model of health transition (redrawn from Hayward and Warner 2005). (*Source*: WHO 2001)

A and C represent person-years lived with a chronic condition (morbidity and disability). These survival curves make it possible to more precisely distinguish changes in health at advanced ages, and such measures of population health are important for a number of reasons (Lamb and Siegel 2004). First, they make it possible to compare the health of one population with the health of another population. Second, they make it possible to compare the health of the same population at different points in time and to identify and quantify health inequalities within populations. At the policy level they can thus be used to inform debates on priorities for health service delivery and planning and to analyze the benefits of health interventions for use in cost-effectiveness analyses. Moreover, it should be noted that disability is not necessarily a permanent condition or one that inevitably precedes death in the older population (Hayward and Warner 2005). It is also associated with both fatal and nonfatal chronic conditions, so it does not necessarily represent the final stage of poor health prior to death.

Basic life table model

Disability-free life expectancy is a summary measure of a population composed of two sets of partial measures: age-specific death rates and age-specific rates of population morbidity, disability, or health-related quality of life that accounts for the morbidity. The model, described below, is similar to a multistate life table in which the individuals alive at age x of the life table are disaggregated into two (or more) states, healthy or unhealthy (Land and Rogers 1982). The method to estimate healthy life expectancy (HLE), the expected number of years in good health, is based on the Sullivan method (Sullivan 1971),

$$\text{DFLE}_x = \frac{1}{l_x} \sum_{y=x}^{\omega} (1 - {}_n\pi_y)\, {}_nL_y \tag{8.1}$$

where DFLE_x denotes disability-free life expectancy at age x, or the number of remaining years of healthy life for persons who have reached age x; l_x is the number of survivors at age x; ${}_n\pi_x$ is the fraction of individuals age x to $x+n$ who are unhealthy; $(1 - {}_n\pi_x)$ is the age-specific rate of being healthy; ${}_nL_x$ is the total number of years lived by a cohort in the age interval x to $x+n$; and ω is the oldest age.

Disabled life expectancy at age x, denoted DLE_x, is given as

$$\text{DLE}_x = \frac{1}{l_x} \sum_{y=x}^{\omega} (1 - {}_n\pi_y)\, {}_nL_y \tag{8.2}$$

Thus, total life expectancy at age x, e_x, disaggregated into disabled and healthy components, is given as

$$e_x = \text{DFLE}_x + \text{DLE}_x = \frac{1}{l_x} \sum_{y=x}^{\omega} {}_nL_y \tag{8.3}$$

Worked problem

A worked example showing calculations for age-specific disability-free life expectancy ($DFLE_x$) is given in table 8.2 using data and methods outlined by Molla and his colleagues (2001). The expected years of life (e_0) and of healthy life ($DFLE_0$) at birth (age 0) were computed as 79.5 and 69.8 years, respectively. The expectation of unhealthy life or of disability life expectancy at age 0 (DLE_0) is then the difference between these two values, or 9.7 years. This means that a newborn white woman, subject to the 1995 US mortality and morbidity rates, would expect to live 79.5 years of which 9.7 ($\approx 12\%$) would be unhealthy. Similarly, at age 65 the average newborn white woman, subject to these rates, would be expected to live an additional 18.9 years of which 5.1 years ($\approx 27\%$), or more than one-quarter, would be unhealthy.

Column 1: Age classes x to $x + 5$.

Column 2: Fraction of the original cohort surviving to age x, l_x. This is the same as in an ordinary cohort table.

Column 3: $_nL_x$ gives the number of life-years lived in the interval x to $x + 5$.

Column 4: $_5\pi_x$ is the fraction of the population in the interval from x to $x + 5$ who are disabled. This fraction is typically estimated from cross-sectional surveys.

Column 5: $(1 - _5\pi_x)$ is the fraction of the population in the interval from x to $x + 5$ who are healthy (i.e., not disabled).

Column 6: $(1 - 5\pi x)_5L_x$ is the number of *healthy life-years* lived in the interval x to $x + 5$.

Column 7: Time spent free of disability after age x, which is the sum of values in column 6 from x to the end of the column.

Column 8: Expectation of time to be spent free of disability, which is computed by dividing column 7 by column 2.

Health Demography II: The Multiple-Decrement Life Table

Background

The multiple-decrement life tables are used widely in human actuarial studies to address questions concerning the frequency of occurrence for causes of death and how life expectancy might change if certain causes were eliminated. The conventional single-decrement life table shows the probability of survivorship of an individual subject to one undifferentiated hazard of death. In multiple-decrement tables the individual is subject to a number of mutually exclusive hazards, such as disease, predators, or parasites, which means that there is more than one way of exiting (Carey 1989; Preston et al. 2001).

Two probabilities, and hence two kinds of tables, are commonly recognized in the study of cause of death. One is the probability of dying of a certain cause in the presence of other causes, which uses the "proper" multiple-decrement table. The other is the probability of dying of a certain cause in the absence of other causes, which is associated with the single-decrement table (Preston et al. 1972). The assumption of the multiple-decrement life table is that multiple causes of death act independently,

Table 8.2. Calculation of age-specific disability-free life expectancy (DFLE$_x$) for white females by Sullivan's method using an abridged US life table in 1995

Age (x)	Number alive at the beginning of each interval l_x	Number of years lived in age interval $_5L_x$	Proportion of persons in age interval in state considered unhealthy $_5\pi_x$	Proportion of persons in age interval healthy state $(1-_5\pi_x)$	Number of healthy years lived in age interval $(1-_5\pi_x)_5L_x$	Number of years lived in healthy state in this and all subsequent intervals $_5T_x$	Number of years in healthy state remaining at beginning of age interval DFLE$_x$
(1)	(2)	(3)	(4)	(5)	(6)	(7)	(8)
0	100,000	497,211	0.0185	0.9815	488,012	6,981,686	69.8
5	99,321	496,412	0.0196	0.9804	486,682	6,493,674	65.4
10	99,247	496,020	0.0189	0.9811	486,645	6,006,992	60.5
15	99,156	495,294	0.0435	0.9565	473,749	5,520,347	55.7
20	98,938	494,163	0.0490	0.9510	469,949	5,046,598	51.0
25	98,720	492,962	0.0617	0.9383	462,546	4,576,649	46.4
30	98,455	491,441	0.0614	0.9386	461,266	4,114,103	41.8
35	98,094	489,247	0.0773	0.9227	451,428	3,652,837	37.2
40	97,580	486,191	0.0890	0.9110	442,920	3,201,409	32.8
45	96,861	481,715	0.1094	0.8906	429,015	2,758,489	28.5
50	95,764	474,612	0.1506	0.8494	403,136	2,329,473	24.3
55	93,969	463,278	0.1919	0.8081	374,375	1,926,338	20.5
60	91,152	445,546	0.2031	0.7969	355,055	1,551,963	17.0
65	86,772	419,113	0.2257	0.7743	324,520	1,196,908	13.8
70	80,441	381,366	0.2364	0.7636	291,211	872,388	10.8
75	71,408	328,775	0.2782	0.7218	237,310	581,177	8.1
80	59,051	257,187	0.3298	0.6702	172,367	343,867	5.8
85+	42,880	255,399	0.3285	0.6715	171,501	171,501	4.0

Source: Molla et al. (2001). Note: The expectations of life at ages 0 and 65 are 79.5 and 18.9 years, respectively.

and it is concerned with the probability that an individual will die of a certain cause in the presence of other causes. The concept itself stems from reliability theory in operations research. Keyfitz (1985) uses an example of a watch that can operate only as long as all its parts are functioning and each part has its own life table. The probability that an individual (i.e., the watch) will survive to a given age is the product of the independent probabilities that each of its components will "survive" to that age. The same notion of probabilities applied to internal components causing the death of a system can be applied to external components, such as disease and accidents in humans or predation and parasitism in nonhuman species. The concept here is that the probability of an individual surviving to a certain age (or stage) is the product of all independent risk probabilities.

In general, multiple-decrement theory is concerned with three questions (Jordan 1967; Elandt-Johnson 1980; Namboodiri and Suchindran 1987):

(1) What is the age (stage) distribution of deaths from different causes acting simultaneously in a given population?

(2) What is the probability that a newborn individual will die after a given age or stage from a specified cause?

(3) How might the mortality pattern or expectation of life change if certain causes were eliminated?

The first two questions are concerned with evaluating patterns and rates of mortality, while the last question is concerned with "competing risk analysis." In all cases the analyses are based on three assumptions. First, each death is due to a single cause. Second, every individual in a population has exactly the same probability of dying from any of the causes operating in the population. Third, the probability of dying from any given cause is independent of the probability of dying from any other cause of death.

Worked problem

Cause-of-death data

In table 8.3 we present the data on cause of death by age group for US females from 1999 to 2001. With an initial number of 10 million individuals, the number of deaths between x and x+n by cause is denoted $_nD_x^i$, and the total number of deaths in this age interval is $_nD_x$. Note that there was a total of approximately 1.8, 6.3, 0.34, and 1.6 million deaths due to neoplasms, cardiovascular and heart-related, accidents and suicides, and "other," respectively. The highest number of deaths by age group was cardiovascular and heart-related in 85- to 90-year-olds. Accidents were the primary cause of death in young children 1 to 5 years old and for deaths in 15- to 20-year-olds.

The number of deaths by cause i,

$$_nd_x^i = \frac{_nD_x^i}{D} \tag{8.4}$$

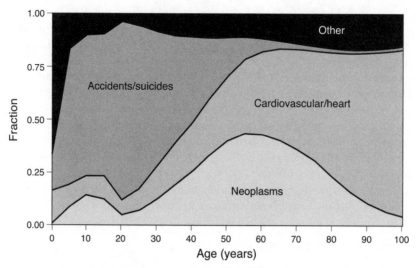

FIGURE 8.14. Frequency distribution (area chart) of three main categories of death and "other" by age in a human population plotted from data presented in table 8.4. (*Source*: Arias et al. 2013)

where D denotes all deaths. The fraction of all deaths in the interval x to x + n is thus the sum of deaths over all causes given as

$$_nd_x = \sum_{i=1}^{4} {}_nd_x \tag{8.5}$$

Plotting deaths as a fraction by cause (fig. 8.14) reveals several age patterns. First, "other" is extremely high in infants due to neonatal and post-neonatal causes. Then, from 10 to 40 years accidents/suicides account for over two-thirds of all deaths. At around age 50, there is a major transition when neoplasms and cardiovascular and heart diseases account for around 75% of all deaths; finally, at the older ages, 7 of 10 deaths are heart-related.

Probability of dying of cause i

The following general formula is used to calculate the probability that a person aged x will eventually die from cause i between ages y and y + s (Arias et al. 2013*):*

$$_\infty q_x = \frac{_\infty d_x^i}{l_x} \tag{8.6}$$

The number of life table deaths by cause due to the ith cause of death, $_nd_x^i$, is estimated by means of the approximation

$$_{n}r_{x}^{i} = \frac{_{n}D_{x}^{i}}{_{n}D_{x}} \qquad (8.7)$$

This formula is applied to $_{n}d_{x}$ for the age categories 0–1 year and 1–5 years, and by 5-year age intervals for ages 0 and over (see tables 8.4 and 8.5).

Competing risk estimation

Competing risk estimation is based on the total probability law of statistics for the geometric mean. Let g denote the proportion of deaths due to cause A. Its complement, $1-g$, then gives the proportion of deaths due to cause B (or all other causes). If p denotes survival probability of surviving both causes as competing risks, it follows that

$$\ln(p) = g\ln(p) + (1-g)\ln(p) \qquad (8.8)$$

and

$$p = p^{g}p^{(1-g)} \qquad (8.9)$$

The independent risks are then computed as

$$p^{g} = \text{risk due cause A} \qquad (8.10)$$

$$p^{(1-g)} = \text{risk due cause AB} \qquad (8.11)$$

For example, if $p = 0.75$ and $g = 0.8$, then the independent risk for cause A is

$$p^{g} = 0.75^{0.8} = 0.7944$$

and for cause B the independent risk is

$$p^{(1-g)} = 0.75^{0.2} = 0.7899$$

Note the verification:

$$0.75 = 0.7899 \times 0.7944$$

Elimination of cause tables

For ages from 1 to 40 years, accidents and suicides were the predominant causes of death, with 40% to over 80% occurring in this group (table 8.6). Around 60%–80% of all deaths between ages 40 and 65 were due to neoplasms and cardiovascular combined in roughly equal proportions. For the oldest ages, 65 through 100 years, cardiovascular and heart-related deaths accounted for most deaths, starting at around 50% at age 65 and increasing to 80% by 100 years of age.

The period survival probabilities with cause i eliminated (table 8.7) are used to compute the cause-eliminated counterparts' survival $l_x^{(-i)}$ (table 8.8) and life expectancy $e_x^{(-i)}$ (table 8.9). Elimination of cardiovascular and heart-related causes of death has the most significant impact on survival and life expectancy. Whereas eliminating neoplasms increases life expectancy by 2.7 years, eliminating cardiovascular and heart-related diseases increases life expectancy by 7.1 years.

Table 8.3. Cause of death by age category in white US females, 1999–2001

| Age (years) | Number surviving | Number of deaths by cause i, $_nD_x^i$ | | | | Total deaths |
		Neoplasms $_nD_x^1$	Cardiovascular/ heart-related $_nD_x^2$	Accidents/ suicides $_nD_x^3$	Other $_nD_x^4$	$_nD_x$
0–1	10,000,000	104	1,861	2,130	7,989	12,084
1–5	9,987,916	572	697	4,316	1,097	6,682
5–10	9,981,233	654	412	3,056	461	4,583
10–15	9,976,650	649	567	3,579	508	5,303
15–20	9,971,347	849	1,244	14,817	627	17,537
20–25	9,953,810	1,250	1,863	13,766	995	17,874
25–30	9,935,936	2,303	2,893	12,070	1,575	18,841
30–35	9,917,095	4,900	5,134	13,267	2,699	26,000
35–40	9,891,095	10,705	9,717	17,021	4,598	42,041
40–45	9,849,053	21,302	17,208	18,586	7,272	64,368
45–50	9,784,685	37,606	28,782	17,683	10,712	94,783
50–55	9,689,902	65,592	53,556	15,831	16,595	151,574
55–60	9,538,328	107,295	98,276	15,596	29,392	250,559
60–65	9,287,769	159,744	172,774	15,175	50,667	398,360
65–70	8,889,409	209,197	278,354	16,030	81,989	585,570
70–75	8,303,840	261,799	451,294	19,233	131,645	863,971
75–80	7,439,870	294,531	757,774	27,080	210,723	1,290,108
80–85	6,149,763	272,558	1,138,773	34,342	294,997	1,740,670
85–90	4,409,092	197,506	1,392,993	36,594	327,968	1,955,061
90–95	2,454,031	103,294	1,227,040	28,781	260,680	1,619,795
95–100	834,237	35,353	657,775	13,913	127,196	834,237
100–105	0	—	—	—	—	—
		1,787,764	6,298,985	342,865	1,570,386	10,000,000

Source: Arias et al. (2013).

Table 8.4. Fraction of all deaths between x and x+n due to cause i, $_nD_x^i$

| Age interval (years) | Cohort survival | Cause of death (i) | | | | Fraction of total deaths |
| | | Neoplasms (i=1) | Cardio-vascular (i=2) | Accidents/suicides (i=3) | Other (i=4) | |
x to x+n	l_x	$_nD_x^1$	$_nD_x^2$	$_nD_x^3$	$_nD_x^4$	$_nd_x$
0–1	1.0000	0.0000	0.0002	0.0002	0.0008	0.0012
1–5	0.9988	0.0001	0.0001	0.0004	0.0001	0.0007
5–10	0.9981	0.0001	0.0000	0.0003	0.0000	0.0005
10–15	0.9977	0.0001	0.0001	0.0004	0.0001	0.0005
15–20	0.9971	0.0001	0.0001	0.0015	0.0001	0.0018
20–25	0.9954	0.0001	0.0002	0.0014	0.0001	0.0018
25–30	0.9936	0.0002	0.0003	0.0012	0.0002	0.0019
30–35	0.9917	0.0005	0.0005	0.0013	0.0003	0.0026
35–40	0.9891	0.0011	0.0010	0.0017	0.0005	0.0042
40–45	0.9849	0.0021	0.0017	0.0019	0.0007	0.0064
45–50	0.9785	0.0038	0.0029	0.0018	0.0011	0.0095
50–55	0.9690	0.0066	0.0054	0.0016	0.0017	0.0152
55–60	0.9538	0.0107	0.0098	0.0016	0.0029	0.0251
60–65	0.9288	0.0160	0.0173	0.0015	0.0051	0.0398
65–70	0.8889	0.0209	0.0278	0.0016	0.0082	0.0586
70–75	0.8304	0.0262	0.0451	0.0019	0.0132	0.0864
75–80	0.7440	0.0295	0.0758	0.0027	0.0211	0.1290
80–85	0.6150	0.0273	0.1139	0.0034	0.0295	0.1741
85–90	0.4409	0.0198	0.1393	0.0037	0.0328	0.1955
90–95	0.2454	0.0103	0.1227	0.0029	0.0261	0.1620
95–100	0.0834	0.0035	0.0658	0.0014	0.0127	0.0834
100–105	0.0000					
		0.17878	0.62990	0.03429	0.15704	1.00000

Source: Arias et al. (2013).

Note: The sum of each column of causes of death equals the fraction of all deaths in the life course attributable to cause i.

Family Demography

Family demography is a subfield of demography and, in the broadest sense, is the study of households and family units, including their structure and the processes that shape them. Families and households can be extraordinarily complex due to the demographic processes or family-related events such as marriage, divorce, and childbearing. Changes in the timing, number, and sequences of these events transform family and household composition. As Watkins and her coworkers note (1987), every individual is at some time a member of a family. And every society defines family roles, the dynamics of which are based on family memberships as well as on the particularities of age, sex, and marital status that specify family positions or statuses. Because longer life means that people can spend more time in the roles of child, parent, grandparent, and spouse, longer life thus alters the demographic foundations of family roles.

Table 8.5. Probability of eventually dying of cause i at age x, $_\infty q_x^i$

Age interval (years)	Neoplasms (i=1)	Cardiovascular (i=2)	Accidents/ suicides (i=3)	Other (i=4)
x to x+n	$_\infty q_x^1$	$_\infty q_x^2$	$_\infty q_x^3$	$_\infty q_x^4$
0–1	0.1788	0.6299	0.0343	0.1570
1–5	0.1790	0.6305	0.0341	0.1564
5–10	0.1790	0.6308	0.0337	0.1564
10–15	0.1791	0.6311	0.0334	0.1564
15–20	0.1791	0.6314	0.0331	0.1565
20–25	0.1793	0.6323	0.0316	0.1567
25–30	0.1795	0.6333	0.0303	0.1569
30–35	0.1796	0.6342	0.0292	0.1570
35–40	0.1796	0.6354	0.0279	0.1572
40–45	0.1793	0.6371	0.0263	0.1574
45–50	0.1783	0.6395	0.0246	0.1577
50–55	0.1761	0.6428	0.0230	0.1581
55–60	0.1721	0.6474	0.0217	0.1589
60–65	0.1652	0.6543	0.0206	0.1600
65–70	0.1546	0.6642	0.0198	0.1615
70–75	0.1403	0.6775	0.0193	0.1630
75–80	0.1214	0.6955	0.0189	0.1642
80–85	0.0990	0.7182	0.0185	0.1644
85–90	0.0762	0.7434	0.0180	0.1624
90–95	0.0565	0.7680	0.0174	0.1581
95–100	0.0424	0.7885	0.0167	0.1525
100–105				

Source: Arias et al. (2013).

Understanding family demography is important for several reasons (Höhn 1987; Ruggles 2012). First, family-level demography situates a specific birth rate into a family context of parents, offspring, and siblings. This, in turn, sheds important light on the within-family sex ratio of children and siblings, including information on sex preselection and stopping rules. For example, if a couple desires one child of each sex or one son, then they might keep trying to have children if they don't have their target number. Second, family demography provides information on both potential and real caregiving arrangements, including both parents caring for children or for their elderly parents. Additionally, family demography sheds light on statistics regarding widowhood and widowerhood, including the period of each or the time from the death of a spouse to remarriage.

Family life cycle

The classic model that is used as a baseline for the family life cycle consists of six stages starting with marriage and ending with the death of the last spouse, and the stages are separated by a specific event where the end of one stage is the beginning of the next (table 8.10). Note that the model can be collapsed into fewer stages if the couple is childless (in which case the family is never actually formed so the model is

Table 8.6. Fraction of deaths due to cause i that occur in the age interval x to x+n

Age interval (years)	Neoplasms (i=1)	Cardiovascular (i=2)	Accidents/ suicides (i=3)	Other (i=4)
x to x+n	$_nd_x^1$	$_nd_x^2$	$_nd_x^3$	$_nd_x^4$
0–1	0.0086	0.1540	0.1763	0.6611
1–5	0.0856	0.1043	0.6459	0.1642
5–10	0.1427	0.0899	0.6668	0.1006
10–15	0.1224	0.1069	0.6749	0.0958
15–20	0.0484	0.0709	0.8449	0.0358
20–25	0.0699	0.1042	0.7702	0.0557
25–30	0.1222	0.1535	0.6406	0.0836
30–35	0.1885	0.1975	0.5103	0.1038
35–40	0.2546	0.2311	0.4049	0.1094
40–45	0.3309	0.2673	0.2887	0.1130
45–50	0.3968	0.3037	0.1866	0.1130
50–55	0.4327	0.3533	0.1044	0.1095
55–60	0.4282	0.3922	0.0622	0.1173
60–65	0.4010	0.4337	0.0381	0.1272
65–70	0.3573	0.4754	0.0274	0.1400
70–75	0.3030	0.5223	0.0223	0.1524
75–80	0.2283	0.5874	0.0210	0.1633
80–85	0.1566	0.6542	0.0197	0.1695
85–90	0.1010	0.7125	0.0187	0.1678
90–95	0.0638	0.7575	0.0178	0.1609
95–100	0.0424	0.7885	0.0167	0.1525
100–105				

Note: The sum of each row is 1.0.
Source: Arias et al. (2013).

not relevant) or if the couple has a single child, in which case stages II and III are the same because extension to one child is also the completed extension. This classic model may apply only to a minority of families since there are innumerable variations due to divorce and remarriage, including the emergence of blended families, adoption of older children, splitting of children due to either preference or separate custody arrangements, and so forth.

Example: Charles Darwin's family

Although most scientists are aware of the seminal contributions made by the giants to their respective fields, few are familiar with details of the domestic lives of these great scholars, including the particulars of their marital status and children. With several notable exceptions for those who chose not to marry or have children—including Gregory Mendel, the Augustinian friar considered the father of genetics, and the English mathematician Sir Isaac Newton—the majority of the pioneers in demography and biology married and had children. For example, "Bills of Mortality" author John Graunt had one child, French naturalist Comte de Buffon had two children, and English cleric Thomas Malthus (*Essay on Population*) and British naturalist Henry

Table 8.7. Period survival from x to x+n when cause i is eliminated

Age interval (years)	Neoplasms (i=1)	Cardiovascular (i=2)	Accidents/ suicides (i=3)	Other (i=4)
x to x+n	$_nP_x^{(-1)}$	$_nP_x^{(-2)}$	$_nP_x^{(-3)}$	$_nP_x^{(-4)}$
0–1	0.9988	0.9990	0.9990	0.9996
1–5	0.9994	0.9994	0.9998	0.9994
5–10	0.9996	0.9996	0.9998	0.9996
10–15	0.9995	0.9995	0.9998	0.9995
15–20	0.9983	0.9984	0.9997	0.9983
20–25	0.9983	0.9984	0.9996	0.9983
25–30	0.9983	0.9984	0.9993	0.9983
30–35	0.9979	0.9979	0.9987	0.9977
35–40	0.9968	0.9967	0.9975	0.9962
40–45	0.9956	0.9952	0.9953	0.9942
45–50	0.9941	0.9932	0.9921	0.9914
50–55	0.9911	0.9899	0.9860	0.9861
55–60	0.9849	0.9840	0.9753	0.9768
60–65	0.9741	0.9755	0.9587	0.9625
65–70	0.9571	0.9649	0.9359	0.9431
70–75	0.9263	0.9489	0.8981	0.9111
75–80	0.8633	0.9244	0.8299	0.8527
80–85	0.7553	0.8913	0.7217	0.7585
85–90	0.5905	0.8450	0.5627	0.6141
90–95	0.3642	0.7698	0.3465	0.4044
95–100	0.0000	0.0000	0.0000	0.0000
100–105				

Source: Arias et al. (2013).

Wallace both had three children. Indeed, it appears that most giants of science had relatively small families. This was not the case for the person widely considered to be the world's greatest biologist—Charles Darwin. He was not only the man who introduced the concept of evolution by natural selection, he was also the married father of 10 children. We examine the dynamics of his family using the marriage model.

A brief background will provide important context. Born in 1809 Darwin was the fifth of six children; his father was a wealthy society doctor and his mother was the granddaughter of Josiah Wedgwood, who had made his fortune in pottery. As a recent graduate of Cambridge University, Darwin embarked on his around-the-world voyage on the *Beagle* on December 27, 1831. Six weeks into the trip he celebrated his twenty-third birthday. When he returned on October 2, 1836, he was four months short of his twenty-eighth birthday. He presumably had no marriage prospects because he had been traveling around the world for five years; however, in fewer than two years after his return he was married (January 29, 1839) to his older first cousin, Emma Wedgwood. In less than three years they were parents of their first-born (William Erasmas Darwin).

Table 8.8. Cohort survival from birth to age x with cause i eliminated

Age interval (years)	Cohort survival	Cohort survival when cause of death (i) is eliminated $p_x^{(-i)}$			
		Neoplasms (i=1)	Cardiovascular (i=2)	Accidents/ suicides (i=3)	Other (i=4)
x to x+n	l_x	$l_x^{(-1)}$	$l_x^{(-2)}$	$l_x^{(-3)}$	$l_x^{(-4)}$
0–1	1.0000	1.0000	1.0000	1.0000	1.0000
1–5	0.9988	0.9988	0.9990	0.9990	0.9996
5–10	0.9981	0.9982	0.9984	0.9988	0.9990
10–15	0.9977	0.9978	0.9980	0.9986	0.9986
15–20	0.9971	0.9973	0.9975	0.9984	0.9981
20–25	0.9954	0.9957	0.9959	0.9982	0.9964
25–30	0.9936	0.9940	0.9943	0.9978	0.9948
30–35	0.9917	0.9923	0.9927	0.9971	0.9930
35–40	0.9891	0.9902	0.9906	0.9958	0.9907
40–45	0.9849	0.9871	0.9873	0.9933	0.9869
45–50	0.9785	0.9828	0.9826	0.9887	0.9812
50–55	0.9690	0.9770	0.9760	0.9809	0.9728
55–60	0.9538	0.9683	0.9661	0.9671	0.9592
60–65	0.9288	0.9537	0.9506	0.9433	0.9369
65–70	0.8889	0.9290	0.9273	0.9043	0.9018
70–75	0.8304	0.8892	0.8947	0.8463	0.8504
75–80	0.7440	0.8236	0.8489	0.7601	0.7748
80–85	0.6150	0.7111	0.7848	0.6308	0.6607
85–90	0.4409	0.5371	0.6995	0.4553	0.5012
90–95	0.2454	0.3171	0.5911	0.2562	0.3078
95–100	0.0834	0.1155	0.4550	0.0888	0.1245
100–105		0.0000	0.0000	0.0000	0.0000
		0.17878	0.62990	0.03429	0.15704

Source: Arias et al. (2013).

A number of interesting aspects of Darwin's life emerge from the information contained in tables 8.11 and 8.12. First, both Charles and Emma were 30 years old or older when they married. This was older than the average marital age in that era and was a late start for creating a family consisting of 10 children. (For women's average marital age see Crafts 1978). Second, the couple added children over a span of 17 years with an average birth interval of around 20 months. Third, Charles was 50 years old when his *Origin of Species* was published (1859), which was the same year that his first child (William Erasmus Darwin) was leaving the family home to attend Cambridge University. Fourth, his children ranged in age from 22 to 34 years old when he died at 73. Finally, Emma was 74 when Charles died and thus a widow for 14 years prior to her own death in 1896 at 88 years. The Charles Darwin "family" thus lasted 57 years. See Costa (2017) and Nicholls (2017) for interesting perspectives on the role of Darwin's house, home, and family in his science.

Clark (2014) notes that the seven surviving children of Charles Darwin produced only nine grandchildren, an average of 1.3/child, who, in turn, produced only

Table 8.9. Person-years lived in interval x to x+n with cause i eliminated

Age interval (years)	Cohort survival	Cause of death (i)			
		Neoplasms (i=1)	Cardiovascular (i=2)	Accidents/ suicides (i=3)	Other (i=4)
x to x+n	$_nL_x$	$_nL_x^{(-1)}$	$_nL_x^{(-2)}$	$_nL_x^{(-3)}$	$_nL_x^{(-4)}$
0–1	0.9994	0.9994	0.9995	0.9995	0.9998
1–5	4.9923	4.9925	4.9934	4.9944	4.9966
5–10	4.9895	4.9900	4.9909	4.9935	4.9941
10–15	4.9870	4.9878	4.9886	4.9926	4.9919
15–20	4.9813	4.9825	4.9834	4.9915	4.9865
20–25	4.9724	4.9742	4.9753	4.9898	4.9780
25–30	4.9633	4.9659	4.9673	4.9871	4.9695
30–35	4.9520	4.9564	4.9581	4.9822	4.9593
35–40	4.9350	4.9433	4.9448	4.9727	4.9441
40–45	4.9084	4.9247	4.9248	4.9548	4.9204
45–50	4.8686	4.8995	4.8964	4.9238	4.8850
50–55	4.8071	4.8634	4.8551	4.8699	4.8300
55–60	4.7065	4.8050	4.7916	4.7759	4.7404
60–65	4.5443	4.7067	4.6945	4.6189	4.5968
65–70	4.2983	4.5454	4.5548	4.3766	4.3805
70–75	3.9359	4.2820	4.3591	4.0161	4.0632
75–80	3.3974	3.8367	4.0843	3.4774	3.5888
80–85	2.6397	3.1203	3.7107	2.7152	2.9047
85–90	1.7158	2.1355	3.2264	1.7786	2.0223
90–95	0.8221	1.0816	2.6151	0.8624	1.0805
95–100	0.2086	0.2887	1.1375	0.2219	0.3111
100–105	0.0000	0.0000	0.0000	0.0000	0.0000
	81.6	84.3	88.7	82.5	83.1

Source: Arias et al. (2013).

Table 8.10. Stages of the classic family life cycle model

Stages	Bounding events	
	Beginning	End
I. Formation	Marriage	Birth of first child
II. Extension	Birth of first child	Birth of last child
III. Completed extension	Birth of last child	First child leaves home
IV. Contraction	First child leaves home	Last child leaves home
V. Completed contraction	Last child leaves home	First spouse dies
VI. Dissolution	First spouse dies	Last spouse dies

Source: Table 4.1 in Höhn (1987).

Table 8.11. Family life cycle table for Charles Darwin

Family name		Parents	Name	Birth	Death	Age at death
DARWIN	Father	Charles Robert Darwin	12 February 1809	19 April 1882	73
	Mother	Emma Wedgwood	2 May 1808	7 October 1896	88

	Age (years)			Children										
	Parents			William Erasmus	Anne Elizabeth	Mary Eleanor	Henrietta Mary	George Howard	Elizabeth	Francis	Leonard	Horace	Charles Waring	Children average age
Year	Mother	Father	Family	1	2	3	4	5	6	7	8	9	10	
1839	31	30	0	0										0.0
1840	32	31	1	1	0									1.0
1841	33	32	2	2	0									1.0
1842	34	33	3	3	1	0								1.3
1843	35	34	4	4	2	—	0							2.0
1844	36	35	5	5	3	—	1							3.0
1845	37	36	6	6	4	—	2	0						3.0
1846	38	37	7	7	5	—	3	1						4.0
1847	39	38	8	8	6	—	4	2	0					4.0
1848	40	39	9	9	7	—	5	3	1	0				4.2
1849	41	40	10	10	8	—	6	4	2	1				5.2
1850	42	41	11	11	9	—	7	5	3	2	0			5.3
1851	43	42	12	12	10	—	8	6	4	3	1	0		5.5
1852	44	43	13	13	—	—	9	7	5	4	2	1		5.9
1853	45	44	14	14	—	—	10	8	6	5	3	2		6.9
1854	46	45	15	15	—	—	11	9	7	6	4	3		7.9
1855	47	46	16	16	—	—	12	10	8	7	5	4		8.9

(continued)

Table 8.11. (*continued*)

Family name	Parents	Name	Birth	Death	Age at death
DARWIN	Father	Charles Robert Darwin	12 February 1809	19 April 1882	73
	Mother	Emma Wedgwood	2 May 1808	7 October 1896	88

| | Age (years) | | | Children | | | | | | | | | | Children average age |
| | Parents | | | | | | | | | | | | | |
Year	Mother	Father	Family	William Erasmus 1	Anne Elizabeth 2	Mary Eleanor 3	Henrietta Mary 4	George Howard 5	Elizabeth 6	Francis 7	Leonard 8	Horace 9	Charles Waring 10	
1856	48	47	17	17	—	—	13	11	9	8	6	5	0	8.6
1857	49	48	18	18	—	—	14	12	10	9	7	6	1	9.6
1858	50	49	19	19	—	—	15	13	11	10	8	7	2	10.6
1859[a]	51	50	20	20	—	—	16	14	12	11	9	8	—	12.9
1860	52	51	21	21	—	—	17	15	13	12	10	9	—	13.9
1861	53	52	22	22	—	—	18	16	14	13	11	10	—	14.9
1862	54	53	23	23	—	—	19	17	15	14	12	11	—	15.9
1863	55	54	24	24	—	—	20	18	16	15	13	12	—	16.9
1864	56	55	25	25	—	—	21	19	17	16	14	13	—	17.9
1865	57	56	26	26	—	—	22	20	18	17	15	14	—	18.9
1866	58	57	27	27	—	—	23	21	19	18	16	15	—	19.9
1867	59	58	28	28	—	—	24	22	20	19	17	16	—	20.9
1868	60	59	29	29	—	—	25	23	21	20	18	17	—	21.9
1869	61	60	30	30	—	—	26	24	22	21	19	18	—	22.9
1870	62	61	31	31	—	—	27	25	23	22	20	19	—	23.9
1871	63	62	32	32	—	—	28	26	24	23	21	20	—	24.9
1872	64	63	33	33	—	—	29	27	25	24	22	21	—	25.9
1873	65	64	34	34	—	—	30	28	26	25	23	22	—	26.9
1874	66	65	35	35	—	—	31	29	27	26	24	23	—	27.9

Year													
1875	67	66	36	—	—	32	30	28	27	25	24	—	28.9
1876	68	67	37	—	—	33	31	29	28	26	25	—	29.9
1877	69	68	38	—	—	34	32	30	29	27	26	—	30.9
1878	70	69	39	—	—	35	33	31	30	28	27	—	31.9
1879	71	70	40	—	—	36	34	32	31	29	28	—	32.9
1880	72	71	41	—	—	37	35	33	32	30	29	—	33.9
1881	73	72	42	—	—	38	36	34	33	31	30	—	34.9
1882[b]	**74**	**73**	**43**	—	—	**39**	**37**	**35**	**34**	**32**	**31**	—	**35.9**
1883	75	—	44	—	—	40	38	36	35	33	32	—	36.9
1884	76	—	45	—	—	41	39	37	36	34	33	—	37.9
1885	77	—	46	—	—	42	40	38	37	35	34	—	38.9
1886	78	—	47	—	—	43	41	39	38	36	35	—	39.9
1887	79	—	48	—	—	44	42	40	39	37	36	—	40.9
1888	80	—	49	—	—	45	43	41	40	38	37	—	41.9
1889	81	—	50	—	—	46	44	42	41	39	38	—	42.9
1890	82	—	51	—	—	47	45	43	42	40	39	—	43.9
1891	83	—	52	—	—	48	46	44	43	41	40	—	44.9
1892	84	—	53	—	—	49	47	45	44	42	41	—	45.9
1893	85	—	54	—	—	50	48	46	45	43	42	—	46.9
1894	86	—	55	—	—	51	49	47	46	44	43	—	47.9
1895	87	—	56	—	—	52	50	48	47	45	44	—	48.9
1896[c]	**88**	—	**57**	—	—	**53**	**51**	**49**	**48**	**46**	**45**	—	**49.9**
Age of death	88	73	75	10	0	84	67	79	77	93	77	2	79.2[d]

[a]Year *Origin of Species* was published
[b]Year Charles Darwin died
[c]Year Emma died
[d]Average age of Charles and Emma's children who lived beyond 10 years

Table 8.12. Summary of Charles and Emma Darwin's family life cycle

Stage	Year		Duration (years)		Parent age (years)	
	Begin	End	Stage	Cumulative	Emma	Charles
I. Formation	1839	1839	<1	<1	31	30
II. Extension	1839	1856	17	17	48	47
III. Completed extension	1856	1858	2	19	50	49
IV. Contraction	1858	1870	12	31	62	61
V. Completed contraction	1870	1882	12	43	74	73
VI. Dissolution	1882	1896	14	57	88	—

20 great-grandchildren, which is 2.2/grandchild, and who, in turn again, produced 28 great-great-grandchildren, 1.4 each. By the time the last generation was born, around 1918, the average family size for this elite group was less than replacement fertility. But, with respect to social mobility, the 27 adult great-great-grandchildren of Charles Darwin, born nearly 150 years after Darwin, are still a distinguished cohort with 11 notable enough to have Wikipedia pages or to appear in the *Times* obituaries. These include six university professors, four authors, a painter, three medical doctors, a well-known conservationist, and a film director.

Kinship

Concepts

Kinship is an extension of genetic or ethnic groupings that can be further divided into relatedness categories, such as tribes, clans, or families. Individuals within family groups are typically organized around a central person (whom we designate as Ego) in one of two ways: either as a common progenitor (e.g., great-grandfather) or as a descendant (e.g., son; grandson). The focus in both cases is on a person's genealogy (direct lineage for either descendant or progenitor) and not on the more complete case of their consanguinity (kinship). That is, everyone is part of a *nuclear family* (parents and children), *stem family* (all direct ancestors and descendants), and *extended family* (nuclear and stem family plus assortment of other kin, such as uncles/aunts, nephews/nieces, and cousins).

Types of Kin

Although kin are classified differently in different societies, Western societies generally classify kin using Ego as the central figure, where *lineal kin* are Ego's direct descendants (children, grandchildren, etc.) and progenitors (parents, grandparents, etc.) and *collaterals* are all of Ego's other kin. These kin, in turn, are divided into two types, the first of which are Ego's *colineal* kin, including aunts/uncles, siblings, and nephews/nieces. These are Ego's siblings and the children of his/her lineal kin. For example, the siblings of Ego's parents are Ego's aunts/uncles. And the children of Ego's siblings are Ego's nephews/nieces. The second group of Ego's collateral kin are his/her *ablineal* kin, otherwise known as Ego's cousins. This group of Ego's relatives are the children of Ego's parents' siblings, in other words, first cousins. Then second cousins are the children of these first cousins, third cousins the children of the second cousins, fourth cousins are the children of the third cousins, and so forth.

Cousins are cross-classified both by *degree* (i.e., first cousin, second cousin) and by *generation removed* (i.e., once removed, twice removed). Ego can only have one set of, say, first cousins in his/her own generation but may have two sets of first cousins once removed; for example, Ego's parents' first cousins are his/her first cousins once removed and Ego's own first cousins' children are also his/her first cousins once removed. In the former case these first cousins are once removed back a generation and in the latter case they are once removed forward a generation.

Examples of relationships

Examples of kinship relations include the following:

(1) The *parents* of Ego's *great-grandmother* are his/her great-great-grandparents and Ego is a great-great-grandson/daughter to them. (2) The *children* of Ego's *grandnephews* are Ego's great-grandnephews and Ego is the great-granduncle/aunt to them.

(3) Ego's *father's brother* is Ego's *uncle* and Ego is the nephew/niece to them.

(4) The *children* of Ego's first cousins are second cousins and he/she is first cousin once removed from them and they are first cousin once removed from Ego.

(5) The *children* of Ego's second cousins are Ego's second cousins once removed and Ego is the same to them.

(6) The *grandchildren* of Ego's *second cousins* are his/her *second cousins twice removed* and Ego is the same to them.

(7) Ego's *grandfather's brothers* are his/her *great-uncles*, their *children* are Ego's *first cousins once removed* and also his/her *father's first cousins*, and their *grandchildren* are Ego's *second cousins* (or Ego's second cousins are the children of Ego's parents' first cousins).

Biological pathways

An important aspect in understanding all kinship relations involves the biological pathways by which someone came to be your kin. This is especially true for all collaterals, but it is also important for lineal kin. For example, most people have eight great-grandparents representing four surnames (traditionally) or lines of descent. Two of these four lines (or surnames) are associated with the mother's side and one of these four is associated with the father's mother's side. In identifying grandparents, it is important to keep the paths straight so that collaterals can be organized and traced accordingly; for example, first cousins may be derived from either the father's side or the mother's side. Additionally, second cousins may be derived from four sources that are the siblings—that is, either the paternal grandfather, the paternal grandmother, the maternal grandfather, or the maternal grandmother.

Pedigree charts

Lineal kin

Identification of *lineal kin* in a pedigree chart (fig. 8.15) involves the straightforward process of connecting parents with their children and, in turn, with their grandchildren, great-grandchildren, and so forth, or the reverse—great-grandparents to grandparents back to children. For example, individuals 6, 13, and 24 in the figure are the children, grandchildren, and great-grandchildren of the patriarch, 1, and matriarch, 2, respectively. And individuals 23, 33, and 40 are, respectively, the great-grandmother, grandmother, and mother of individual 45. Equivalencies of relatedness among some of the common kin are shown in fig. 8.16.

Colineal kin

Identification of *colineal kin* involves the reciprocal relationships of siblings across two or more generations. For example, individual 12 in fig. 8.15 is the sister of individual 13 who, in turn, is the mother of individual 23. Therefore, individual 12 is the aunt of individual 23, who is the niece of individual 12. Similarly, individual 31 is the grandniece of individual 12, who is the great-aunt of individual 31.

Ablineal kin—same generation

Identification of *ablineal kin* involves the relationship of offspring relative to aunts and uncles (i.e., parents' siblings). For same-generation cousins (e.g., first cousin, second cousin), the first step is to identify first cousins (children of siblings) and then identify the children of these first cousins as second cousins. Then the children of these second cousins are third cousins and so forth.

Ablineal kin—different generation

For cousinship relationships that are in different generations, the first step to identify ablineal kin is to determine the cousinship relations of the progenitor of the person in the later generation. Then count the number of generations separating these two individuals, and this information identifies the cousins—for example, first cousins once removed, third cousins twice removed, and so forth (fig. 8.16).

Pedigree collapse

If a recent newborn kept multiplying his/her progenitors by two every generation—doubling his/her parents, their parents, and so on—to the time of Charlemagne (ca. 800 AD), he/she would have between 4 and 17 billion progenitors. *Pedigree collapse* prevents this theoretical population implosion from taking place. In other words, the intentional mating between close cousins and random mating between distant cousins, who do not even know they are related, causes a duplication in their descendants'

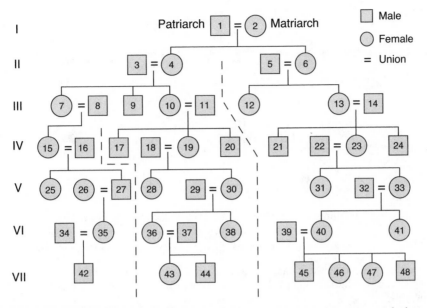

FIGURE 8.15. Hypothetical pedigree chart depicting seven generations starting with the patriarch and matriarch in generation I through descendants 42 through 48 in generation VII. The dashed lines separate family lines in subsequent generations, thus grouping lineal and collateral kin.

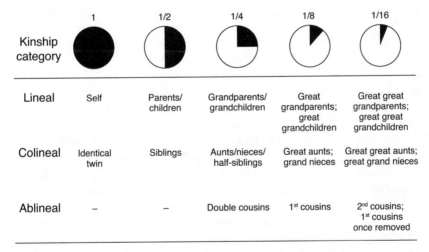

FIGURE 8.16. Consanguinity and equivalences of genetic relatedness across common kinship relationships.

pedigrees because cousins already occupy a slot there (see fig. 8.17). The farther back one traces any person's genealogy the greater the rate of duplication grows, until finally when there is more cousin intermarriage than input from new people, the shape of one's pedigree stops expanding and begins to narrow. Each person's complete family tree, in other words, is shaped like a diamond. In the beginning it expands

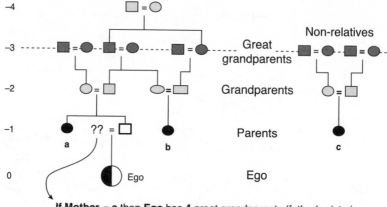

If **Mother** = **a** then **Ego** has <u>**4** great grandparents</u> (father's sister)
            = **b** then **Ego** has <u>**6** great grandparents</u> (father's 1st cousin)
            = **c** then **Ego** has <u>**8** great grandparents</u> (non-relative of father)

FIGURE 8.17. Schematic showing great-grandparent numbers relative to three different mating scenarios—Ego's mother is her father's sister, her father's first cousin, or a nonrelative.

upward as an inverted triangle, but at some point, hundreds of years back, the rate of expansion peaks, and the base of the inverted triangle is reached and overwhelmed by "collapse" when the pedigree starts to narrow again. Eventually, the triangle comes to a point at a theoretical first couple, Adam and Eve (Shoumatoff 1985).

Consider Further

Most demographic concepts and models were first introduced to the literature in the context of human populations. It thus follows that much of the content in this chapter on human demography is directly or indirectly connected to the content in nearly all the other chapters in this book. This includes concepts on rates, the life course, and the Lexis plane in chapter 1 (Basics), and virtually all content on life tables, mortality, reproduction, and stable theory in chapters 2 (Life Tables), 3 (Mortality), 4 (Reproduction), and 5 (Population I: Basic Models), respectively. Subsections in each of chapters 6 (Population II: Stage Models) and 7 (Population III: Extensions to Stable Theory) contain content relevant to human demography, including age/stage and increment/decrement concepts in the former and two-sex, stochastic, and multiregional models in the latter. The majority of the entries on demography in chapter 11 (Biodemography Shorts) pertain to humans at virtually all levels, from individuals to cohorts to family to populations.

The list of excellent textbooks and source books that outline demography methods is too extensive to name, but a small fraction include the textbook *Demography: Measuring and Modeling Populations* (Preston et al. 2001), the edited book *Methods and Materials of Demography* (Siegel and Swanson 2004), the four-volume *Demography: Analysis and Synthesis—A Treatise in Population* (Caselli et al. 2006a), and *Handbook of Population* (Poston and Micklin 2005).

The paper by Colchero and his colleagues (2016) establishes deep evolutionary links to human evolutionary history by connecting life expectancy and life equality (life span variation relative to average) between humans and a wide range of nonhuman primate species. Doblhammer's demography monograph (2004) contains important information on the effects of early life experience on late-life health and longevity in humans. The paper by Metcalf and Pavard (2007) contains important new perspectives on biodemography as an emerging area situated at the interface of population biology and mainstream demography. The book by Hertz and Nelson (2019) titled *Random Families* discusses the emergence of entirely new categories of kin and families resulting from the ready availability of donated sperm and eggs. Primary sources of demographic information include articles in the journals *Demography*, *Population Development and Review*, *Population Studies*, *Population*, *Genus*, and *Biodemography and Social Biology* among many others.

Applied Demography I:
Estimating Parameters

As is well known, the celebrated Malthus has established as a principle
that the human population tends to increase in geometric progress so as to
double after a certain period. . . . In fact, all other things equal, if one thousand
souls have become two thousand after twenty-five years, these two thousand
will become four thousand after the same lapse of time. Pierre-Francois
Verhulst (1838),

Correspondance Mathematique et Physique Publiee 10, 113

Applied demography refers to the application of basic demographic concepts and
tools to practical problems concerned with individuals and cohorts in populations.
For human demography, Murdoch and Ellis (1991) distinguish basic from applied
demography across five dimensions: (1) *scientific goals,* where basic demography is
concerned largely with explanation and applied demography with prediction; (2) *time
referent,* where basic demography is concerned with the past and applied demography
with the present and future; (3) *geographic focus,* where basic demography is concerned with international or national patterns and applied demography with aggregate
data for small areas; (4) *purpose of analysis,* where basic demography is concerned
with the advance of scientific knowledge and applied demography with the application
of knowledge to discern the consequences; and (5) *intended use,* where basic demography is concerned with the advance of knowledge and sharing, including the use of
research results to inform decision making among nondemographers, and applied
demography is concerned with more specific and often more local decisions.

In the applied biological sciences, such as wildlife, fisheries, conservation, invasion
biology, pest management, epidemiology, and environmental sciences, demographic
methods are used widely in the contexts of both cohort analyses, such as life tables,
and the manipulation of population growth rates, including both age- and stage-
structured population models (Caughley 1977; Krebs 1999; Amstrup et al. 2005).
In this chapter we make the transition from basic to applied demography and focus
on how the parameters we have described in earlier chapters can be estimated. We
first introduce techniques for estimating population numbers, N_x. We then discuss the
basic foundations of mark-recapture models to estimate survival, look at methods
for estimating population growth rates under conditions of discrete and continuous
growth, and present methods to estimate population age and stage structure. At the
end of this chapter we consider the special case where these demographic parameters
are estimated from captive cohorts that have been moved from the wild to the lab.

Estimating Population Numbers

Obtaining an accurate estimate of the number of individuals in a cohort or population is clearly critical to biodemographic analysis, but obtaining these estimates can be difficult. For species in their natural habitats, there are a number of different methods to estimate population sizes, examples of which are described below.

Complete counts

Complete counts refers to the straightforward concept in which every member in a population is counted. This is often based on photographs, for example, waterfowl on lakes, seals on beaches, or elephants in open areas. Other methods, such as deer drives, where all individuals must pass through a narrow area for counting, also provide a complete count of nearly all individuals. In studies with plants, complete counts are possible in species where an "individual" can be clearly defined; in other words, when the definition is not obfuscated due to clonal growth. In all cases where individuals can be easily identified and marked, longitudinal studies, as discussed in chapter 1, are the gold standard for demographic studies.

Incomplete counts

An incomplete count involves counting subpopulations within one or more quadrats and then extrapolating to the entire population. Strip censuses, roadside counts, and flushing counts are examples of incomplete count methods. The basic model for estimating population size is based on the assumption that the ratio of the population number in a quadrat to the quadrat's area equals the ratio of the total population to the total area. That is,

$$\frac{P_T}{A_T} = \frac{C_q}{A_q} \tag{9.1}$$

$$P_T = \frac{A_T C_q}{A_q} \tag{9.2}$$

where P_T, A_T, C_q, and A_q denote the total population, total area, quadrat count, and quadrat area, respectively. A basic assumption of this method is that individuals are scattered randomly across the study area (Burnham et al. 1980).

Indirect counts

Indirect counts use indirect signs of the number of animals present as indices of relative abundance. This method does not provide estimates of the number of individuals in the population but rather can be used to show population trends (increase; decrease; stationary). Examples of indirect counts include counting scats (fecal pellets) or the number of nests/dens in a given area.

Mark-recapture

The use of marked individuals that are captured, uniquely marked, and then recaptured is a major technique that is used to monitor biological populations. Here we briefly focus on how this technique can be used to estimate population size; later in this chapter we provide more detail on marking techniques and how this approach can be used to estimate survival. Mark-recapture methods, first developed by Petersen (1896) to study European plaice (a flatfish) in the Baltic Sea and later proposed by Lincoln (1930) to estimate numbers of ducks, is often referred to as the Lincoln-Petersen index (although it is used to estimate actual population sizes). The method involves capturing a number of animals, marking them, releasing them back into the population, and then determining the ratio of marked to unmarked animals in the population in the next sampling event. The total population, P_T, is estimated by the formula

$$\frac{P_T}{M} = \frac{C}{R} \tag{9.3}$$

$$P_T = \frac{M \times C}{R} \tag{9.4}$$

where M is the number of animals marked in the first trapping session, C is the number of animals captured in a second trapping session, and R is the number of marked animals recaptured in the second trapping session.

Assumptions for this method of population estimation are that the proportion of marked animals captured in the second trapping session is the same as the proportion of total marked animals in the total population, that each trapping session captures a representative sample of various age and sex categories from within the population, and that marked individuals do not lose their markings. Other assumptions are that the capture rate, and the mortality rate, for marked and unmarked individuals is identical, that marked animals mix randomly with unmarked animals, and that the population has no significant recruitment, or ingress (births or immigration) or egress (deaths or emigration).

As an example, suppose 10 rabbits are captured and marked on day 1 (M = 10) and on day 2 the total captured is 20 (C = 20), 5 of which are marked (R = 5). Then the estimated total population is

$$P = \frac{MC}{R}$$

$$P = \frac{10 \times 20}{5}$$

$$P = 40 \text{ rabbits in total population}$$

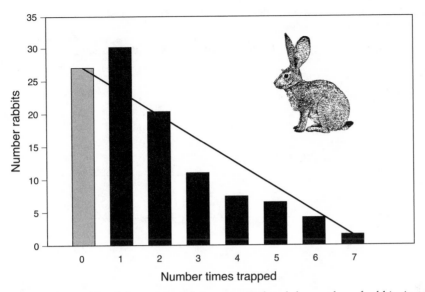

FIGURE 9.1. Relationship of the number of times trapped and the number of rabbits in a hypothetical population. The line is used to estimate the number of rabbits that were *never* captured, which is then added to the total number that were captured.

If recapture data are available over multiple days, the frequency of capture method developed by Eberhardt (1969) can be used. This concept is based on plotting the number of times individuals are captured versus the number of individuals captured each time.

As an example, suppose over a one-month period the number of marked individuals that are captured 1, 2, 3, 4, 5, 6, and 7 times is 30, 20, 10, 6, 5, 3, and 1, respectively. These data are then fit to a statistical distribution to determine how many rabbits were never trapped even though they were present. In this case, if we assume that rabbits were captured at random, the estimate of the number never trapped is between 25 and 30 (fig. 9.1). The number never trapped at all but present is added to the total number captured to give the *frequency of capture* estimate. In this example, the population size estimate is slightly over 100 individuals (i.e., the sum of value for all bars in fig. 9.1).

Hunting effort

A method related to the frequency of capture method is referred to as the DeLury method (DeLury 1954); it is based on the concept that the number of animals killed per unit of hunting time is proportional to the population density. By plotting kill rate (animals killed per day), versus total kill, it is possible to get an estimate of the total population by extrapolating the line to the x-axis where the intersection is the estimate of the total population (table 9.1; fig. 9.2). Assumptions of this method include that the population has no immigration or emigration, that there is an identical hunting risk for all animals, that variable hunter skills can be quantified and averaged, and that hunting is done individually.

Table 9.1. Example application of the use of data on animals killed per hours hunted to estimate total population prehunt

Day	Animals killed	Hours hunted	Kill/gun-hours	Cumulative kill
1	235	940	0.250	235
2	212	881	0.240	447
3	190	964	0.198	637
4	172	952	0.180	808
5	155	917	0.169	964
6	139	905	0.153	1102
7	125	928	0.134	1227

Source: DeLury (1954).

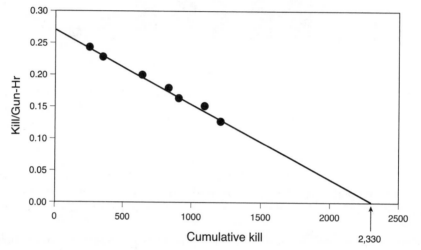

FIGURE 9.2. Regression line (Y = 0.2815 − 0.00012X) is fitted to kill per gun-hour relative to cumulative kill. The regression line is extended to the x-axis to determine the prehunting period population level of N = 2,330.

Change-in-ratio method

Originally developed by Kelker (1940) and applied to deer populations, the change-in-ratio method is used to estimate the size of a population when a population can be divided into two classes and when the ratio of these two classes changes due to a selective removal from the population. The classes within the population can be based, for example, on sex, size, age, or maturity status (Dawe et al. 1993). An example application using sex status is given in Pierce et al. (2012), where N_1 and N_2 denote the pre- and posthunt populations, T denotes the total kill (both sexes), F the total females killed, and P_1 and P_2 the proportion of females pre- and posthunt, respectively. The formulas for estimating both the pre- and posthunt populations are, then,

$$N_1 = \frac{P_2 T - F}{(P_2 - P_1)} \tag{9.5}$$

$$N_2 = N_1 - T_1 \qquad\qquad (9.6)$$

Inserting numbers for each of the four parameters (T, F, P_1, P_2) into these formulas to obtain N_1 and N_2 is simple and straightforward.

Estimating Survival: Mark-Recapture

Accurate estimates of age- or size-dependent survival, and transitions in state, are critical to demographic models (see chapters 5–7). The best data—in other words, the data with minimal error—are obtained when 100% of the marked individuals can be accounted for at each census interval and the fate (survival and state) of each individual can be determined unambiguously. However, for many mobile species in complex habitats some portion of the marked individuals are not accounted for at a given census, even if they are still alive. Error is then introduced into estimates of total population size that can complicate demographic accounting. Methods have been developed to estimate the survival and transition probabilities from data sets where marked individuals are not recaptured at each census.

Methods

Methods for estimating survival, using mark-recapture data, were originally outlined by Cormack (1964), Jolly (1965), and Seber (1965, 1970), and they have been subsequently extended by others (Pollack et al. 1990; Lebreton et al. 1992; Lebreton et al. 2009; Barbour et al. 2013; Morehouse and Boyce 2016). These models all begin with a set of individuals who are marked at time $t = 1$, and then each is seen, or not, at each subsequent census interval. Individual sighting histories can then be used to estimate survival. Here we first outline methods for marking and tagging individuals and then we present the foundations of this approach. A detailed discussion of the method for parameter estimation from these models is beyond the scope of this book, but with a worked example and two examples of alternative visualizations of these approaches, the basic idea, and powerful breadth of this approach, will be clear.

Marking and tagging individuals

The methods for marking individuals fall into three broad categories. *Natural marks* and/or tags (fig. 9.3a–c) refer to the unique, natural markings or properties of individuals. Animals' natural markings can be used for some snakes (ventral patterns); lizards (dorsal and throat patterns); salamanders (spot patterns); swans (bill patterns); ospreys (head markings); felids, including tigers, lions, cheetahs, and leopards (unique coat patterns); giraffes (coat patterns); cetaceans (fluke patterns and notches); and rhinoceroses (horn shape and wrinkle patterns). Camera trapping, DNA sampling using "hair snares," and acoustic sampling (e.g., birds; bats; whales) are also considered forms of natural, noninvasive techniques for identifying individual animals (Royle et al. 2014). *Noninvasive markings* (fig. 9.3d–f) range from neck collars (ungulates), leg, arm, and wing bands (amphibians, reptiles, birds, small rodents, bats),

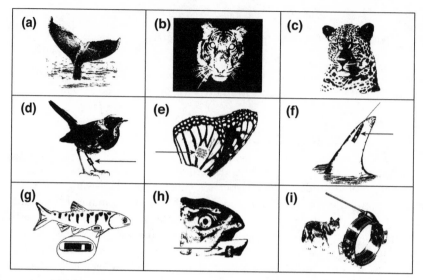

Figure 9.3. Examples of approaches used in mark-recapture studies on free-ranging animals. *Natural marking*: (a) fluke patterns in cetaceans; (b) camera trap image of tiger; (c) spot patterns in leopard. *Noninvasive marking*: (d) leg bands on songbird (e) butterfly wing tag; (f) fin transponder on hammerhead shark. *Invasive marking methods*: (g) internal transponder in trout; (h) jaw tag in salmon; (i) wolf with collar. Arrows show mark or transponder.

and backpacks, to trailing devices, tapes, streamers, and external dyes and paints. These techniques can also include chemical markers; transponders; tattoos; ear, jaw, and wing tags; hot-iron, freeze, and chemical branding; feather and toenail clipping; and shell notching (Royle et al. 2014). *Invasive markings* (fig. 9.3g–i) can include internal transponders and tags. Silvy et al. (2012) provide references to the literature and URLs for websites describing various marking methods for all major categories of animals, including amphibians, reptiles, birds, and mammals.

Incomplete data

Both the concept and the limitations of mark-recapture studies are illustrated in the schematic diagram shown in fig. 9.4. There are always some individuals that are never captured due to either chance or their behavior, such as being (so-called) trap shy (Marked 1), thus not all individuals are brought into the study once they are marked and released even though they may live through several sampling periods. The data in this figure provide an illustration of imperfect detection because not all living individuals were detected during each sampling occasion across all census periods (i.e., Marked 2 through 5). In other words, the data points of "0" are sampling zeros rather than structural zeros. This means that individuals still alive at a given sampling period often need to be verified as alive based on captures at later sampling periods. Additionally, the majority of individuals experience a postcaptive period prior to death since capture efficiencies for most studies are typically far less than 100% (Marked 2, 4, 5). This means that the age (or date) of last capture is not an accurate proxy for oldest age at death. Mark-recapture may not always yield accurate survival estimates.

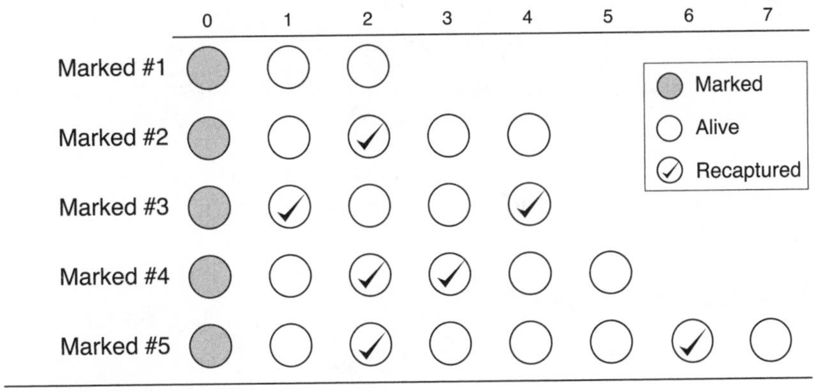

Number living	0	1	2	3	4	5	6	7
Confirmed	5	4	4	3	2	1	1	0
Actual	5	5	4	4	4	2	1	1

FIGURE 9.4. Schematic illustration of mark-recapture survival method in which five hypothetical individuals are released and subject to recapture. Note that if an individual is recaptured at a late sampling occasion, then it is confirmed to have been alive at all previous sampling occasions.

For example, for the data in fig. 9.4, whereas actual survival to sampling occasion #3 was 100%, the verified survival was only 60% (3 of 5 recaptures). The actual survival to sampling occasion #5 was twice that of the verified survival.

Models

Cormack-Jolly-Seber model

Methods for estimating survival from mark-recapture methods of data analysis are based on some version of the Cormack-Jolly-Seber (CJS) approaches (Cormack 1964; Jolly 1965; Seber 1970) and involve data on captures, or noncaptures, of marked individuals. Survival analysis, and thus mark-recapture methods, involve estimating the number of individuals that died over a given period relative to the total number who were at risk of dying. The challenge is essentially the bookkeeping required to compute the number of individuals at risk versus the number who have died.

The general concept of the CJS mark-recapture method of survival estimation is based on the equivalency of the ratios A and B, that is,

$$A = B \tag{9.7}$$

where

$$A = \frac{\text{number unobserved at time j but confirmed alive later}}{\text{number unobserved but subject to risk from j} - 1 \text{ to j}} \tag{9.8}$$

$$B = \frac{\text{number verified alive at time j}}{\text{number verified alive at time j}-1} \tag{9.9}$$

The formal concept on which these mark-recapture survival estimates are based involves equating these two ratios and solving for the total marked population sizes at sampling occasion j, that is,

$$\frac{z_j}{(M_j - m_j)} = \frac{r_j}{R_j} \tag{9.10}$$

where

z_j = the number of members of the marked population not captured at sampling occasion j that are captured again later

M_j = the *marked population size* just before period j

m_j = the number of survivors captured at sampling occasion j that are *marked*

u_j = the number of animals captured at sampling occasion j that are *unmarked*

R_j = the total number of animals captured at sampling occasion j that are *released* (includes both previously marked recaptures and newly marked individuals)

r_j = the number of members of R_j *captured again later*

Note that the denominator for the left-hand ratio in eqn. (9.10) is the difference at time j between the total at-risk marked population, M_j, and the number of the marked population that are confirmed alive, m_j. This difference is the number of the marked population that was unknown at a particular sampling time. The left-hand ratio, then, is the number of individuals that are unknown but later confirmed alive (i.e., the numerator), divided by the number at risk of dying but unknown at sampling time j (i.e., the denominator). And the right-hand ratio is the observed fraction marked and recaptured relative to the total number that were marked and released. This model can be reexpressed as

$$M_j = m_j + \frac{R_j z_j}{r_j} \tag{9.11}$$

The values of M_j, M_{j+1}, and R_j are then used to compute the estimated probability of survival from period j to period j + 1, denoted ϕ_j:

$$\phi_j = \frac{M_{j+1}}{(M_j - m_j) + R_j} \tag{9.12}$$

In words, the estimated probability of surviving from sampling occasion j to j + 1 equals the number of individuals who survived from j to j + 1 divided by the number who were at risk of dying at sampling occasion j. This number at risk, in the denominator, is the sum of the total number captured that were then released and an estimate of the number who were alive at sampling occasion j but were not observed alive through subsequent capture. Note that this is the estimated, or "apparent," probability of survival because it does not account for animals that may have emigrated from the study area and are not available for recapture.

Worked example

An example of this CJS model, for estimating survival over a 7-year period for the passerine bird known as the European dipper, is given in table 9.2 (taken from McDonald et al. 2005, although the original data are from Marzolin 1988). These recapture data with both sexes combined are condensed into what is referred to as an m-array. The m-array, shown in table 9.3, contains values for three parameters, including r_j, m_j, and z_j, all of which are used for computing M_j (eqn. 9.11) and, in turn, ϕ_j (eqn. 9.12). The other key value needed to compute survival is the number released at sampling occasion j, R_j, which is given in table 9.4. The information for R_j, m_j, r_j, and z_j is summarized in table 9.5 and is used to estimate M_j and ϕ_j. For example, using eqns. (9.11) and (9.12) yields

$$M_2 = m_2 + \frac{R_2 z_2}{r_2}$$

$$= 11 + \frac{60 \times 2}{25} \tag{9.13}$$

$$= 15.80$$

and

$$\phi_2 = \frac{M_3}{(M_2 - m_2) + R_2}$$

$$= \frac{28.17}{(15.8 - 11.0) + 60} \tag{9.14}$$

$$= 0.4347$$

Multistate models

The CJS-like models are the most general approach to survival estimation using mark-recapture approaches, but for many situations this model is restrictive because it requires that all individuals have the same probabilities of capture and survival

Table 9.2. Release-recapture sequences for the European dipper (males and females combined) for release periods 1981, 1982, 1983, 1984, 1985, 1986, and 1987

Capture history sequence	Sampling occasion, j							Marked by		
	1	2	3	4	5	6	7	Sequence	Number released	Year
1	0	0	0	0	0	0	1	39	39	1987
2	0	0	0	0	0	1	0	23		
3	0	0	0	0	0	1	1	23	46	1986
4	0	0	0	0	1	0	0	16		
5	0	0	0	0	1	1	0	9	41	1985
6	0	0	0	0	1	1	1	16		
7	0	0	0	1	0	0	0	16		
8	0	0	0	1	0	0	1	2		
9	0	0	0	1	0	1	1	1		
10	0	0	0	1	1	0	0	11	45	1984
11	0	0	0	1	1	1	0	7		
12	0	0	0	1	1	1	1	8		
13	0	0	1	0	0	0	0	29		
14	0	0	1	0	1	1	0	1		
15	0	0	1	1	0	0	0	12		
16	0	0	1	1	1	0	0	6	52	1983
17	0	0	1	1	1	1	0	2		
18	0	0	1	1	1	1	1	2		
19	0	1	0	0	0	0	0	29		
20	0	1	1	0	0	0	0	11		
21	0	1	1	0	1	1	0	1		
22	0	1	1	1	0	0	0	2	49	1982
23	0	1	1	1	1	0	0	3		
24	0	1	1	1	1	1	0	1		
25	0	1	1	1	1	1	1	2		
26	1	0	0	0	0	0	0	9		
27	1	0	1	0	0	0	0	2		
28	1	1	0	0	0	0	0	6		
29	1	1	0	1	1	1	0	1	22	1981
30	1	1	1	1	0	0	0	2		
31	1	1	1	1	1	0	0	1		
32	1	1	1	1	1	1	0	1		
									294	

Source: Table 9.1 in McDonald et al. (2005, 199).

Note: Release numbers are contained in rightmost column ordered bottom to top. Coding indicates whether a bird was captured (1s) or not (0s) in a given sampling occasion, j (e.g., j = 1 for 1981, = 2 for 1982, etc.).

(Cormack 1964; Lebreton et al. 1992). A set of multistate models have been developed to address the increasing complexity of field studies, and these models closely correspond to the stage-classified matrix models discussed in chapter 6. Multistate mark-recapture models deal with animals in different states, where "state" may be, for example, physiology, breeding status, or disease. Location can also be considered

Table 9.3. Observed m-arrays for the European dipper data in table 9.2

Release occasion (j)	Number released (R_j)	m-matrix (m_{jj})						Total recaptured r_j
		j'= 2	3	4	5	6	7	
1	22	11	2	0	0	0	0	13
2	60		24	1	0	0	0	25
3	78			34	2	0	0	36
4	80				45	1	2	48
5	88					51	0	51
6	98						52	52
7	93							
$m_j =$		11	26	35	47	52	54	

Source: Table 9.2 in McDonald et al. (2005, 200).

Note: R_j are the number of birds captured and released on occasion j; in the matrix, $m_{jj'}$ are the number of birds released on occasion j who were first recaptured on subsequent occasion j'. The values for z_j, the marked individuals not captured at j that were captured later, are the row sums of numbers to the right of the subdiagonal shaded line; and the values for m_j, the number of marked survivors captured at time j, are the column sums at the bottom of the table.

Table 9.4. Computation of the number of marked European dippers released at each sampling occasion (R_j) based on coding information contained in table 9.2

Sampling occasion (j)	Newly marked and released (u_j)	Marked from previous time and recaptured						Previously marked and re-released (m_j)	Total released [$R_j = (u_j + m_j)$]
		1	2	3	4	5	6		
1	22	—	—	—	—	—	—	0	22
2	49	11	—	—	—	—	—	11	60
3	52	6	20	—	—	—	—	26	78
4	45	5	8	22	—	—	—	35	80
5	41	3	7	11	26	—	—	47	88
6	46	2	4	5	16	25	—	52	98
7	39	0	2	2	11	16	23	54	93

Note: Each cell value for the number of marked individuals in one of the previous sampling occasions is the sum total of individuals in each sampling sequence that were recaptured at time j. For example, there were 1, 7, and 8 (sum = 16) individuals with sequence histories 9, 11, and 12, respectively, that were recaptured during sampling occasion 6 that had been released in sampling period 4.

a state, which connects these approaches to the metapopulation analysis introduced in chapter 6. The state of an animal may then change from one census time to the next and states can have different probabilities of survival and capture. Advanced models can also address issues such as environmental covariates, mortality causes, and recaptures in continuous time rather than discrete census intervals. A comprehensive description of the foundation of these multistate models can be found in Lebreton et al. (2009). Multistate mark-recapture models have broad applications that can address many demographic questions in population biology, conservation, and metapopulation dynamics. Comprehensive software for these models is available, including MARK (White and Burnham 1999) and M-Surge (Choquet et al. 2009). Both of these packages allow for a wide range of models to be fit. Models can

Table 9.5. Estimates of survival (ϕ_j in rightmost column) for the European dipper example

Occasion	Number marked that are released	Number captured that are marked	Number marked seen after j	Number seen before j, not seen at j, and seen after j	Estimated number of marks	Survival estimate
j	R_j	m_j	r_j	z_j	M_j	ϕ_j
1	22	0	13	—	0.0	0.7182
2	60	11	25	2	15.8	0.4347
3	78	26	36	1	28.2	0.4782
4	80	35	48	2	38.3	0.6261
5	88	47	51	3	52.2	0.5985
6	98	52	52	2	55.8	—
7	93	54	—	—	—	—

Source: Table 9.3 in McDonald et al. (2005, 202).

also be fit in a Bayesian state-space framework, including models developed by King (2012) and by Cohchero and Clark (2012).

Mark-recapture as cohort analysis

An alternative approach for survival analysis using mark-recapture data is cohort analysis, which is based on the concept that each release group is considered a cohort whose long-term (longitudinal) survival can be estimated from the recapture data in subsequent years. The annual survival of each cohort is not only compared with survival in other released cohorts but is also used for computing the mean survival across all cohorts (i.e., both cross-sectional and longitudinal survival). Again, using the European dipper data (Marzolin 1988; McDonald et al. 2005) in table 9.2, we visualize this approach as a Lexis diagram (fig. 9.5), where the recapture history of each released cohort is tracked along a diagonal line in the time-year (age-period) plane. These numbers are then arranged by year of release in table 9.6 along with survival rates for both real (longitudinal) and synthetic (cross-sectional) cohorts.

At least two aspects of this approach merit comment. First, the Lexis diagram (fig. 9.5) serves as an important aid not only for visualizing the recapture data but also for understanding the larger problem in the time-of-release by year-of-recapture plane. It is extremely difficult to see both the data and the larger problem from the near-raw coded data (0s and 1s) in table 9.2. Both longitudinal and cross-sectional perspectives provide important insights into between- and within-year survival rates. This includes estimation of some rates that are based on extremely small numbers due to the dwindling number of survivors within released cohorts. For example, in the first year after initial releases for 1981 and 1982 cohorts, the between-cohort annual survival rates varied by 20% to 25%. The variation in overall survival rates for each released cohort then varies by nearly this amount; for example, the average (geometric) survival in the 1981 cohort was around 0.62, whereas the average annual survival in the 1983 release cohort was only 0.44—a difference of 18%. Note also that the variance in the differences in postrelease survival rates was relatively small for the cross-sectional rates (e.g., 0.6133 at $t=2$ versus 0.5157 at $t=3$). These are

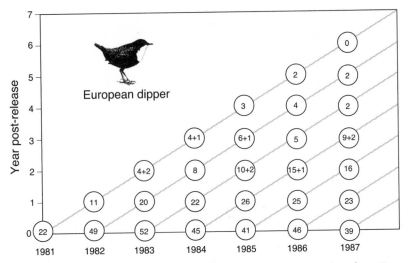

FIGURE 9.5. Lexis diagram depicting European dipper mark-recapture data for a 7-year period starting in 1981. Each diagonal line represents a new mark-recapture release group with the release numbers at the diagonal base (including newly marked captures and previously marked captures that were recaptured) and incremented diagonally by year. The second number in a circle indicates the number of marked individuals that were captured that year within a cohort that had not been detected previously. For example, 4 + 1 in the 1981 cohort for the 1984 captures (circles vertically arranged above 1984) indicates that a total of 5 individuals were captured from the 1981 cohort, 1 individual of which was not captured previously in either 1982 or 1983. (*Source*: Marzolin 1988)

thus a better reflection of the average long-term postrelease survival rates. Second, estimates of average annual survival across all released cohorts and all years were relatively small, differing by only 0.0028 (i.e., 0.5547 and 0.5518) depending on whether the computations were based on longitudinal (columns) or cross-sectional (rows) rates. Note that these estimates differed by less than 1% from the average annual survival in the CJS model contained in table 9.5. This cohort approach thus provides a complementary if not alternative approach to the CJS mark-recapture methodology.

Visualization

The survival estimation problem with mark-recapture is a demographic problem that can be visualized to increase clarity and illustrate its properties. For example, we created a hypothetical cohort with 1,000 individual butterflies that survived in the wild according to a Gompertz mortality model with a and b parameter values of 0.0005 and 0.02, respectively. This cohort has an expectation of life, $e_0 = 45.8$ days, with a maximum life span of 100 days. We then simulated a sampling of each individual using a 10% daily probability of capture (and re-release). The results of this simulation are illustrated in fig. 9.6 and table 9.7. Several general properties of mark-recapture models are illustrated in these results. First, a total of 63 individuals were never captured prior to their death. The average life span of those never-captured

Table 9.6. Arrangement of European dipper mark-recapture data by release cohort showing both longitudinal and cross-sectional survival estimates, p(t), where p(t)=N(t+1)/N(t)

| Time | 1981 | | 1982 | | 1983 | | 1984 | | 1985 | | 1986 | | Geometric |
t	N(t)	p(t)	N(t)	p(t)	N(t)	p(t)	N(t)	p(t)	N(t)	p(t)	N(t)	p(t)	means
0	22	0.6364	49	0.4082	52	0.4615	45	0.6444	41	0.6098	46	0.5000	0.5354
1	14	0.5000	20	0.4500	24	0.5000	29	0.6207	25	0.6400	23		0.5371
2	7	0.7143	9	0.7778	12	0.4167	18	0.6111	16				0.6133
3	5	0.6000	7	0.5714	5	0.4000	11						0.5157
4	3	0.6667	4	0.5000	2								0.5774
5	2		2										**0.5547**
	0	0.6190		0.5274		0.4429		0.6253		0.6247		0.5000	**0.5518**

Note: The numbers captured, N(t), were adjusted according to later captures as shown in fig. 9.5. For example, 3 individuals from the released cohort of 22 individuals in 1981 were first captured in either 1983 or 1984, thus verifying that they were alive in 1982. These 3 were thus added to the 11 captured that year, bringing to 14 the number of individuals verified as alive. Bold values are geometric means.

individuals was 13.2 days but ranged from a few days up to a remarkable 81 days for 1 never-captured individual. This illustrates that by chance alone it is possible to completely miss some very old individuals, but more importantly, there was a significant fraction of individuals who were never recaptured. Second, around 80% of the life-days lived by the cohort were recorded due to the initial capture and recaptures throughout the lives of the hypothetical individual butterflies. If the age of last capture is used as the age of death for an individual, this means that life expectancy was underestimated by 20%. Third, postcapture life-days constituted nearly 18% of the total life-days lived by the cohort. In other words, nearly 1 of 5 days was lived by this cohort "outside" of the study due to a combination of never-captures and postcapture days. Finally, the average duration of the postcapture period was longer for those that died at younger ages relative to those that died at older ages because mortality at the older ages is higher and therefore individuals are more likely to die before they are recaptured.

We know the probability of recapture was 10% daily because this was built into the simulation, but if this was not known it could be approximated using the life-days as

$$\text{Capture probability} = \frac{\#\text{ marks} - \left[1,000 - (\text{non} - \text{captures})\right]}{\text{capture-days}}$$

$$= \frac{4,391 - 937}{35,809}$$

$$= 0.0965$$

Given that the average long-term interval between captures is the inverse of the probability of capture, this interval can be estimated as 10.4 days ($=0.0965^{-1}$). This information can be used to adjust the age of last capture by generating a random capture add-on duration to each age of capture, including individuals that were never captured. The results of this adjustment in comparison to the actual survival curve and the survival curve based on oldest age captured is shown in fig. 9.7. Note the close approximation of the adjusted survival curve to the original. Any departure of

Table 9.7. Summary of life-days in hypothetical butterfly mark-release-recapture cohort

	n	Total life-days	Average duration (days)
Noncaptures	63	830	13.2
Captures	937		
Confirmed		35,809	38.2
Postcapture		7,615	8.1
		44,200	

Note: Cohort is disaggregated into life-days for butterflies never detected prior to their death, life-days confirmed through capture once released up to the last capture, and postcapture life-days.

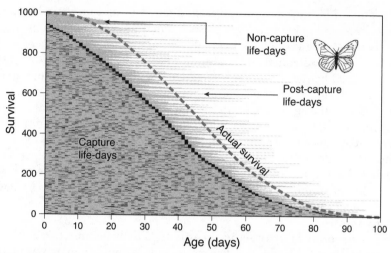

FIGURE 9.6. Results of a computer-generated mark-recapture study of a hypothetical butterfly population subject to the Gompertz mortality model in the wild ($a = 0.0005$; $b = 0.02$) and a 10% daily probability of capture/recapture. The life course of 1,000 individuals is shown here depicted as horizontal life-lines arranged top-to-bottom from shortest to longest precapture age; black cells depict recapture ages. The arrow at the top points to the life-days lived by individuals that were never captured before they died. The arrow labeled "postcapture life-days" points to the days lived by individuals after their last capture before dying. See table 9.7 for partitioning of life-days.

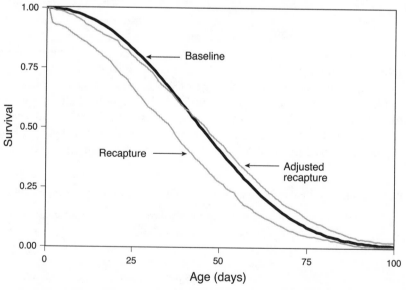

FIGURE 9.7. Comparison of three survival curves for the hypothetical butterfly cohort, including the original baseline, the age of last recapture, and the adjusted recapture curve based on the addition of a randomly generated postcapture life span.

these curves is due to the actual longer-than-average postcapture life spans at young ages and the shorter-than-average life spans at older ages.

Estimating Population Growth Rates

Population growth rate is the change in the number of individuals in a population over a specified time period. In chapters 5 and 6 we introduced the concept of population growth rate and noted that it can be expressed as the finite rate of increase, λ, or the continuous growth rate, r. Here we will consider how growth rate can be estimated either through changes in population numbers or through changes in population age structure.

Growth rate from population number

Both the finite and continuous rate of change in the sequence of numbers in a hypothetical population are presented in table 9.8, where

$$\lambda = \frac{N_{t+1}}{N_t} \qquad (9.15)$$

$$r = \frac{\ln \dfrac{N(t+1)}{N(t)}}{\Delta t} \qquad (9.16)$$

For example,

$$\lambda = \frac{N_3}{N_2} = \frac{159}{137} \qquad (9.17)$$

$$= 1.16$$

and

$$r = \ln\left[\frac{\left(\dfrac{159}{137}\right)}{137}\right] \qquad (9.18)$$

$$= 0.149$$

The value for λ applies to the model

Table 9.8. Change in number with time in a hypothetical population expressed in both finite (λ) and continuous (r) rates

Time (t)	0	1	2	3	4
N(t)	100	121	137	159	183
λ		1.21	1.13	1.16	1.15
r		0.191	0.124	0.149	0.141

Table 9.9. Relationship of numbers in a population with three age classes that are doubling each time step

	Time (t)		
Number age x	0	1	2
N_0	12	24	48
N_1	6	12	24
N_2	3	6	12
TOTAL	21	42	84

$$N_{t+1} = \lambda N_t \qquad (9.19)$$

and the value for r applies to the model

$$N(t+1) = N(t)e^{rt} \qquad (9.20)$$

These two models correspond to the discrete and continuous versions, respectively.

Growth rate from age structure

The underlying concept for estimating population growth rate from age structure is illustrated in table 9.9. Suppose the counts in this table were observed in a hypothetical population over three time steps with three age classes. There is no mortality until the last age (age class 2) when all individuals die. Note that the ratio of any adjacent age class at any time is 2:1. In other words, there is a twofold difference between the number of individuals in age classes 0 and 1 and between the number in age classes 1 and 2. These differences in numbers are strictly due to the effect of growth rate on age structure. Therefore, if we are certain that there is no mortality in the population between ages 1 and 3, and we only know the age structure of the population, we could state that the population would grow by a factor of two. This relationship is made more explicit with

$$c_x = \frac{n_x}{N} \tag{9.21}$$

$$c_y = \frac{n_y}{N} \tag{9.22}$$

where n_x and n_y are the number in age classes x and y, c_x and c_y are the fraction of the total population in age classes x and y, and N is the total number in the population. Then

$$\frac{c_x}{c_y} = \frac{n_x}{n_y} \tag{9.23}$$

That is, the ratio of fractions in each of the two age classes is equal to the ratio of the numbers in each age class. The formulas for the fractions of the total stable population at ages x and y are

$$c_x = be^{-rl_x} \tag{9.24}$$

and

$$c_y = be^{-rl_y} \tag{9.25}$$

where r denotes the intrinsic rate of increase, b the intrinsic birth rate, and l_x and l_y survival to ages x and y, respectively. Therefore, the ratio of c_x to c_y is

$$\frac{c_x}{c_y} = \frac{be^{-rl_x}}{be^{-rl_y}} \tag{9.26}$$

$$= \frac{e^{-rl_x}}{e^{-rl_y}} \tag{9.27}$$

Since $\dfrac{c_x}{c_y} = \dfrac{n_x}{n_y}$, then

$$\frac{n_x}{n_y} = \frac{e^{-rl_x}}{e^{-rl_y}} \tag{9.28}$$

Table 9.10. Number of egg and immature stages in a hypothetical arthropod population and estimated growth rates

Sample #	1	2	3	4	5
Number in egg stage	2,311	1,150	2,048	3,654	2,145
Number in immature stage	824	596	541	5,820	4,643
Estimated r	0.206	0.131	0.266	−0.093	−0.154

Note: Estimates of population growth rates are based on durations and assumptions given in text.

Rearranging and taking logs yields a solution for r,

$$r = \frac{\ln \dfrac{N(t_2)}{N(t_1)}}{t_2 - t_1} \qquad (9.29)$$

which yields the same equation for growth rate as given in eqn. (9.16).

An example of the application of this technique is given in table 9.10 for five samples of a hypothetical arthropod population where the mean ages of the egg ($=x$) and immature ($=y$) stages are 2.5 and 7.5 days, respectively, and no between-stage mortality occurs. Therefore, $(y - x) = 5$ days, $l_x = l_y = 1.0$, and the growth rate at sample 1 is computed as $r = \frac{1}{5} \ln \frac{2311}{824} = 0.206$. The growth rate, r, for sample 2 is computed as $r = \frac{1}{5} \ln \frac{1150}{596} = 0.131$, and so forth. Note that the negative estimates for r in samples 4 and 5 suggest that the population is decreasing at these sample times.

Estimating Population Structure

Importance of age data

Without age information, human demography would be unimaginable in some types of studies and impossible in others—for example, constructing Lexis diagrams, disaggregating age-period-cohort effects, tabulating actuarial rates, predicting future births and deaths, analyzing migration trends, projecting population numbers, or developing population policies (Carey et al. 2018). Indeed, demographers concerned primarily with human populations consider age as central to and as inextricable from their discipline as the concept of supply and demand is to economists, Darwinian selection is to evolutionary biologists, and differential calculus is to mechanical engineers. Without age data, the field of demography would be reduced to a shadow of its current self at best and completely disappear at worst. Aside from population studies in a few subspecialties in human demography (e.g., remote indigenous peoples), the absence of age data in human population studies is the rare exception.

The profundity of not having information on individual and population age in studies of nonhuman species is often not fully appreciated by mainstream demographers because of their exclusive focus on humans. But this lack is deeply frustrating to the majority of population biologists and applied ecologists. This is because the absence of information on age and age structure in populations of nonhuman species severely limits the scope and depth of demographic analysis and modeling in several important respects. First, the majority of the most sophisticated demographic models in the literature are developed for and concerned with human populations. These methods both assume and require information on individual age and population age structure. Therefore, without age data on nonhuman species, most of the classical demographic models, including cohort life tables and age-structured population models, apply mostly in theoretical and laboratory contexts rather than in the natural settings where they are the most relevant. Second, age is a major source of risk, which, as a general concept, underlies the quantification of various age-specific forces of transition, for example, in sexual maturation, marriage and divorce, reproduction, disease acquisition, disablement, retirement, and death. Because force-of-transition concepts apply to changes of state in species across the Tree of Life (Jones et al. 2013), the lack of age information limits demographic analysis. Third, the results of demographic studies in the laboratory are of marginal value without the availability of age data for cohorts and populations in the field. These limitations preclude opportunities to refine, adapt, and expand powerful demographic tools for use in the analysis of populations of nonhuman species. They also restrict the range of possibilities for creating new demographic concepts and building new models based on the treasure trove of life history (and thus demographic) characteristics observed across the Tree of Life.

Age structure

Age estimation for preadults of most nonhuman species is relatively straightforward because of extensive documentation of the duration of well-defined developmental stages (Lyons et al. 2012). Whereas age estimates for preadults of many invertebrate groups (e.g., nematodes) as well as for two vertebrate groups (i.e., fish and reptiles) are based largely on size, in other groups age is estimated by developmental stage, such as for insects (e.g., larvae; wingless preadults) and amphibians (e.g., tadpoles). The preadult stages for both avian and mammalian species are also relatively distinct and the duration is well documented. For example, the young of passerine birds are classified as hatchling, nestling, and fledgling with known developmental times (and thus age intervals) for each. The replacement and growth of primary feathers indicates later developmental stages of preadulthood for many species, with feather coloration often distinguishing juvenile birds that are transitioning to adulthood. Preadult stages of mammals are typically based on size but often also on coloration patterns (e.g., spotted deer fawns), head features (short head morphology in ungulate calves), or tooth development, eruption, and wear (e.g., felids; rodents).

Whereas the methods of age estimation in preadults is relatively straightforward, for most species approximating the age of adult individuals, particularly individuals at middle and advanced ages, is challenging at best and intractable at worst. Many concepts and methods for estimating the age of individuals sampled from the wild are based on what is referred to as the "recording structure" (Klevezal 1996). This

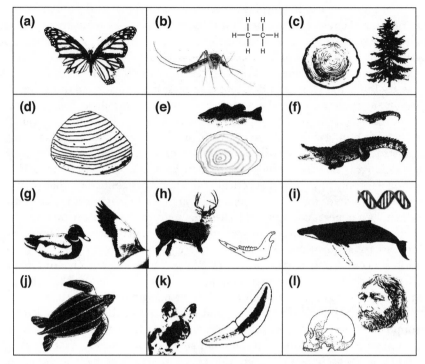

FIGURE 9.8. Overview of aging methods for selected species: (a) wear and tear in butterflies (Molleman et al. 2007); (b) cuticular hydrocarbon analysis in mosquitoes (Gerade et al. 2004); (c) tree ring dating (Baillie 1982); (d) shell thin sectioning in freshwater mussels (Neves and Moyer 1988); (e) otolith structure in fish (Campana and Thorrold 2001); (f) size in crocodiles (Zug 1993); (g) primary feathers in waterfowl (Lyons et al. 2012); (h) tooth wear and eruption in white-tailed deer (Gee et al. 2002); (i) epigenetic estimation in humpback whales (Polanowski et al. 2014); (j) skeletochronological analysis in leatherback sea turtles (Avens et al. 2009); (k) counting cementum annuli in wild dogs (Mbizah et al. 2016); (l) cranial suture closure in humans (Key et al. 1994).

term refers to animal structures, such as the shells of mollusks; scales, otoliths, and bones of fish; bones of amphibians, reptiles, and mammals; dentin and cementum of mammalian teeth; and the horny substance of claws and of some mammals. Recording structures are thus morphological structures that respond to changes of physiological condition of an individual as they grow (Klevezal 1996). A requirement of all recording structures is that these morphologically changed features must persist for a long time. Examples of the use of recording structure concepts as well as other approaches for estimating individual age are highlighted in fig. 9.8.

Methods for estimating the age of adult insects include wear and tear and follicular relics (Tyndale-Biscoe 1984), transcriptional profiling (Cook et al. 2006, 2008; Cook and Sinkins 2010), pteridine density in eye capsules (Lehane 1985; Krafsur et al. 1995), and hydrocarbon layering in insect cuticles (Desena et al. 1999; Moore et al. 2017). Although all of these methods can distinguish young individuals from old ones, few can accurately classify mid-aged individuals, and none can differentiate the age of individuals within the more advanced age groups. A comprehensive overview of methods for estimating adult age in birds and mammals is presented in Lyons et al. (2012). Methods for aging fish using otolith layering are given in the papers by Campana and Limburg (Campana and Thorrold 2001; Limburg et al. 2013), and for estimating tree age using methods of tree ring analysis are given in Baillie (2015) and Speer (2010). Haussmann and Vleck (2002) suggest that telomere length can be used to age animals (e.g., finches) of unknown age captured in the wild.

Stage structure

Stage-frequency analysis is the average time required for arthropods and other cold-blooded organisms to attain various stages in their development, and their survival to those stages (Pontius et al. 1989). This analysis has received considerable attention because of its importance in applied ecology. The definitive method for determining individual-level developmental and survival rates is to monitor the survival and developmental rates of newborn individuals maintained in solitary confinement; however, for logistical reasons, or because some species subgroups, such as solitary insects, cannot be maintained in solitary confinement, this approach may not be possible. Data collection thus necessarily involves information on the development and survival outcomes of individuals maintained in groups.

For example, suppose 100 newly laid spider mite eggs are placed in solitary confinement on a single leaf disc in a petri dish lined with saturated cotton to keep the disc fresh. Each day the number in each stage (i.e., egg, immature, adult) is counted until all individuals have either completed development or died. On any given sampling date, more than one of the stages is present because there is variation in the developmental rates for each stage (fig. 9.9). The objective is to determine the average duration of each of the first two stages.

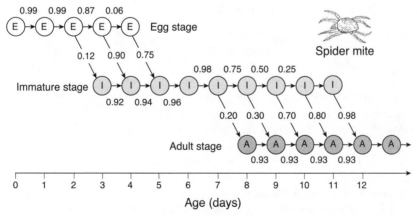

FIGURE 9.9. Intra- and interstage transition rates for spider mite (*T. urticae*) preadult development from newly laid eggs through adulthood (Carey and Bradley 1982). Note the variance within and between stages: the first-to-last egg hatch ranges from 3 to 5 days; the first-to-last immature stage occurs from 3 to 12 days. The numbers aligned with the horizontal arrows correspond to the age-specific within-stage transition probabilities, and the numbers aligned with the downward-pointing arrows correspond to the between-stage transition probabilities.

Table 9.11. Multistage life table framework for estimating stage-specific developmental and survival probabilities for the spider mite (*Tetranychus urticae*) from egg to adult

| | Number alive by stage | | | | Surviving | |
Age (x)	Egg (i=1)	Immature (i=2)	Adult (i=3)	Time (t)	unhatched eggs	Surviving preadults
(1)	(2)	(3)	(4)	(5)	(6)	(7)
0	100			0	1.0000	1.0000
1	99			1	0.9900	0.9900
2	98			2	0.9800	0.9800
3	85	12		3	0.8526	0.9702
4	5	88		4	0.0490	0.9245
5	0	86		5	0.0000	0.8598
6		83		6	—	0.8254
7		81		7	—	0.8089
8		61	16	8	—	0.6066
9		30	33	9	—	0.3033
10		8	52	10	—	0.0758
11		1	55	11	—	0.0076
12		0	52	12	—	0.0000
					3.37	7.85

Source: Carey and Bradley (1982).

There are three ways for an individual to exit an age class: (1) surviving to the next age within a stage, (2) surviving to the next age and into the next stage, or (3) dying. Thus each individual is subject to two competing transition risks, including the force of mortality and the force of (stage) transition. Using the general formula given in eqn. (9.30) for the multistage life tables contained in table 9.11,

$$N_{x+1}^i = \left(N_x^{i-1}\right)\left(p_x^{(i-1)\,\text{to}\,i}\right) + \left(N_x^i\right)\left(p_x^i\right) \tag{9.30}$$

The transitions shown for the immature stage at age 3 days can then be described as the number of immatures (i=2) attaining age 4 (x=3) and denoted as (N_4^2). This is equal to the sum of the number of immatures that survive to day 4 (x=4) while remaining in the immature stage (= $N_3^2 p_3^2$), and the product of the number of eggs age 3 (N_3^1) and the probability of transitioning from the egg to the immature stage ($p_3^{(1\text{to}2)}$). The results (table 9.11) show that the mean durations for the egg and immature stages are 3.37 and 4.48, respectively. The duration of the immature stage is determined by subtracting the egg duration from the total duration.

Extracting Parameters from Captive Cohorts

In this chapter we have thus far focused on estimating demographic parameters from populations in situ, but there is another set of models that have been developed for *captive cohorts* that use data gathered on the remaining life spans of captured individuals to estimate the age structure of the wild population from which they are sampled (Müller et al. 2004; Carey, Papadopoulos, et al. 2012). The underlying concept is that a population's age structure and its death distribution are uniquely interconnected, and with certain simplifying assumptions the age structure can be estimated from the death distribution using demographic models and reference life tables. The theoretical foundations for the captive cohort method (Müller et al. 2007; Vaupel 2009; Carey et al. 2018) have been extended to estimate the age structure in a wild Medfly population (Carey, Papadopoulos, et al. 2008; Carey 2011), to explore the implications of postcapture patterns of reproduction (Kouloussis et al. 2011) and age bias in sampling (Kouloussis et al. 2009), to estimate shifts in mean age (Carey, Papadopoulos, et al. 2012), and to identify physiological changes as populations (of mosquitoes) prepare for hibernation in late fall (Papadopoulos et al. 2016). In this section we illustrate the analysis of postcapture information using methods commonly used in cohort studies (e.g., life tables), and then, with simplifying assumptions, we show how this information can be used to estimate population age structure.

Captive cohort life table

Life table concepts and methods can be brought to bear on live-captured insects by considering them as a "population cohort"—a group of individuals (a cohort) experiencing the same event (their capture) at the same time, in other words, a captive cohort. Survival and mortality of these individuals is monitored (typically in the laboratory) from capture through the death of the last individual. Standard life table methods are used for cohort data analysis, but age is confounded with other sources of frailty, for example, host or reproductive status. Thus the interpretation of differences in life table parameters between two population samples (e.g., cohort survival or period mortality) is that differences in the postcapture age patterns of survival imply differences in their overall frailty of which chronological (or biological) age may be the primary source of the frailty differences.

A summary of postcapture longevity for Medflies captured during July and September in Greece by Carey and his colleagues (Carey, Papadopoulos, et al. 2008) is given as an abridged life table in table 9.12 and graphically in fig. 9.10. Average longevity of captured Medflies differed by two weeks (56 days in September versus 42 days in July). The September captures were less frail, as shown with the high early postcapture survival, and also much longer lived, with around a 40-day difference in the last fly dying between the two months of capture. The inset in fig. 9.10 shows the differences in life-days over the postcapture life course of the sampled flies.

Mortality equivalencies

Mortality equivalences for the different postcapture life expectancies can be generated by scaling an original mortality schedule. Let $\hat{e}_0^A$ and $\hat{e}_0^B$ denote new expectations of life at birth for cohorts A and B based on modified age-specific mortality schedules $\hat{q}_x^A$ and $\hat{q}_x^B$, where

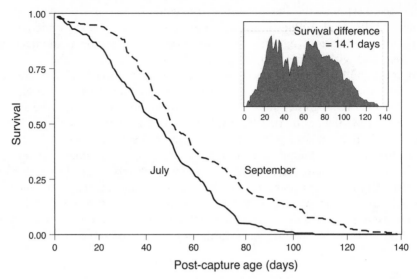

FIGURE 9.10. Medfly postcapture survival for individuals trapped in Chios Island, Greece, during July ($e_0 = 42.1$ days) and September ($e_0 = 56.0$ days) of 2005. Inset shows the survival differences between cohorts that sum to the 14.1 day difference in average post-capture longevity. (*Source*: Carey, Papadopoulos, et al. 2008)

$$\hat{q}_x^A = (1+\delta)q_x^A \tag{9.31}$$

$$\hat{q}_x^B = (1+\delta)q_x^B \tag{9.32}$$

and where q_x^A and q_x^B denote the original mortality schedules for cohorts A and B, respectively (Carey 1993). The odds ratios (OR) for the original schedule relative to the new schedule is then scaled to generate an equivalent life expectancy that can be computed as

$$OR = \frac{q_x^B}{\hat{q}_x^B} \tag{9.33}$$

Results shown in fig. 9.11 reveal that the highest mortality occurred in early July, and the lowest in September and October when mortality differed by nearly eightfold—that is, from a δ-value of nearly 4 to a δ-value of around 0.5.

Survival curve analysis

Robine (2001) and Cheung, Robine, and their colleagues (Cheung et al. 2005) introduced the concept of three dimensions of the survival curve: (1) *horizontalization*, which corresponds to how long a cohort and how many survivors can live before aging-related deaths significantly decrease the proportion of survivors; (2) *verticalization*, which corresponds to how concentrated aging-related deaths are around the modal age of death; and (3) *longevity extension*, which corresponds to how far the

Table 9.12. Abridged (5-day interval) life tables for Medfly captive cohorts sampled in July (n = 195) and September (n = 268)

Age	July captures					September captures				
x	N_x	l_x	q_x	d_x	e_x	N_x	l_x	q_x	d_x	e_x
0	195	1.0000	0.0359	0.0359	42.1	268	1.0000	0.0261	0.0261	56.0
5	188	0.9641	0.0426	0.0410	38.6	261	0.9739	0.0077	0.0075	52.4
10	180	0.9231	0.0444	0.0410	35.2	259	0.9664	0.0039	0.0037	47.8
15	172	0.8821	0.0523	0.0462	31.7	258	0.9627	0.0155	0.0149	43.0
20	163	0.8359	0.1227	0.1026	28.4	254	0.9478	0.0354	0.0336	38.6
25	143	0.7333	0.0839	0.0615	27.0	245	0.9142	0.0898	0.0821	35.0
30	131	0.6718	0.1069	0.0718	24.2	223	0.8321	0.0987	0.0821	33.2
35	117	0.6000	0.1111	0.0667	21.8	201	0.7500	0.1443	0.1082	31.5
40	104	0.5333	0.1635	0.0872	19.2	172	0.6418	0.0930	0.0597	31.4
45	87	0.4462	0.2069	0.0923	17.5	156	0.5821	0.1795	0.1045	29.4
50	69	0.3538	0.1449	0.0513	16.4	128	0.4776	0.1328	0.0634	30.3
55	59	0.3026	0.2034	0.0615	13.8	111	0.4142	0.1441	0.0597	29.5
60	47	0.2410	0.3617	0.0872	11.6	95	0.3545	0.0632	0.0224	29.1
65	30	0.1538	0.2333	0.0359	11.8	89	0.3321	0.1348	0.0448	25.9
70	23	0.1179	0.5652	0.0667	9.7	77	0.2873	0.1948	0.0560	24.5
75	10	0.0513	0.1000	0.0051	14.0	62	0.2313	0.1774	0.0410	24.8
80	9	0.0462	0.3333	0.0154	9.3	51	0.1903	0.0980	0.0187	24.7
85	6	0.0308	0.3333	0.0103	9.2	46	0.1716	0.1304	0.0224	22.1
90	4	0.0205	0.5000	0.0103	7.5	40	0.1493	0.0750	0.0112	20.0
95	2	0.0103	0.5000	0.0051	7.5	37	0.1381	0.1622	0.0224	16.4
100	1	0.0051	0.0000	0.0000	7.5	31	0.1157	0.3226	0.0373	14.1
105	1	0.0051	1.0000	0.0051	2.5	21	0.0784	0.0952	0.0075	14.6
110	0	0.0000		0.0000	0.0	19	0.0709	0.2632	0.0187	9.9
115						14	0.0522	0.5000	0.0261	8.9
120						7	0.0261	0.2857	0.0075	9.4
125						5	0.0187	0.4000	0.0075	8.5
130						3	0.0112	0.0000	0.0000	7.5
135						3	0.0112	1.0000	0.0112	2.5
140						0	0.0000		0.0000	0.0

Source: Carey, Papadopoulos, et al. (2008).

highest normal life durations can exceed the modal age of death. These measures were developed primarily for characterizing historical changes in the degree of rectangularization ("squaring the curve") of human survival.

One of the major differences between longevity databases for humans versus those for nonhuman species is sample size. Whereas there are typically records for tens, or hundreds, of thousands of deaths by age and sex for humans, the longevity records involving studies of nonhuman species often are in the range of a few score (primates), a few hundred (rodents), or a few thousand (insects, worms). Although a modal age of death can be computed with any data set, a statistically significant modal age is only identifiable for data sets containing deaths of a few thousand individuals. This is especially true for the use of the captive cohort method, where the sample sizes involve a few score or a few hundred individuals.

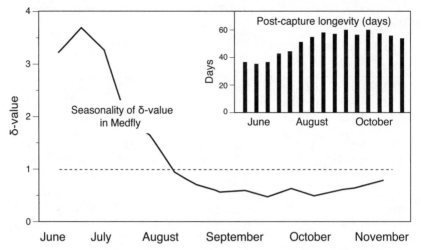

FIGURE 9.11. Seasonal changes in the value of δ in estimating seasonal changes in Medfly frailty based on equivalencies in the scale of mortality relative to an overall seasonal mean. Inset shows the changes in the post-capture longevity through the summer and fall months.

Given this constraint on estimating the modal ages for the death distribution for small data sets from captive cohort studies, we introduce a simplified approach here that is based on the slope of the survival curve at early, middle, and advanced ages. We start with the concept presented in fig. 9.12 of a stylized survival curve in which three slopes are computed using regression methods, including the following (modified from Cheung et al. 2005):

(1) *horizontalization*—survival slope between age 0 and $l_x = 0.90$

(2) *verticalization*—survival slope between $l_x = 0.90$ and $l_x = 0.10$

(3) *longevity extension*—survival slope between $l_x = 0.10$ and the age of the last individual to die $(x = \omega)$

Let S_H, S_V, and S_{LE} denote the slopes of the straight lines defining horizontalization, verticalization, and longevity extension, respectively. Simple algebra yields slopes where A, B, and C in fig. 9.12 specify survival at birth and survival to the ages when 10% and 90% are dead, respectively, and D is the age of the oldest individual.

$$S_H = \frac{(l_0 - l_{x_B}^B)}{x_B} \tag{9.34}$$

$$S_V = \frac{(l_{x_B} - l_{x_C})}{(x_C - x_B)} \tag{9.35}$$

$$S_{LE} = \frac{(l_{x_C} - l_{x_\omega})}{(\omega - x_C)} \tag{9.36}$$

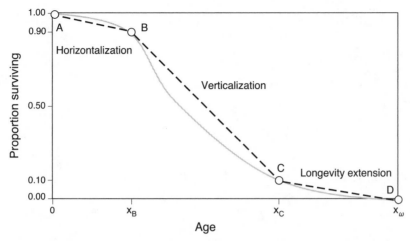

FIGURE 9.12. Schematic showing the three dimensions of the survival curve, including horizontalization, verticalization, and longevity extension. The gray curve represents observed survival, and the black lines represent the straight-line fits between points labeled A, B, C, and D that correspond to survival rates of 1.0, 0.9, 0.1 and 0.0, respectively. The slopes of these three straight lines for A-B, B-C, and C-D specify the three survival curve dimensions.

For example, using data from the Medfly postcapture survival curves in July and September (fig. 9.13), we can do a comparison of these three dimensions across the two capture periods. *Horizontalization,* the rate of decline for this component of the survival curves, was over 2-fold greater in the July captures relative to those captured in September. Whereas the first 10% of deaths occurred after two weeks for the July Medfly captures, it occurred after around a month for the September captures—that is, twice as long. For *verticalization,* there was a nearly 1.5-fold difference in the slopes of the postcapture survival curves between the July versus the September captures. The period that elapsed from 90% alive to 10% alive was 58 and 75 days for the July and September captures, respectively. For *longevity extension,* the tails of the survival curves for Medflies captured on the two sampling dates were very similar with similar slopes, to nearly four decimal places, and durations differing by only 3 days—that is, 34 days for July captures versus 38 days for September captures.

Population structure

Frailty structure

In the analysis of postcapture survival, the "cohort" of individuals may be heterogeneous, and this can pose unique issues for the estimation of survival. In populations of insects and other invertebrates, individuals within a captured cohort are a single age class for species where there is a synchronous emergence of adults, such as mayflies. In most other plant and animal species, a captive cohort consists of individuals of mixed ages and the exact composition is unknown. Frailty, and thus the probability of death, varies with age; thus the time trajectory of mortality for a sample of individuals captured in the wild is determined by the age composition of

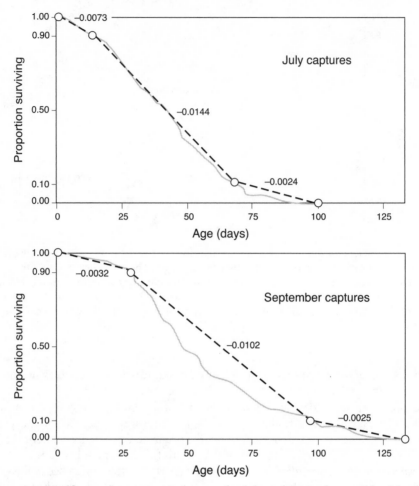

FIGURE 9.13. Medfly postcapture survival curves for July and September samples. Numbers in dashed lines are linear regression slopes for the three dimensions of the curves. See legend in fig. 9.12.

the captured cohort. This age composition may change during the season or between locations. We address the consequences of age on the captured cohort's survival under four different scenarios.

CASE I

For the first case we consider single cohorts at different capture ages. The survival curves shown in fig. 9.14a correspond to the hypothetical situation in which individuals from the wild are collected from a single wild Medfly cohort immediately after emerging (young only) and again at 50 days (middle age) and 90 days (old age). Survival of young individuals is extremely high (flat slope) until around 50 days at which time they begin dying off more rapidly. Note that some individuals survive beyond 125 days. For individuals collected at 50 days initial survival is moderate, but it is substantially less than survival for young individuals during their first days postcapture. Survival then drops off rapidly and no individuals live beyond 75 days.

Survival for 90-day-old individuals is relatively short with an extremely steep decline immediately after capture and no individuals live beyond 30 days postcapture.

CASE II
In the second case we consider a cohort with a mixture of young and middle-aged subcohorts. At the time of sampling this hypothetical population consists of a mixture of newborn (age 0) and 50-day-old individuals (fig. 9.14b). The earliest deaths of the middle-aged subcohort drives the overall cohort survival downward immediately. The steepness of the decline depends on the relative initial number of middle-aged individuals in the cohort. As the (originally) middle-age subcohort dies off, the higher survival of the (original) newborn subcohort (though now older) becomes a larger proportion of the total. As a result, the decrease in survival slows into the intermediate postcaptive ages until the time when the original newborn cohort reaches more advanced ages. At this point the steepness of the decline in survival increases drastically. Maximum postcapture longevity is 140 days due to the presence of the newborn age class in the original cohort.

CASE III
In the third case we consider a cohort with a mixture of young and old sub-cohorts. At the time of sampling this hypothetical population consists of a mixture of newborn (age 0) and 90-day-old individuals (fig. 9.14c). The much higher death rates of the older individuals immediately drive the overall cohort survival steeply downward. At middle captive ages, the depth and steepness of the decline depends on the relative proportion of the older individuals in the cohort. As the older subcohort completely dies off, after around three weeks, the overall cohort survival is determined solely by survival in the only remaining subcohort, the original newborns. This change in the composition of the cohort results in improved survival in the intermediate postcaptive ages until which time the original newborn cohort reaches the more advanced ages. Maximum postcapture longevity is 140 days due to the presence of newborns in the original cohort.

CASE IV
In the fourth case we consider a cohort with a mixture of middle-aged and old subcohorts. At the time of sampling this hypothetical population consists of a mixture of middle-age (50-day) and old-age (90 day) subcohorts (fig. 9.14d). Because the average age of the overall cohort is high (because there are no newborns present), survival begins declining immediately after capture with virtually no slowing at any time through the maximum postcapture age of 75 days.

Age structure

The general concept of age estimation using captive cohort methods is illustrated in fig. 9.15, in which the postcapture death distributions of population samples differ markedly depending on population age and age structure. Here we introduce the model from Müller and his colleagues (2007) that was applied to field experiments with Medflies (Carey, Papadopoulos, et al. 2008). The model includes three cohorts: C, the population of wild-caught captive cohorts; R, the reference cohorts that are age homogeneous and are raised in the laboratory from birth; and W, the wild cohorts. Here x^* denotes the time elapsed since capture of the captive cohort. Since the ages

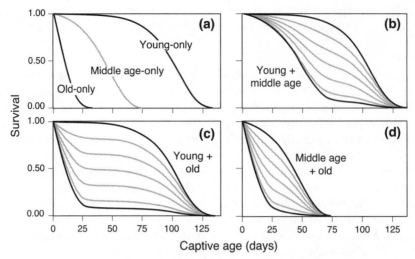

FIGURE 9.14. Captive cohort survival patterns for hypothetical Medfly populations based on (a) single cohorts at different ages when captured, including young only (0 days old), middle age only (50 days old), and old only (90 days old); and mixtures of (b) young and middle age, (c) young and old, and (d) middle age and old. The solid black curves at the top and bottom correspond to the mixture extremes, with the lighter-shaded curves depicting progressively greater proportions of the most frail subcohort. The original cohort was created using a Gompertz model with a and b parameters equal to 0.0001 and 0.072, respectively. Expectations of life (e_0) at capture were 83.0, 35.4, and 8.6 days, respectively, for young, middle-age, and old cohorts.

of individuals in this cohort are unknown, x^* does not correspond to any individual's actual age. Age x denotes the actual age of individuals in the wild, the population distribution of which the model aims to infer. With these conventions, eqn. (3) in Müller et al. (2007)—which provides a basic relationship between the survival function of the captive population, the survival function of the reference cohort, and the density of the age distribution of the wild population—is found to have the following discrete counterpart in demographic notation:

$$l_s(x^*) = \sum_x l_x(x + x^*) \frac{c_w(x)}{l_x(x)} \qquad (9.37)$$

This is obtained from the continuous version by replacing the survival functions of the reference and captive cohorts by their discrete counterparts, l_x and l_R, and replacing the (continuous) age density in the wild by c_W. Taking differences in eqn. (9.37) simultaneously on both sides of the equation (corresponding to differentiation of both sides of eqn. (3) in Müller et al. 2007) yields

$$c_s(x^*) = \sum_x c_x(x + x^*) \frac{c_w(x)}{l_x(x)} \qquad (9.38)$$

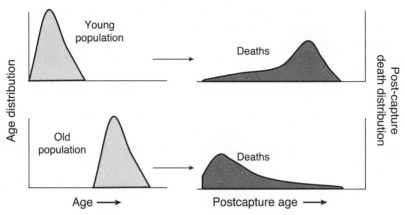

FIGURE 9.15. Schematic depicting hypothetical age distributions of younger and older populations and their corresponding postcapture death distributions.

Given that the quantities c_S and c_R can be determined from the observed survival in captive and reference cohorts, eqn. (9.38) can be used to determine the unknown target quantity c_W by numerical deconvolution. This deconvolution is not entirely straightforward, as the quantity of interest, c_W, is nested in the sum on the right-hand side of the equation. Müller et al. (2007) provide an analysis of the assumptions and the implementation of this somewhat complex procedure.

Conceptually, the basic assumption for estimating the age structure using the captive cohort method with a reference life table is that, once removed from the wild and reared from that moment forward in the laboratory, individuals in the captive cohort will be subject to the age-specific mortality rates as reference individuals of known age. Thus, the survival of all individuals age 10 days at capture will, regardless of their past history, die at the same rate as 10-day-old individuals that have always been maintained in the laboratory. That this assumption is certainly violated does not invalidate the broader concept if age is generalized as a frailty proxy and the results of the study are considered as an outcome of the frailty structure of the population.

For example, consider a population sample consisting of two age classes, 10 and 15 days, where $N_{10} = 800$ and $N_{15} = 200$ denote the number in the sample for these respective age classes (table 9.13). If these flies are subject to the age-specific survival rates, p_x, as given in table 9.13, then

$$N_s(0) = 1{,}000$$

$$N_s(1) = p_{10} N_w(10) + p_{15} N_w(15) = 900$$

$$N_s(2) = p_{10} p_{11} N_w(10) + p_{15} p_{16} N_w(15) = 775$$

$$N_s(3) = p_{10} p_{11} p_{12} N_w(10)\ p_{15} p_{16} p_{17} N_w(15) = 636$$

$$N_s(4) = p_{10} p_{11} p_{12} p_{13} N_w(10) + p_{15} p_{16} p_{17} p_{18} N_w(15) = 495$$

$$N_s(5) = p_{10} p_{11} p_{12} p_{13} p_{14} N_w(10) + p_{15} p_{16} p_{17} p_{18} p_{19} N_w(15) = 364$$

Survival of the mixed age cohort of 1,000 newly captured individuals through the first iteration, denoted $N_s(1)$, will equal the sum total of survival of each captured age class through one time step. For example, the fraction of all wild individuals in age class 0, $N_w(0)$, that survive one step is calculated from the reference life table as $\frac{l_R(1)}{l_R(0)}$ (i.e., the probability of surviving from age class 0 to age class 1). It follows that the fraction of all wild individuals in age class 1, $N_w(1)$, that survive one step is calculated as $\frac{l_R(2)}{l_R(1)}$ (i.e., the probability of surviving from age class 1 to age class 2). The same analytics apply to the fraction of individuals in age class 2 that survive to age class 3. These relationships are given as

$$N_s(1) = N_w(0)\frac{l_R(1)}{l_R(0)} + N_w(1)\frac{l_R(2)}{l_R(1)} + N_w(2)\frac{l_R(3)}{l_R(2)} \qquad (9.39)$$

Thus $N_s(1)$ is computed as

$$= 300 \times \frac{0.75}{1.00} + 125 \times \frac{0.50}{0.75} + 225 \times \frac{0.25}{0.50}$$

$$N_s(1) = 421$$

This process is repeated for the number surviving 2 days:

$$N_s(2) = N_w(0)\frac{l_R(2)}{l_R(0)} + N_w(1)\frac{l_R(3)}{l_R(1)} \qquad (9.40)$$

The fraction of all wild individuals in age class 0, $N_w(0)$, that survive two time steps is then calculated from the reference life table as $\frac{l_R(2)}{l_R(0)}$ (i.e., the probability of surviving from age class 0 to age class 2). And the fraction of individuals who are age 1 that survive two captive time steps is calculated as $\frac{l_R(3)}{l_R(1)}$. Thus

$$= 300 \times \frac{0.50}{1.00} + 125 \times \frac{0.25}{0.75}$$

$$N_s(2) = 192$$

And the same is true for the number in each age class surviving for 3 days:

$$N_s(3) = N_w(0) \frac{l_R(3)}{l_R(0)}$$

$$= 300 \times \frac{0.25}{1.00} \qquad\qquad 9.41$$

$$= 75$$

The formulas for Drosophila captive cohort survival, $N_S(x)$, in 10-day intervals for a wild population, the relationship of age distribution in the wild and death distribution in the laboratory, and the parameters in a nonstationary hypothetical wild population are contained in tables 9.14, 9.15, and 9.16, respectively.

Consider Further

The foundational concepts and models on which much of the content in this chapter rests are contained in the earlier chapters (e.g., demography basics; life tables; stable population theory), and a number of the entries in chapter 11 (Biodemography Shorts) involve applications of demographic concepts and principles (e.g., the Group 4 subgroup concerned with animal biodemography). A starting point for readers interested in delving deeper into applied demography is Swanson et al. (1996) and Burch (2018), both of which are concerned with the major scope of applied demography, and the books by Keyfitz (Keyfitz and Beekman 1984; Keyfitz and Caswell 2010) and Caswell (2001) for perspectives on human demography and population biology, respectively.

Other excellent sources of more in-depth demographic content include methods for Bayesian estimation of age-specific survival from incomplete mark–recapture/

Table 9.13. Example of postcapture survival in a sample of flies consisting of two subcohorts of 10- and 15-day-old individuals

Captive Age x*	Flies of unknown age in mixed-age population sample								Captive cohort	
	Captured 10-day-old flies				Captured 15-day-old flies				Survival	Deaths
	x	N(x)	p(x)	d(x)	x	N(x)	p(x)	d(x)	N(x*)	d(x*)
0	10	800	0.95	40	15	200	0.70	60	1000	100
1	11	760	0.90	76	16	140	0.65	49	900	125
2	12	684	0.85	103	17	91	0.60	36	775	139
3	13	581	0.80	116	18	55	0.55	25	636	141
4	14	465	0.75	116	19	30	0.50	15	495	131
5	15	349	—		20	15	—		364	

Table 9.14. Formulas for captive cohort survival, $N_S(x)$, in 10-day intervals for a wild *D. melanogaster* population

Age class	$N_S(x)$	Formulas
0	$1,000 =$	Radix
10	$N_S(10) =$	$N_w(0)\dfrac{l_R(10)}{l_R(0)} + N_w(10)\dfrac{l_R(20)}{l_R(10)} + \cdots + N_w(70)\dfrac{l_R(80)}{l_R(70)}$
20	$N_S(20) =$	$N_w(0)\dfrac{l_R(20)}{l_R(0)} + N_w(10)\dfrac{l_R(30)}{l_R(10)} + \cdots + N_w(60)\dfrac{l_R(80)}{l_R(60)}$
30	$N_S(30) =$	$N_w(0)\dfrac{l_R(30)}{l_R(0)} + N_w(10)\dfrac{l_R(40)}{l_R(10)} + \cdots + N_w(50)\dfrac{l_R(80)}{l_R(50)}$
40	$N_S(40) =$	$N_w(0)\dfrac{l_R(40)}{l_R(0)} + N_w(10)\dfrac{l_R(50)}{l_R(10)} + \cdots + N_w(40)\dfrac{l_R(80)}{l_R(40)}$
50	$N_S(50) =$	$N_w(0)\dfrac{l_R(50)}{l_R(0)} + N_w(10)\dfrac{l_R(60)}{l_R(10)} + \cdots + N_w(30)\dfrac{l_R(80)}{l_R(30)}$
60	$N_S(60) =$	$N_w(0)\dfrac{l_R(60)}{l_R(0)} + N_w(10)\dfrac{l_R(70)}{l_R(10)} + N_w(20)\dfrac{l_R(80)}{l_R(20)}$
70	$N_S(70) =$	$N_w(0)\dfrac{l_R(70)}{l_R(0)} + N_w(10)\dfrac{l_R(80)}{l_R(10)}$
80	$N_S(80) =$	$N_w(0)\dfrac{l_R(80)}{l_R(0)}$
90	0	

Note: Based on a radix of 1,000 captured individuals of unknown age and life table rates in a reference cohort, $l_R(x)$.

recovery data (Colchero and Clark 2012; Colchero et al. 2012), mark-recapture concepts applied to human populations and in epidemiology (Fienberg 1972; Hook and Regal 1995; vanderHeijden et al. 2009; Coumans et al. 2017), and senescence in the wild when birth and death dates are unknown (Zajitschek et al. 2009). Nearly all the chapters in the seventh edition of the *Wildlife Techniques Manual* (Silvy 2012) and especially the chapter on estimating animal abundance by Pierce et al. (2012) is relevant to this material. The software program MARK provides parameter (e.g., survival) estimates in studies using recaptured or reencountered marked animals. The program and its capabilities are described in White and Burnham (1999) and available at http://www.phidot.org/software/mark/index.html. The US Centers for Disease Control (CDC) posts online lessons in basic epidemiological concepts, principles, and methods at https://www.cdc.gov/ophss/csels/dsepd/ss1978/index.html. The chapter on wildlife demography by Jean-Dominique Lebreton and Jean-Michel Gaillard (2016) contains sections ranging from the history of wildlife demography research and tools for demographic assessment to an overview of wildlife demography and population growth rate along the slow-fast continuum. A paper

Table 9.15. Relationship of age distribution in the wild and death distribution in the lab subject to reference life table rates for three hypothetical *D. melanogaster* populations with young, middle, and old age structures

Age	Reference	Young population		Middle-age population		Old population	
		Postcapture survival	Wild population age structure	Postcapture survival	Wild population age structure	Postcapture survival	Wild population age structure
		$N_S(x)$	$N_W(x)$	$N_S(x)$	$N_W(x)$	$N_S(x)$	$N_W(x)$
0	1.0000	1000.0	500	1000.0	0	1000.0	0
10	0.9851	912.4	300	663.7	0	263.9	0
20	0.8785	7410.4	200	369.9	200	310.4	0
30	0.6690	533.6	0	165.7	500	2.0	0
40	0.4629	335.1	0	52.8	300	0.0	0
50	0.2553	170.2	0	10.8	0	0.0	200
60	0.1138	610.1	0	0.6	0	0.0	500
70	0.0348	110.2	0	0.0	0	0.0	300
80	0.0025	1.3	0	0.0	0	0.0	0
90	0.0000	0.0	0	0.0	0	0.0	0
		3,787.1		2,262.5		1,304.3	

Table 9.16. Parameters in nonstationary hypothetical wild population for captive cohort analysis

Age or captive age	Wild population	Reference life table	Captive survival	Captive deaths
x or x^*	$N_W(x)$	$l_R(x)$	$N_S(x^*)$	$d_w(x^*)$
(1)	(2)	(3)	(4)	(5)
0	300	1.000	1000	579
1	125	0.750	421	229
2	225	0.500	192	117
3	350	0.250	75	75
4	0	0.000	0	

on insect demography by Carey (2001) includes sections on applied demography in an entomological context. The paper by Hargrove et al. (2011) describes a general mortality model developed for the tsetse fly—an important disease vector in Africa. Additional papers concerned with residual demography include work by Carey, Müller, and their colleagues (Müller et al. 2004, 2007; Carey, Papadopoulos, et al. 2008; Carey, Müller, et al. 2012; Carey 2019).

Applied Demography II:
Evaluating and Managing Populations

> A model is neither true nor false; only the theory similarity to what it represents. A theory is thus a metaphor between a model and data.
> And understanding in science is the feeling of similarity between complicated data and a familiar model.
>
> Julian Jaynes (1976), *The Origin of Consciousness in the Breakdown of the Bicameral Mind*

The primary focus of this chapter is the use of demographic models for evaluating, comparing, and managing populations. In the first section on comparative demography, we draw on concepts introduced in earlier chapters to discuss methods to quantify differences in demographic metrics among populations. In the section on health and health span, we use the concept of population health in a life table context and apply it to such diverse topics as African buffalo populations, biomarkers of aging, a health classification for cities based on "infestation status," and biological control. In the section on population harvesting, we use demographic approaches to determine how best to manipulate populations for mass rearing and population culling. And in the last section on conservation, we develop models to evaluate the impacts of hunting and poaching, and we quantify data from a captive breeding program.

Comparative Demography

In these examples we start by introducing (or reintroducing) techniques to identify and compare the detailed actuarial properties of two or more cohorts in both basic (e.g., aging; population biology) and applied (e.g., conservation biology; population control) contexts. Using cohort life table data for three fruit fly species, we illustrate a collection of comparative concepts and methods, including graphical inspection of survival and mortality, synopses of life table traits, age-specific mortality slopes, Gompertzian parameters, mortality ratio crossovers, proportional differences in overall mortality, and age-specific contributions to differences in expectation of life at birth.

We will compare the actuarial properties of three fruit fly species: two species in the fly family Tephritidae (referred to as the "true" fruit flies), including the Mediterranean fruit fly (*Ceratitis capitata*) and the Mexican fruit fly (*Anastrepha ludens*), and one species in the fly family Drosophilidae (referred to as "small fruit flies"), the vinegar fly (*Drosophila melanogaster*) (Carey et al. 1998a, 2005). The two tephritid species are 5–8 times larger than the vinegar fly (i.e., housefly-size or slightly larger, versus gnat-size). Unlike the vinegar flies that lay their eggs on or in rotting fruit, the two tephritid species lay their eggs in growing fruit that are approaching market ripeness. These ecological differences between the species have shaped the evolution of their survival and mortality patterns.

Basic metrics

This first step of demographic comparison builds on concepts and graphical analysis presented in chapters 2 and 3, where we introduced survival and mortality curves, life table metrics, and the Gompertz model of mortality.

Survival and mortality curves

One of the first steps in any comparative actuarial analysis is to inspect and graphically compare the survival (l_x) and mortality (q_x) schedules of cohorts. Several similarities and differences are evident in the fruit fly survival and mortality schedules (fig. 10.1). First, the Mexfly's survival is different from the survival in the Medfly and the Drosophila. Survival patterns are similar in these two latter species for their first 20–30 days after which their patterns diverge slightly. These survival differences are reflected in the much lower mortality for the Mexfly relative to the other species (fig. 10.1, right) and the similarity in mortality for the other two species through their first 20+ days. Note also that the mortality in all species diverge at ≈ 25 days, and thus they have different actuarial aging rates (slopes); but at an age of slightly over 60 days, the mortality rates of all three species converge.

Life table metrics

A complementary and usually simultaneous step in comparative actuarial analysis is to summarize and evaluate the key metrics from the life tables of, for this case, each of the three fruit fly species. There are three subcategories to this analysis: central tendencies, cohort age survival, and survival age spread (table 10.1).

Central tendencies refer to the mean, median, and mode. The mean longevity in the Mexfly (48.7 days) exceeds that for the Medfly (34.1 days) by around two weeks, and

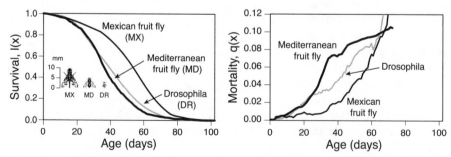

FIGURE 10.1. Female age-specific survival (left panel) and mortality (right panel) for the Medfly, Mexfly, and *Drosophila melanogaster*, based on smoothed data for each species. Note relative size differences between species (left-panel inset).

Table 10.1. Summaries of key metrics (in days) from life table analyses of female cohorts from three fruit fly species

Parameter/metric (in days)	Fruit fly species		
	Mexican	Mediterranean	Drosophila
Central tendencies			
Mean (e_0)	48.7	34.1	38.0
Median ($l_x = 0.5$)	52	35	42
Mode ($d_x = $ max)	60	35	36
Cohort survival age (x)			
$l_x = 0.75$	39	27	26
$= 0.25$	62	45	50
$= 0.10$	71	56	62
$= 0.01$	87	81	75
Survival durations			
75–25% (central half)	23	18	24
10–1% (tail)	16	25	13

it is greater than the Drosophila (38.0 days) by approximately 11 days. The median survival, defined as the age when only 50% of the cohort is still living, exceeds the mean longevity for all three species; but the relative age of modal death—the most frequent age of death—varies across species with the mode for the Mexfly exceeding, for the Medfly equaling, and for the Drosophila lower than, their respective survival means.

To compare the *cohort age survival,* we can compare the three species at the same points on their survival curves. For example, the point where cohort survival was 75% occurred by 40 days in the Mexfly, but only 26–27 days for the other two species. At approximately 60 days, 25% of the original Mexfly cohort was still alive, but 10% or less of the cohorts of the other two species were still alive. The original cohorts were not reduced to the last 1% in the Mexfly for nearly three months, but this survival level occurred 1–2 weeks earlier in the Medfly and Drosophila.

The survival age spread, with respect to life expectancy, did not differ across species. Although life expectancies differed by over 10 days for the Mexfly and Drosophila, the time required for cohort survival in both species to drop from 75% to 25% differed by only 1 day (i.e., 23 and 24 days, respectively). Interestingly, the Medfly had the shortest life span but the longest survival tail, as reflected in the time required for survival in the original cohort to decrease from 10% to 1%.

Gompertzian analysis

The Gompertz model [$\mu(x) = ae^{bx}$] is useful in comparative contexts because fitting the model to data reveals differences in mortality intercept (parameter a), slope (parameter b), and doubling time $DT = \dfrac{\ln 2}{b}$. The results of model fitting for the mortality schedules of the three fruit fly species for the first 30 days are given in table 10.2. The initial mortality levels for the 30-day Gompertz fits of species-specific mortality are generally similar, and extremely low, ranging from around 0.002 (Drosophila) to 0.004 (Mexfly). These low and relatively similar levels of initial mortality

Table 10.2. Comparison of the Gompertz model parameters and doubling time for three fruit fly species

Gompertz parameter or metric	Fruit fly species		
	Mexfly	Medfly	Drosophila
a (initial mortality)	0.0037	0.0034	0.0019
b (slope)	0.0310	0.10047	0.1180
Mortality doubling time (days)	22.4	6.9	5.9

Note: Linear least squares regressions of age versus $\ln[\mu(x)]$ for the first 30 days were used to estimate model parameters.

clearly had minimal impact on the between-species longevity differences. The main differences in 30-day survival probabilities across species were due to the rate of change in mortality with age, in other words, the Gompertz slope parameter, b. Note that b-values were over threefold greater in the Medfly and Drosophila compared to the Mexfly. This is reflected in the large differences in mortality doubling times, which were greater than 3 weeks for the Mexfly and less than 1 week for both of the other two species.

Comparative mortality dynamics

Mortality slopes

The life table aging rate (LAR) (Horiuchi and Coale 1990), discussed in chapter 2, is denoted k_x and is defined as the rate of change with age in the central death rate, m_x, based on the formula

$$k_x = \ln(m_{x+1}) - \ln(m_x) \tag{10.1}$$

The LARs for each of the species shown in fig. 10.2 reveal important across-species differences in changes in mortality schedules with age and in the magnitudes of these changes. The LAR for the Mexfly varies widely through age 40, ranging from a mortality slope of over 0.6 (rapid increase) to around −0.2 (moderate decrease). Mortality in this species from around 40 to 70 days increases at a modest slope of 0.2 after which it rapidly decreases and then rapidly increases. The overall LAR patterns for the Medfly and the Drosophila are roughly similar to each other with high initial values gradually decreasing through ages 30 to 40 at which time they generally stabilize. These more constant levels are positive but very low for the Medfly and positive at 0.01 to 0.02 for the Drosophila.

Mortality ratios

The mortality ratio (MR) of two age-specific schedules A and B is computed as

$$MR = \frac{q_x^A}{q_x^B} \tag{10.2}$$

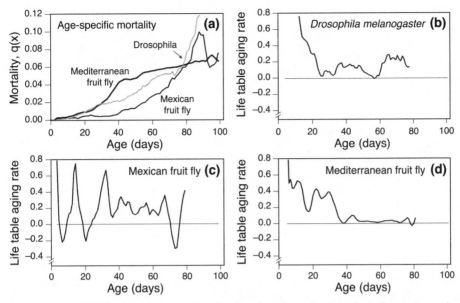

FIGURE 10.2. Life table aging rates (LARs) for female cohorts in three fruit fly species and the species-specific mortality schedules from which the LARs were derived (panel (a)).

where q_x^A and q_x^B denote mortality rates at age x for cohorts A and B, respectively. These ratios provide perspectives on the relative differences in mortality over the entire life course. For example, a ratio greater than unity (1.00) over all ages would indicate that mortality is greater in cohort A than in cohort B across their entire life courses. However, a mortality crossover occurs if respective mortality schedules change such that the schedules converge. A mortality crossover (or convergence) in humans can be described as an attribute of the relative rate of change, and the level of age-specific mortality rates in two population groups, where one group is "advantaged" (i.e., lower relative mortality) and the other is "disadvantaged" (i.e., higher relative mortality) (Manton and Stallard 1984). For a crossover, the disadvantaged population must have age-specific mortality rates markedly higher than the advantaged population through middle age at which time the rates change. Mortality crossovers thus occur due to differences in rates of aging at the individual level and due to demographic selection where individuals with high mortality are selected out of the population and only the robust individuals survive to the older ages (Manton and Stallard 1984). This is also referred to as the "cohort-inversion model," which is based on the concept that cohorts experiencing particularly hard or good times early in life will respond inversely later in life (Hobcraft et al. 1982). Mortality ratios provide insights into the magnitude of mortality differences between two cohorts at each age, but they do not quantify the extent to which these differences contribute to differences in life expectancy. Although not illustrated here, the age ratios of other functions, including life expectancy and survival, are also useful in identifying important relative patterns of the cumulative consequences of mortality between cohorts (Carey 1993).

The ratio of the Drosophila and Medfly mortality schedules, given in fig. 10.3, show several important differences. The first is that the relative magnitude of the

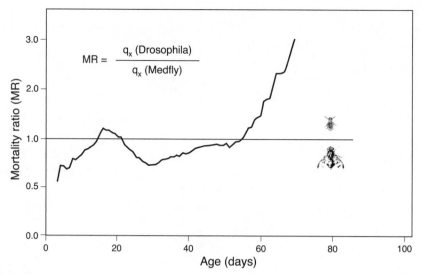

FIGURE 10.3. Mortality ratio for the Drosophila and the Medfly. Note the three mortality crossovers at ages when their ratios are 1.0.

differences in the mortality levels between the two species is modest. It was not until age 60 days that Drosophila mortality exceeded Medfly mortality, when it was two- to threefold greater, and before that time the schedules differed only slightly. The second major difference between these two mortality schedules is that Drosophila mortality was lower than the Medfly for nearly two full months (with a brief exception at 18–20 days). The third difference is that there were multiple mortality crossovers, twice in the age interval 18 to 20 days and then again at 60 days. These crossovers reveal the divergence and convergence patterns of the between-species mortality patterns (i.e., relative waxing and waning of differences in actuarial aging).

Mortality contributions to e_0

Life expectancies of the Mexfly and the Drosophila differed by 10.7 days (i.e., 48.7 days and 38.0 days, respectively) (see table 10.1). The question of which age group accounts for these differences can be addressed using the formula for decomposing life expectancy differences between cohorts, as given in Carey (1993) (but also see Jdanov et al. 2017; Chisumpa and Odimegwu 2018). Let Δ_x denote the change in expectation of life between two cohorts from age x to x + 1, and let l_x^A and l_x^B denote survival to age x for cohorts A and B, respectively, and e_x^A and e_x^B denote expectation of life for cohorts A and B, respectively. Then

$$\Delta_x = A - B \tag{10.3}$$

where

$$A = \left(e_x^A - e_x^B\right)\left[\frac{1_x^A + 1_x^B}{2}\right] \tag{10.4}$$

$$B = \left(e_{x+n}^A - e_{x+n}^B\right)\left[\frac{1_{x+n}^A + 1_{x+n}^B}{2}\right] \tag{10.5}$$

The contributions of mortality differences by age for the Drosophila versus the Mexfly are shown in fig. 10.4. The net sum of all bars (daily intervals) in this graphic equals the difference in the life expectancies of these two species (i.e., 10.7 days). A negative number indicates a contribution in favor of the Drosophila (leftmost group of bars below line), and a positive number indicates a contribution in favor of the Mexfly (rightmost group of bars above line).

The distribution of the daily contributions of mortality differences reveals two important points. First, the tiny mortality advantage of the Drosophila in the first week relative to the Mexfly made a disproportionately large offsetting contribution to the differences in life expectancies. This is due to the effects at early ages because more Drosophila survived early ages and thus could contribute more life-days to life expectancy than otherwise would have occurred. Second, the majority of the differences in contributions occurred at early-middle age in both species, a time when mortality rates were in the earlier stages of convergence. Thus the effects on e_0 of earlier mortality in both cases had the most substantial impact on e_0 differences, even though both the relative and absolute mortality differences between these two species was much greater at older ages.

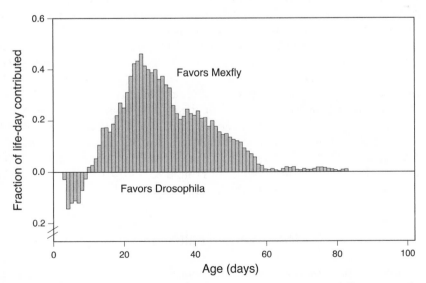

FIGURE 10.4. Contributions of mortality differences to life expectancy differences in the Mexfly versus the Drosophila.

Mortality proportional equivalences

Although the life expectancies of two cohorts may differ by, for example, 10%, it does not follow that the differences in their mortality schedules differ by this same percentage. Indeed, a number of demographers have asked this question: How much do differences in mortality rates determine differences in life expectancies? (see Pollard 1982; Keyfitz 1985; Vaupel 1986). Adopting methods from these demographers, this question can be addressed by letting $\hat{e}_0^A$ and $\hat{e}_0^B$ denote new expectations of life at birth for cohorts A and B based on modified age-specific mortality schedules $\hat{q}_x^A$ and $\hat{q}_x^B$,

$$\hat{q}_x^A = (1+\delta)q_x^A \qquad (10.6)$$

$$\hat{q}_x^B = (1+\delta)q_x^B \qquad (10.7)$$

where q_x^A and q_x^B denote the original mortality schedules for cohorts A and B, respectively (Carey 1993).

The results of uniform scaling of mortality from $\delta = -0.5$ to $\delta = 1.0$ for the mortality schedules for the three fruit fly species is shown in fig. 10.5. The interpretation of the case when $\delta = -0.4$ indicates that if the Mexfly age-specific mortality schedule is multiplied by a factor that reduces the rates uniformly across all age classes by 40% (i.e., 60% of the original since $1 + \delta = 0.6$ when $\delta = -0.4$), then the expectation of life for a newborn cohort of this species is increased from its original value of just under 49 days (at open circle labeled A in fig. 10.5) to approximately 60 days. Thus, a 40% uniform decrease in mortality across all ages yields a 20% increase in life expectancy for this species. Similarly, a doubling of mortality across all Mexfly ages occurs when $\delta = 1.0$ (i.e., $1 + \delta = 1 + 1 = 2$). Thus Mexfly life expectancy is computed as ≈ 30 days when age-specific mortality across all ages is doubled; in other words, a 40% decrease in life expectancy has an outcome resulting from a 100% uniform increase in mortality.

These same concepts can be used to interpret the level of mortality changes needed in the species-specific life expectancies for equivalencies (fig. 10.5). Specifically, (1) an overall 50% reduction in mortality ($\delta = -0.5$) in the Mediterranean and Drosophila fruit fly mortality schedules is required for their e_0's to equal that of the Mexican fruit fly (from A to Â); (2) a 20% reduction ($\delta = -0.2$) in Medfly mortality is required for its e_0 to equal that of the Drosophila (from B to B̂); (3) an increase in Mexican fruit fly mortality of 100% ($\delta = 1.0$) is needed for its e_0 to equal that of the Medfly (from B to B̌); and (4) an increase in mortality of around 35% ($\delta = 0.35$) for e_0 in the Drosophila is required to equal that for the Medfly (from C to Ĉ).

Health and Health Span

In his perspective piece on development of genetically tractable models to assess health span in animal models, Tatar (2009) argues that understanding health span is the most relevant clinical, social, and economic feature of aging research and that the

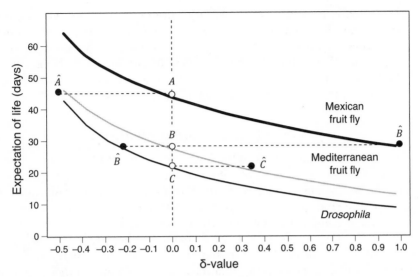

FIGURE 10.5. Change in expectation of life at age 0 for three fruit fly species relative to uniform proportional changes in their mortality (δ-value). The open circles represent observed species-specific e_0's labeled A, B, and C, and the closed circles correspond to the δ-values required for equivalencies across species. For example, A corresponds to the observed e_0 of the Mexican fruit fly and Â corresponds to a reduction of 0.5 (δ = −0.5) in the overall mortality of both the Medfly and Drosophila that is needed for each of their e_0's to equal that of the Mexfly.

model systems of worms, flies, and mice are potentially powerful tools to achieve this aim. Most importantly, in order to maximize the use of these nonhuman systems to address health aspects of human aging, health span needs to be made an operational metric.

There are at least four reasons why basic health span studies on nonhuman species are important (Grotewiel et al. 2005; Kirkland and Peterson 2009; Hamerman 2010; Murphy et al. 2011).

(1) *Identification of system(s) failure(s)*. Functional senescence studies will identify key organ systems that fail with age, some of which may be directly involved in survival or tied to impaired functional status but not mortality.

(2) *Establishing connections between life span and health span*. Studies of functional senescence can shed important light on the mechanistic connections between life span and health span. For example, it is often not clear whether manipulations that extend life span affect systems equally or selectively.

(3) *Biomarker identification*. Studies of health span have the potential to identify biomarkers for aging and health. Such biomarkers will allow researchers to assess whether an intervention aimed at extending life span or health span is effective.

(4) *Intervention concepts and techniques*. Given that older adults often prioritize late-life functionality over life extension, studying functional senescence will help identify the pathophysiological changes associated with aging. This will further our understanding of how these changes affect both health span and life span in order to ultimately identify interventions that lessen the adverse effects of age on health status.

Condition scoring: African buffalo

It is well known in human demography that the quantity of life (individual longevity) generally follows from the quality of life (individual health). Health statistics for human populations are used to interpret trends in mortality and the basis for assumptions for projecting mortality into the future (Lamb and Siegel 2004). The conceptualization of human population health includes not only life expectancy and infant mortality but also the prevalence of morbid conditions, healthy life expectancy, and health-related quality of life (Manton and Stallard 1991). Summary measures of health have a number of potential applications, including (among others) comparing the health of one population with the health of another population or of the same population at two different times. These measures provide insights into the effects of nonfatal health conditions on overall population health and inform debates on priorities for health research (Murray et al. 1999).

Many of these same concepts and arguments apply to populations of nonhuman species and are especially relevant to species that are threatened or critically endangered or to species that are seasonally stressed. In this section we present as a case study an overview of assessment and scoring methods for health in the African buffalo, a species whose populations experience drought stress virtually every year and severe stress every several decades when dearth conditions become extreme as occurs in the African savanna.

Condition indices

Rather than focusing on individual-level measures, conservation biologists have traditionally approached health in nonhuman species in the wild based on population change, where increasing populations are considered healthy and decreasing populations unhealthy (Stevenson and Woods Jr. 2006). However, in an attempt to identify the mechanisms underlying these population changes, the focus can be shifted to assessing the health of wild animals that are based, at least in part, on the health literature in the veterinary sciences (Edmonson et al. 1989). For example, studies of many species have shown that annual survival is highly correlated with body mass, particularly during the winter (Haramis et al. 1986).

Health multistate life table

Criteria for describing the health state and risk of death are presented in table 10.3. The life cycle schematic and the corresponding transition matrix for projecting the development and health stage transitions in the African Cape buffalo are shown in fig. 10.6, and a multistate life table for this species is given in table 10.4. (See chapter 6 for more background on multistage models).

Several aspects of fig. 10.6a, b, and table 10.4 merit comment. First, these illustrate differences in the ages at which individuals may transition between stages. For example, not all individuals become sexually mature at the same age, nor do individuals transition to different health states at the same rates. Second, death rates may differ drastically among individuals of the same age who are in different health states. For example, the hypothetical multistage life table shows that individuals between 14 and 18 years old are distributed across all four health stages and their risk is less determined by their chronological age per se and more by their health state. Some

Table 10.3. Description of assessment criteria for individual components of body condition scoring in the African buffalo

Number	Stage	Adult health description	Risk
1	Newborn/ juvenile	—	High to moderate
2	Mature adult (Health I)	Ribs not visible; fatty layer on and between ribs; spine bones not visible; convex; smooth rear; hip bones not visually apparent; tail base sits in depression; coat glossy over entire body; walking and cantering fully enabled; social rank high; dominance and threat displays common.	Low
3	Mature adult (Health II)	Few to some ribs visible; spine bones not visible; spine feels flat; hip bones can be seen, round, smooth appearance; tail base on level with surrounding tissue; thin coat covering entire body; few bald patches; walking fully enabled; minor trouble with cantering; social rank intact; minor changes in frequency of dominance and threat displays.	Elevated
4	Old adult (Health III)	Ribs clearly visible in center of rib cage; abdomen ridged; spine palpable as slightly elevated bony center line; points of hips distinctly visible; hip bones easy to feel; tail base protrudes slightly; some bald patches of coat behind shoulders or flanks; some problems walking and running; social rank moderately reduced; dominance displays rare; increasing frequency of submissive displays.	High
5	Senile adult (Health IV)	Ribs clearly visible with deep depressions; vertebrae distinguishable by sight; hip bones protrude beyond the hip point; emaciated rear; tissue surrounding tail base forms round hollow; majority of body area bald or very sparsely coated; major struggles walking and running; social rank low; location in herd or subherd in rear; submissive displays common with no dominance threats or displays.	Severe

Source: Abbreviated version from Table 1 in Ezenwa et al. (2009).

Note: We used *The Behavioral Guide to African Mammals* (Estes 1991) as a guide for inferring behavioral changes that likely occur with (or perhaps cause) changes in health state. Also see Hafez and Schein (1962), Sinclair (1977), Mloszewski (1983), Jolles et al. (2008), and Peterson and Ferro (2012) for ecological conditions that reduce health state and increase risk.

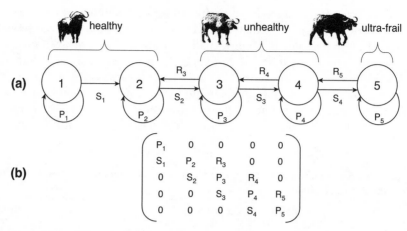

FIGURE 10.6. African Cape buffalo health models: (a) life cycle schematic depicting within- and between-stage transitions for a hypothetical cohort of African Cape buffalo; (b) corresponding transition matrix for a five-state model or multistate life table (see table 10.4). Note that adding arrows depicting reproduction in (a) and corresponding elements in the matrix shown in (b) would transform this life table model into a population model.

individuals in this age range are subject to low risk whereas others are subject to acute or severe risk. Third, it is likely that individuals, cohorts, and populations of buffalo can transition both forward and backward across multiple health stages as environmental conditions change. Individuals experience ever-increasing stress, and thus heightened health and mortality risk, as the herds move across increasing distances between vegetation for grazing and water sources (Dublin and Ogutu 2015; vanWyk 2017).

One take-home message for the use of health demography methods in research in wildlife management and conservation biology is that the tools and concepts used in human health demography can be adapted for use in the systematic study and analysis of health in populations of nonhuman species (Meine et al. 2006; Mills 2013). With respect to African buffalo, only around 2% of populations are disease-free, and incorporating this information into a broader analysis of health dynamics in wild populations may shed important light on their mortality dynamics (vanWyk 2017). Just as in human populations, different diseases in wild populations of non-human species may have little impact on mortality risk, but under certain conditions (e.g., severe drought) they could have dire mortality consequences. Similarly, some diseases may serve as *underlying* causes of death in the wild, other diseases may be *contributing* causes of death, and still other diseases may be *intervening* causes of death. Across all of these hypothetical cases it is likely that predation would be the *immediate* cause of death, which for the African buffalo would be predation by the African lion.

These same concepts, regarding health assessment, apply to a wide range of other nonhuman species and situations; for example, for fishing-gear-related injury in marine birds (Dau et al. 2009), motor boat injuries in terrapins (Lester et al. 2013), automobile-related injuries in reptilians (Rivas et al. 2014), disability in free-ranging baboons (Beamish and O'Riain 2014), injuries and impairments in several arboreal

Table 10.4. Hypothetical multistage health life table for the African buffalo

Stage	x	Number alive at age x N(x)	Survival to age x l(x)	Preadult stage 1	Adult stage 2	3	4	5
Newborn	0	1000	1.000	1000				
Juvenile	1	740	0.740	740				
	2	629	0.629	629				
	3	535	0.535	535				
	4	454	0.454	364	90			
Mature adult	5	386	0.386	221	165			
	6	328	0.328	102	226			
	7	318	0.318		318			
	8	309	0.309		308	1		
	9	300	0.300		298	2		
	10	291	0.291		290	1		
	11	282	0.282		279	3		
	12	273	0.273		269	3	1	
	13	265	0.265		249	11	5	
	14	257	0.257		231	12	13	1
Old/senescent adult	15	250	0.250		100	99	48	3
	16	242	0.242		68	115	55	4
	17	177	0.177		8	32	131	6
	18	129	0.129		1	12	112	4
	19	94	0.094			2	87	5
	20	69	0.069			1	53	15
	21	50	0.050				28	22
	22	37	0.037				13	24
	23	27	0.027				8	19
	24	24	0.024				3	21
	25	14	0.014				1	13
	26	8	0.008					8
	27	6	0.006					6
	28	4	0.004					4
	29	1	0.001					1
	30	0	0.000					0
			7.50	3590.3	2901.4	294.0	558.0	156.0
		Fraction life-years		(0.4787)	(0.3869)	(0.0392)	(0.0744)	(0.0208)

Note: Values of age-specific survival are from Table 2 in Jolles (2007), adult health stages are from table 10.3 (this section), and between-stage transition rates were created ad hoc for illustrative and heuristic purposes. Note the fraction of life-years in each stage in bottom row. Approximately 90% of the life-years are in the newborn, juvenile, and mature adult stages.

primate species (Arlet et al. 2009), anthropogenic-induced injury in wildlife (Schenk and Souza 2014), and wear and tear in fruit-feeding butterflies (Molleman et al. 2007).

Active life expectancy: Fruit flies

Health concepts apply to all animal species, including invertebrates such as insects. In this section we bring the life table methods to bear on fruit fly data (using concepts and approaches introduced in chapter 8 on Human Demography), where we use behavioral criteria to dichotomize the life course of individual flies as either healthy or unhealthy.

Biomarker of aging

A *biomarker of aging* is a behavioral or biological parameter of an organism that will predict future functional capacity or mortality risk better than using chronological age as a predictor (Markowska and Breckler 1999). Behavior biomarkers of aging are important, because behavior changes with aging and thus behavior itself can be used as an index of aging. Moreover, any intervention to alter the chronological course of aging must be assessed behaviorally to evaluate the impact of the innovation on the quality of life. While monitoring the behavior of male Medflies throughout their lives, Papadopoulos and his colleagues (2002) discovered a behavioral trait that is unique to older, geriatric flies in general, and particularly to individuals that are gradually approaching death. This trait was termed "supine behavior" in accordance with the upside-down position of the temporarily immobile flies. Supine males lie on their backs at the bottom of their cage appearing dead or moribund; but when these flies right themselves, either spontaneously or after gentle prodding, it is clear that they are very much alive and moderately robust when they show walking, eating, and wing-fanning behaviors, much of which are indistinguishable from similar behavior in normal (nonsupine) flies. The question is, Is this behavior a biomarker of aging?

Event history chart

An event history chart (see appendix I) showing the age patterns of the supine behavior for the 200+ males (fig. 10.7) reveals a distinct association between individual life span and both the age of onset and the intensity of this behavior. Supine behavior in many flies begins to occur about two to three weeks prior to death, and therefore the period of its occurrence closely follows the cohort survival (l_x) schedule. Supine behavior seldom occurs in very young flies, where young is defined as less than 25 days, but frequently occurs in flies that are over 50 days old, which are ages at which the mortality rate begins to increase substantially. The event history chart helps identify four general properties of the Medfly supine behavior: it is *persistent* in that it occurs on subsequent days after onset; it is *progressive* because the intensity increases with age; it is *predictive* because the onset and intensity is a strong indicator of impending death; and it is *universal* because nearly all male Medflies exhibit this behavior prior to death. These properties of supine behavior are consistent with the properties of biomarkers of aging (Papadopoulos et al. 2002).

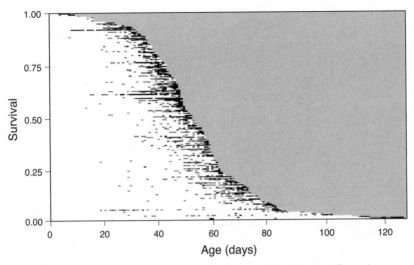

FIGURE 10.7. Event history charts for supine behavior in 203 male Medflies relative to cohort survival. Each individual is depicted as a horizontal "line" proportional to his life span. Each day an individual displayed supine behavior is coded according to the number exhibited—light gray indicates no supine observations; dark gray indicates one to six supine observations; and black indicates seven to twelve supine observations. (*Source*: Carey et al. 1998b)

Evaluation of active health expectancy

We provide a worked example for determining the disability-free, or active, life expectancy at age x using fruit fly supine behavior data. Table 10.5 gives an abridged, prevalence-based life table (also known as the Sullivan method) that shows the proportion of life spent in a particular state. Note that the summation of columns 3, 6, and 7 represents the overall life expectancy (62.8 days), the fly-days in the unhealthy state (11.5 days), and the active (or healthy) life expectancy (51.3 days). This shows that around 18% of the average male Medfly's life for this cohort was spent in the unhealthy state. The unabridged schedules for total and unhealthy life span in these flies are shown in fig. 10.8.

Invasion status as health classification for cities

Background

A generic underlying concept of all actuarial models, including life tables, is that of the risk of a negative occurrence—in other words, death, illness, or damage. Generally, risk can be defined as the probability in a change of state (e.g., single to married; unemployed to employed), and life table concepts can be brought to bear on a wide range of problems involving change of state. One example in which life table methods have not yet been applied but in which the concepts and methods are relevant is invasion biology (Davis 2011; Simberloff and Rejmanek 2011). For this case cities (or more generally regions) can be considered as individuals that are at risk of experiencing a change of health state according to their pest invasion status. For

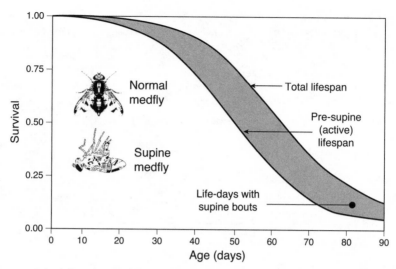

FIGURE 10.8. Schedules of active life and total survival for the male Medfly cohort. The age at which active life for an individual ended is defined as the age at which supine behavior was first observed. Inactive life is defined as all life-days beyond this onset age. The shaded zone shows the difference in life-days between total and active supine life spans.

example, the "health" status of a city that has never experienced a pest fruit fly outbreak would be considered much differently than a city that has experienced multiple fruit fly outbreaks over a series of years, particularly because the city that has had outbreaks may require the long-term use of chemical pesticides for pest control. In this section we bring life table concepts and methods to bear on data from the pest fruit fly invasion of California that started in the middle of the twentieth century and continues into the twenty-first (Papadopoulos et al. 2013). In the mid twentieth century, California began experiencing outbreaks of a number of exotic fruit fly species that are some of the most notorious agricultural pest species in the world. These species, in the dipteran (fly) family Tephritidae, include the oriental fruit fly (*Bactrocera dorsalis*) and two species discussed earlier in this chapter, the Mediterranean fruit fly (*Ceratitis capitata*) and the Mexican fruit fly (*Anastrepha ludens*). These species pose a grave threat to agriculture because, unlike the common vinegar fly (*Drosophila melanogaster*), which feeds on rotting fruit, the tephritid species attack growing (fresh) fruit destined to both national and international markets.

Application of life table methods

We approach the fruit fly invasion problem by considering California cities as a cohort of pristine (healthy) "individuals" subject to the risk of experiencing a fruit fly outbreak (transitioning to the first of several "unhealthy" states). In 1950 no city in California had experienced any fruit fly outbreaks. However, by 2016 a total of 331 of the 478 cities in the state had experienced at least one outbreak. Some of these outbreaks were due to invasions of new species, and others were recurrences of previously detected species. Of the 331 cities that were "infested" between 1950 and 2016, a total of 189, 122, 91, and 60 of the cities experienced two, three, four, and five

Table 10.5. Active health expectancy life table for Medfly males subject to the risk of supine onset

Age interval (5 days)	Number alive at the beginning of each interval	Number of days lived in age interval	Proportion of flies in age interval in state considered unhealthy	Proportion of flies in age interval in healthy state	Number of days lived in unhealthy state in interval	Number of days lived in healthy state in interval	Number of days lived in healthy state in this and all subsequent intervals	Disability-free life expectancy at age x
x	l_x	$_5L_x$	$_5\pi_x$	$(1 - {_5\pi_x})$	$_5L_x \cdot {_5\pi_x}$	$_5L_x (1 - {_5\pi_x})$	$_5T_x$	$DFLE_x$
(1)	(2)	(3)	(4)	(5)	(6)	(7)	(8)	(9)
0	1.0000	4.9989	0.0000	1.0000	0.0000	4.9989	51.2821	51.3
5	0.9996	4.9887	0.0038	0.9962	0.0191	4.9697	46.2832	46.3
10	0.9959	4.9635	0.0092	0.9908	0.0457	4.9178	41.3135	41.5
15	0.9895	4.9272	0.0184	0.9816	0.0909	4.8363	36.3957	36.8
20	0.9814	4.8815	0.0350	0.9650	0.1711	4.7105	31.5594	32.2
25	0.9712	4.8185	0.0592	0.9408	0.2852	4.5333	26.8490	27.6
30	0.9562	4.7151	0.0954	0.9046	0.4499	4.2652	22.3156	23.3
35	0.9298	4.5396	0.1445	0.8555	0.6560	3.8836	18.0505	19.4
40	0.8860	4.2726	0.2081	0.7919	0.8893	3.3833	14.1668	16.0
45	0.8230	3.9046	0.2804	0.7196	1.0950	2.8096	10.7836	13.1
50	0.7388	3.4498	0.3528	0.6472	1.2171	2.2327	7.9740	10.8
55	0.6411	2.9521	0.4181	0.5819	1.2342	1.7179	5.7413	9.0
60	0.5397	2.4352	0.4801	0.5199	1.1691	1.2661	4.0234	7.5
65	0.4344	1.9346	0.5458	0.4542	1.0559	0.8787	2.7573	6.3
70	0.3395	1.5075	0.6127	0.3873	0.9237	0.5838	1.8786	5.5
75	0.2635	1.1670	0.6738	0.3262	0.7863	0.3807	1.2947	4.9
80	0.2033	0.9028	0.6682	0.3318	0.6032	0.2996	0.9141	4.5
85	0.1578	0.6935	0.6483	0.3517	0.4496	0.2439	0.6145	3.9
90	0.1196	0.4990	0.6321	0.3679	0.3154	0.1836	0.3706	3.1
95	0.0800	0.2000	0.0650	0.9350	0.0130	0.1870	0.1870	2.3
100	0.0000	0.0000	0.0000	1.0000	0.0000	0.0000	0.0000	0.0
		62.8			11.5	51.3		51.3

Source: Papadopoulos et al. (2002).

outbreaks, respectively. A total of 44 of the cities experienced 10 or more fruit fly outbreaks, and 3 cities (Los Angeles, Anaheim, and San Diego) experienced 20 or more fruit fly outbreaks. The data for the 331 California cities in which one or more fruit fly outbreaks have occurred are presented in a life table format in table 10.6 by year (col. 1) arranged in two groups of three columns each according to quarantine number (i.e., first, second, or third). Columns 2 to 4 and 5 to 7 contain the numbers and fractions, respectively, of cities that had not experienced either one, two, or three fruit fly outbreaks. Note the analogy here where an outbreak is considered a change of state similar to a health state.

This life table approach given in table 10.6 and illustrated in fig. 10.9 that is used to analyze the timing and spread of an invasive fruit fly merits several comments (Zhao et al. 2019a,b). First, the life table concept provides a framework for characterizing general trends in the invasion. For example, the life table graphic in fig. 10.9 shows the 60-year period between the first and last cities invaded starting in 1954. Second, this figure shows that the rate at which new cities were infested was extremely slow after the first city experienced an outbreak, with only around 60 cities (20% of total) experiencing their first outbreak by 1980. Then the remaining 270 cities experienced their first outbreak after 30 more years. Third, the life table approach encourages comparisons of invasion trajectories. For example, the graphic shows that the time between the first and second outbreaks that occurred in approximately 10% ($n \approx 30$) of the cities was around 11 years. However, the time between the second and third outbreaks in this same subgroup of cities was only around 3 years. Thus a few cities are experiencing a large number of the outbreaks. Fourth, the life table approach reveals "health" transitions. For example, the highest frequency of cities transition-

Table 10.6. Health span life table methods applied to oriental fruit fly invasion status of 331 southern California cities

Year (1)	Number of cities, N_t^j			Fraction of cities, l_t^j		
	$j=1$ (2)	$j=2$ (3)	$j=3$ (4)	$j=1$ (5)	$j=2$ (6)	$j=3$ (7)
1950	331	331	331	1.0000	1.0000	1.0000
1955	330	331	331	0.9970	1.0000	1.0000
1960	328	330	330	0.9909	0.9970	0.9970
1965	328	330	330	0.9909	0.9970	0.9970
1970	321	329	330	0.9698	0.9940	0.9970
1975	300	326	329	0.9063	0.9849	0.9940
1980	266	319	325	0.8036	0.9637	0.9819
1985	206	301	320	0.6224	0.9094	0.9668
1990	126	251	294	0.3807	0.7583	0.8882
1995	76	212	264	0.2296	0.6405	0.7976
2000	41	187	247	0.1239	0.5650	0.7462
2005	18	168	231	0.0544	0.5076	0.6979
2010	3	145	215	0.0091	0.4381	0.6495
2015	0	142	209	0.0000	0.4290	0.6314

Note: N_t^j and l_t^j denote, respectively, the number and fraction of cities at time t that have not experienced j outbreaks.

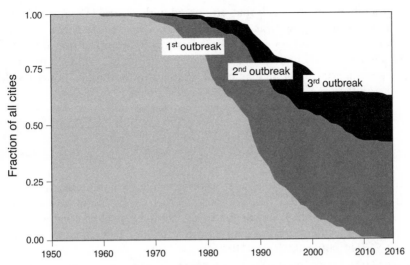

FIGURE 10.9. Life table survival model applied to tephritid fruit fly invasion of 331 California cities starting in 1950, a year when no fruit flies had ever been detected in the state, through 2015. Each survival curve depicts the transition of the cities from never having experienced an outbreak to having experienced one (first detection), from having experienced one outbreak to having experienced two outbreaks (second detection), and so forth.

ing to the less healthy state (from pest-free to pest present) was in the 1980s, with around half of the cities transitioning to "pest-infested" and large fractions of many cities experiencing second and third outbreaks. Finally, the life table approach provides a framework for visualizing, if not projecting, future trends in fruit fly outbreaks. For example, extrapolation suggests that by midcentury there will still be approximately 50 cities that will likely experience annual fruit fly outbreaks.

Biological control as health demography

Biological control (biocontrol) is a method of insect and plant pest control based on the use of natural enemies, including predators, parasitoids, or pathogens. This may involve either introducing natural enemies from world regions where the pest is endemic or using mass-released predators or parasitoids or biopesticides, such as the microbial pesticide *Bacillus thuringiensis*, by conservation methods aimed at increasing populations of naturally occurring predators and parasitoids (Heimpel and Mills 2017). Although the practice of biological control has been around for centuries, it began in earnest in the late nineteenth century and early twentieth century. Publication of the book *Silent Spring* (Carson 1962) brought biological control front and center because of Carson's emphasis on looking for alternatives to the use of chemical pesticides (Gay 2012). In all cases the aim is to decrease the health, and thus the life span, of individuals in pest populations.

Measuring the impact of natural enemies on pest survival and population growth is a central component of studies in biological control. It thus follows that tools are needed to quantify the effects of natural enemies on pest populations. A method used for many biological control investigations in the mid to late twentieth century was key factor analysis, an approach that originated with Morris (1959) and Varley and Gradwell (1960). Their idea was to identify factors that are largely responsible for the observed changes in populations based on the assumption that mortality was caused by a key factor that could be recognized using its k-value, a quantity that changes with time in the same way as the changes in total mortality. They defined k_x as

$$k_x = \log_{10}a_x - \log_{10}a_{x+1}$$

where a_x denotes the number in stage x. For example, if the numbers in stages 0, 1, and 2 are $a_0 = 44,000$, $a_1 = 3,513$, and $a_2 = 2,539$, then the k_x-values would be $k_0 = 1.09$ and $k_1 = 0.15$ (from Table 4.1 in Begon et al. 1996, 150). Thus k-values are simply cause- or stage-specific mortality expressed as the base-10 exponent. For example, $q_0 = 1 - (3,513/44,000) = 0.9202 = (10^{1.09})^{-1}$.

Royama (1996) noted a fundamental problem with the concept of key factor analysis is that the variation in population is inappropriately interpreted as the variance. He contended that, for judging which factor is major, there are multiple and subtle criteria beyond the simplistic idea of key factors. Carey (1989) introduced the multiple-decrement life table to the ecology literature as an alternative to key factor analysis, and later Robert Peterson and his colleagues (2009) built on and applied the multiple-decrement concept in insect pest control with particular emphasis on *irreplaceable mortality*. (For more on the concepts of multiple-decrement life tables, see chapter 8.)

Irreplaceable mortality

The concept of irreplaceable mortality was first introduced in a study concerned with the impact of two parasitoids on the biological control of a pest scale insect, where irreplaceable mortality is defined as "that portion of the parasitism which, if it were lacking, would not be replaced by some other cause of death" (Huffaker and Kennett 1966). Here we define irreplaceable mortality with reference to competing risk and multidecrement life table concepts (see chapter 8) as "the marginal difference between all-cause mortality and mortality due to the elimination of one or more risk factors." For example, if egg, larvae, and pupal mortality rates are 10%, 40%, and 90%, respectively, then out of 1,000 eggs, a total of 900, 540, and 54 individuals will survive to the larval, pupal, and adult stages, respectively. However, if egg mortality is eliminated, then a total of 1,000, 600, and 60 individuals will survive to these respective stages. In other words, six more individuals would survive to adulthood by eliminating egg mortality. Thus the value of irreplaceable mortality due to egg mortality in survival to the adult state is 0.6% (i.e., $100 \times 6/1,000$).

Multiple-decrement life table of a pest

As an illustration, we present a case study of the alfalfa weevil, a damaging insect pest of alfalfa, with the objective of identifying the probability of death in the presence or absence of a combination of multiple causes of mortality. Data from a multiple-decrement life table for this species (table 10.7) is used to determine the probability of cause of death in the absence of other causes (table 10.8). The irreplaceable mortalities are then calculated to identify mortality solutions in the absence of all other factors (table 10.9).

From the tables, irreplaceable mortality caused by the elimination of a single causal factor ranges from 0.2% to 58.9%, with the largest irreplaceable factor being fungal disease. It is also shown that if fungal disease were eliminated as a mortality factor, overall mortality would be reduced by nearly 60% because that percentage cannot be replaced by other factors. Additionally, for the elimination of two factors, values range from 0.7% to 81.8%, with large irreplaceable mortality values for the combination of fungal disease with general establishment failure. In other words, this approach can be used to identify the important components to a biological control plan for this pest species.

Table 10.7. Multiple-decrement life table for the alfalfa weevil, *Hypera postica*

| | | | | Fraction of deaths by cause | | | | |
| | Mortality | Survival | Deaths | Wasp | Infertile | Failure | Fungus | Rain |
Stage (x)	aq_x	al_x	ad_x	ad_{1x}	ad_{2x}	ad_{3x}	ad_{4x}	ad_{5x}
Eggs	0.021	1.000	0.021	0.002	0.019	0.000	0.000	0.000
Early larvae	0.257	0.979	0.252	0.000	0.000	0.252	0.000	0.000
Late larvae	0.853	0.727	0.620	0.000	0.000	0.000	0.606	0.014
Prepupae	0.222	0.107	0.023	0.007	0.000	0.000	0.016	0.000
Late pupae	0.114	0.084	0.010	0.000	0.000	0.000	0.010	0.000
Adults		0.074						
TOTALS			0.926	0.009	0.019	0.252	0.632	0.014

Source: Table 3 in Peterson et al. (2009).

Table 10.8. Elimination of mortality causes for the alfalfa weevil, *H. postica*

| | | | Probability for cause of death in absence of other causes | | | | |
| | Survival | Mortality | Wasp | Infertile | Failure | Fungus | Rain |
Stage (x)	al_x	aq_x	ad_{1x}	ad_{2x}	ad_{3x}	ad_{4x}	ad_{5x}
Eggs	421	0.021	0.002	0.019	0.000	0.000	0.000
Early larvae	412	0.257	0.000	0.000	0.257	0.000	0.000
Late larvae	306	0.853	0.000	0.000	0.000	0.850	0.020
Prepupae	45	0.222	0.070	0.000	0.000	0.164	0.000
Late pupae	35	0.114	0.000	0.000	0.000	0.114	0.000
Survival egg to adult	31		0.072	0.019	0.257	0.889	0.020

Source: Table 4 in Peterson et al. (2009).

Table 10.9. Irreplaceable mortality for the alfalfa weevil, *H. postica*

Combination of causes	% mortality	Eliminated causes	% irreplaceable mortality
All causes (Wasp + infertility + failure + fungus + rain)	92.6	—	—
Wasp + infertility + failure + fungus	92.5	Rain	0.2
Wasp + infertility + failure + rain	33.8	Fungus	58.9
Wasp + infertility + fungus + rain	90.1	Failure	2.6
Wasp + failure + fungus + rain	92.5	Infertility	0.1
Infertility + failure + fungus + rain	92.1	Wasp	0.6
Wasp + infertility + failure	32.4	Fungus, rain	60.2
Wasp + infertility + fungus	89.9	Failure, rain	2.8
Wasp + infertility + rain	10.8	Failure, fungus	81.8
Wasp + failure + rain	32.5	Infertility, fungus	60.2
Wasp + fungus + rain	89.9	Infertility, failure	2.7
Infertility + failure + fungus	91.9	Wasp, rain	0.7
Infertility + failure + rain	28.6	Wasp, fungus	64.0
Infertility + fungus + rain	89.3	Wasp, failure	3.3
Failure + fungus + rain	91.9	Wasp, infertility	0.7

Source: Table 5 in Peterson et al. (2009).

Population Harvesting

From a demographic perspective, harvesting is concerned with cohort-structured populations where the objective is to remove precisely the right number of individuals from the population to achieve zero population growth (Beddington and Taylor 1973; Getz 1984; Connelly et al. 2012). A number of different demographic approaches to harvesting problems have been developed for use by wildlife and fisheries managers to help them determine which age classes to harvest to maintain the highest sustainable yields (Caughley 1977, 1983; Goodman 1978). Here we describe the demographic models that are at the foundation of several different harvesting concepts, including single–age/stage and all–age/stage harvesting and population culling, to control population growth rates. In the next section, on conservation, we discuss hunting and poaching, which may also be considered forms of population harvesting.

Insect mass rearing

The concept of demographic harvesting was applied to mass rearing of insects for biological control by Carey and Vargas (1985). This approach involves identifying the sustainable yield for a cohort subjected to interventions at two key ages: the primary harvest age that is typically pre-reproductive and the secondary harvest (or discard) age that is typically reproductive or post-reproductive. A new survival schedule is then imposed on a harvested population (Beddington and Taylor 1973). As an illustration of population harvesting, we consider an example with the Mediterranean fruit fly that is often reared in massive numbers.

Single–age/stage harvesting

Let θ denote the target age for harvesting and h denote the fraction of the individuals at the target age that are harvested. This leaves the fraction $(1-h)$ for renewal. The rate of harvest must confer zero population growth, so the value of h must be the solution to the equation

$$1 = (1-h)\sum_{\theta}^{\delta} l_x m_x \tag{10.8}$$

$$h = 1 - \frac{1}{(\sum_{\theta}^{\delta} l_x m_x)} \tag{10.9}$$

where δ is the discard age and thus the artificially imposed last day of reproduction. The demographic harvesting scheme is illustrated in fig. 10.10.

The associated stable (and stationary) age distribution for the population is given by

$$c_x = \frac{l_x}{s} \quad \text{for } x < \theta \tag{10.10}$$

$$c_x = \frac{l_x(1-h)}{s} \quad \text{for } x \ge \theta \tag{10.11}$$

where

$$s = \sum_{x=0}^{\theta-1} l_x + (1-h)\sum_{x=\theta}^{\delta} l_x \tag{10.12}$$

If c_θ is the proportion of the target stage at age θ in the colony after harvest, then the expression

$$P = \frac{h}{(1-h)}\left(\frac{c_\theta}{\sum_{x=\alpha}^{\delta} c_x}\right) \tag{10.13}$$

gives the daily per-female yield of the target stage P, where ε denotes the age of eclosion. This number requires doubling (2P) if males need to be accounted for and the assumption of a 1:1 sex ratio is valid. Carey (1993) shows that P can also be computed from the equation

$$P = \frac{l_\theta(\sum_{x=0}^{\delta} l_x m_x) - 1}{\sum_{x=\epsilon}^{\delta} l_x} \qquad (10.14)$$

Using the Medfly example, determination of the harvesting parameters requires two steps (Carey 1993).

Step 1. *Harvest rate, h.* Using a harvest age of $x = 19$ days (late pupae) and discard age of 40 days (mid-aged adults), the fraction harvested, h, is computed as $h = 1 - (1/120) = 0.9917$, where 120 is the net reproduction rate for the discard age of $\delta = 40$ days.

Step 2. *Production rate, P.* Substituting $l_\theta = 0.55$ for survival to the harvest age (i.e., egg to late pupae), 120 for net reproduction to the discard age, and 10.9 days as the denominator in eqn. (10.14) yields a production rate, P, of approximately 12 pupae/female/day.

All-age/stage harvesting

Here we use an example taken from Carey and Krainacker (1988), who derived and applied the model to tetranychid spider mite mass rearing. These acarines are typically mass-reared by infesting flats of young host plants such as snap bean sprouts. Typically, after two weeks, a fraction of all flats are removed for harvesting 100% of the mites on the plants, with the mites on the remaining flats used to inoculate the

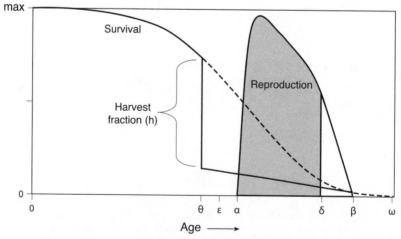

FIGURE 10.10. Single-age harvesting model, where θ, ϵ, α, δ, β, and ω denote age of harvest, age of eclosion, first reproduction, discard age, last reproduction (absent harvesting), and oldest age, respectively.

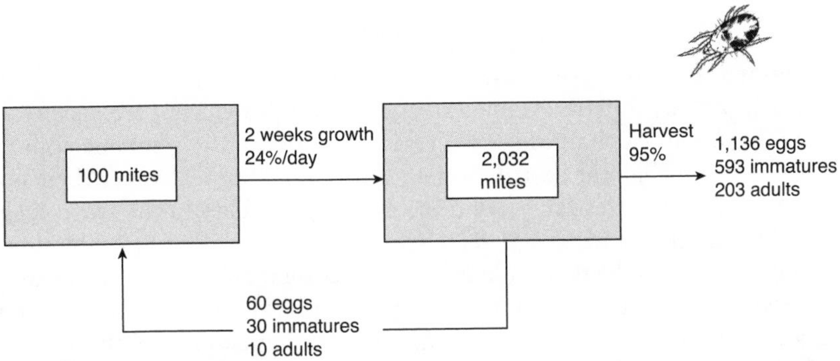

FIGURE 10.11. All-age harvesting model. All ages or stages of entire subpopulations of spider mites at the stable age distribution are harvested (or culled) at once. One or more subpopulations are retained for replenishing the harvested subpopulations.

plants growing in mite-free flats. Then two weeks later this process is repeated. The objective is to harvest the appropriate number to confer population replacement.

Let λ denote the finite rate of increase of the population and h the harvest rate for the entire population; then the fraction of all stages that can be harvested (removed) while maintaining population replacement is the solution to the equation

$$1 = (1 - h)\lambda$$

$$h = \frac{(\lambda - 1)}{\lambda} \tag{10.15}$$

$$h = 1 - \lambda^{-1}$$

Because the same fraction of each age class is removed from the population, the age structure of the factory population will be identical to the age structure of the unconstrained (unharvested) population. The rate of population increase for spider mites is $\lambda = 1.24$, suggesting that $1 - (1.24)^{-1}$, or 0.194 fraction of the entire population (all stages) can be removed daily and still maintain population replacement, or that $h = 1 - 1.24^{-14}$ ($= 0.95$) can be removed every two weeks. A schematic diagram of this mite-harvesting model is shown in fig. 10.11.

Culling: African elephants

The basic age/stage harvesting model presented for the spider mite assumes that populations consist of subpopulations that can be harvested independently and in their entirety. Next we consider harvesting as it applies to culling in social groups, such as in elephants, where the best practice in elephant management is to remove the entire family or subherds because selective killing disrupts the social organization of the subpopulation.

Background

Kruger National Park is a game reserve in northeastern South Africa that covers an area of nearly 20,000 square kilometers and where all the Big Five species reside, including the African elephant. In some regions of Africa, the elephant populations are in decline due to habitat destruction and poaching, but in Kruger their populations are expanding to sizes far beyond the carrying capacity of the park. As a consequence, African elephants in Kruger have the potential to completely alter their surroundings and act as the predominant ecosystem engineer in protected areas (Fazio 2014). Given that an adult elephant can consume 180 kg of vegetation per day and can uproot trees for access to a few choice leaves, elephants have the potential to convert habitats from dense woodlands to sparse shrub, which can have far-reaching ramifications for other plants and animals residing in these parks. Because of these potentially destructive pressures, an elephant culling program, with selective slaughter, was implemented in 1967. Due to public outcry, park managers discontinued this program in 1994 and, as a result, by 2006 the elephant population increased to over 15,000 individuals (Owen-Smith et al. 2006; Fulton 2012).

Three options exist for managing elephant populations (Fulton 2012). The first is *contraception*, which involves either immune contraception (elephant "pill") or sterilization. One problem with the contraceptive option is that population modeling revealed that an effective program would require the treatment of several thousand cows annually at a cost greater than the entire annual management budget for South Africa's national parks. Another problem with this approach is that contraceptive implants create disturbing posttreatment effects and behavioral aberrancies in elephants. Additionally, sterilization by this method is not reversible, and reversing contraception may be important if, at another time, populations are in decline due to a disease epidemic or heavy poaching. The second option for managing elephant populations is *relocation*. There are at least three problems with relocation: (1) moving adult elephants is stressful to the animals, extremely costly, and logistically challenging due to their large size; (2) translocation of juvenile animals is considered inhumane because it removes them from their family unit; and (3) opportunities for translocations are extremely limited as there is little space left into which elephants can be moved. The third option is population harvesting through *culling* or selective slaughter. Because sociality in elephants is highly developed, removal of any one individual is disruptive at the family, subherd, and herd levels. If the removed individual is the matriarch, or even one of the senior subordinate breeding cows, the effect on the group can be devastating because the removal constitutes a loss of leadership, knowledge, and wisdom.

Culling as a management actuarial strategy

Hypothetically, to reduce the size of a population with an initial population of 15,000 elephants by half would require culling elephants of all ages for 10 years (fig. 10.12). Using the population growth equations

$$N_{t+1} = N_t(b-d) \tag{10.16}$$

$$N_{t+1} = N_t \lambda \qquad (10.17)$$

the annual rate at which the population must decline in order to reduce the population of 15,000 by one-half (7,500) in a decade is

$$\lambda^{10} = 0.5$$

$$\lambda = \sqrt[10]{0.5}$$

$$= 0.933$$

In other words, by reducing the elephant population growth rate to a below-replacement rate of 0.933 (i.e., 6.7% less than replacement) each year for 10 years, the population will fall to half the level of its current 15,000 individuals.

Two aspects of this hypothetical culling program should be noted. First, the number of elephants culled becomes progressively smaller through phase I because the fraction to be culled is applied to an ever-shrinking population. The number killed the first year of phase I is over 2,000 animals, and by the last year of phase I it is almost half that number at ≈ 1,100. Second, the ≈ 7% annual growth rate requires culling half as many elephants (i.e., 525 individuals) each year as was required for the last year of phase I. This is still a large number of animals to kill and dispose of,

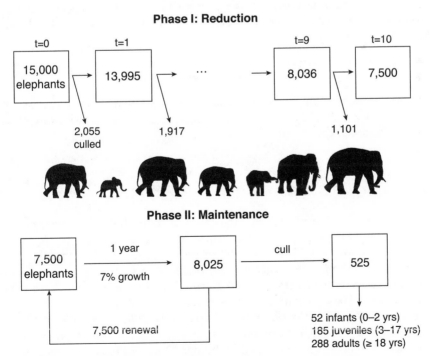

FIGURE 10.12. The demography of African elephant management by culling.

but it is only a quarter of what was required at the start of the program. Finally, it is an understatement to say that these culling programs create large logistical and societal challenges (Caughley 1981, 1983; deVos et al. 1983), and there are many pros and cons of these approaches that need to be considered (Bell 1983).

Conservation

Demography is an inherent component of conservation biology inasmuch as the characteristics of a threatened species' population size (number of individuals), population structure (age or sex), spatial distribution (fragmentation), and population growth are demographic data used to classify conservation status. Here we consider populations threatened by hunting and poaching and how data from a captive breeding program can be quantified.

Hunting

Overexploitation is one of the major threats to populations of mammals, reptiles, and birds, second only to habitat destruction (Weinbaum et al. 2013). Hunting occurs at levels that are often largely unsustainable (Robinson and Redford 1991), and it follows that to avert species extinctions and to maintain acceptable levels of ecosystem health and structure, hunting policies must be informed by models that, at the very least, incorporate the competing risks of natural and hunting mortality (Weinbaum et al. 2013). For example, it may be possible to explore the role of private wildlife ranching and hunting preserves as a conservation tool (Cousins et al. 2008). In this section we use demographic approaches to evaluate competing risks of mortality for large animals.

Age- and season-specific hunting

Total mortality (μ_x) for animals who are subject to hunting include at least two competing causes: the risk of dying "naturally," denoted μ_x^N, and the risk of dying as an outcome of hunting, denoted μ_x^H.

$$\mu_x = \mu_x^N + \mu_x^H \tag{10.18}$$

Hypothetical age patterns showing the baseline (natural) mortality and mortality due to seasonal hunting, poaching, and trophy hunting are depicted in fig. 10.13.

The results of an example application of the outcome of the competing risk of hunting and natural mortality is given in fig. 10.14. We assume a 10% seasonal hunting kill rate, a 3.5-year maturation, and a growth period during which no hunting mortality occurs. The expectation of life with and without hunting was 66.3 months (5.5 years) and 75.1 months (6.3 years), respectively. A total of 75% of all deaths were due to natural (Gompertzian) mortality and 25% due to hunting mortality (i.e., 250 of the 1,000 deer were harvested by hunters). Interestingly, an increase in seasonal hunting pressure from 10% seasonally to 90% only reduces the expectation of

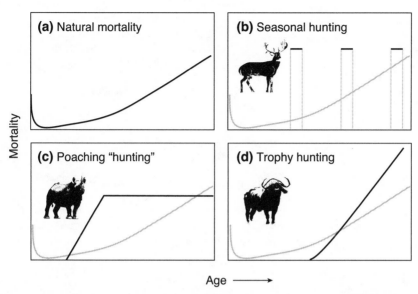

(b) Seasonal hunting

Mortality

Age ⟶

FIGURE 10.13. Schematics of age-specific baseline and hunting mortality as competing risks: (a) baseline natural mortality due to all causes other than hunting; (b) mortality due to seasonal (regulated) hunting; (c) mortality due to poaching and bushmeat (unregulated) hunting; and (d) mortality due to trophy hunting, where the larger (older) individuals are the most prized.

life at birth from 5.5 years to around 4 years. This moderate decrease occurs despite a ninefold increase in hunting pressure for two reasons: (1) the deer are protected from hunting until they are 3.5 years old; and (2) around 20% of the deer are already dead when individuals are subject to mortality due to hunting. This means that additional mortality, whether 10% or 90%, is acting on a reduced fraction of the total cohort.

Rear and release

"Rear and release" harvesting, or release-pulse cycles, refers to game animals that are (typically) reared in captivity and then released into the field for shooting by hunters (e.g., Thacker et al. 2016) or fish that are reared in hatcheries and released into streams and ponds for capture by anglers (e.g., Kientz et al. 2017). A schematic of survival of released individuals, in this case pheasants, for an idealized release program is shown in fig. 10.15, where N_R denotes the number released, C denotes the release interval (cycle length), and N_L denotes the number in the standing population at its lower limit (i.e., just prior to next release). A key concept of rear and release is that the life-days (or life-weeks, life-months, life-years) lived in each of the released cohorts is equal to the same life-quantities lived in each of the release cycles. In table 10.10 the sum of the columns for each of the release numbers (=286) equals the sum of the life-quantities lived for each of the release cycles (=286).

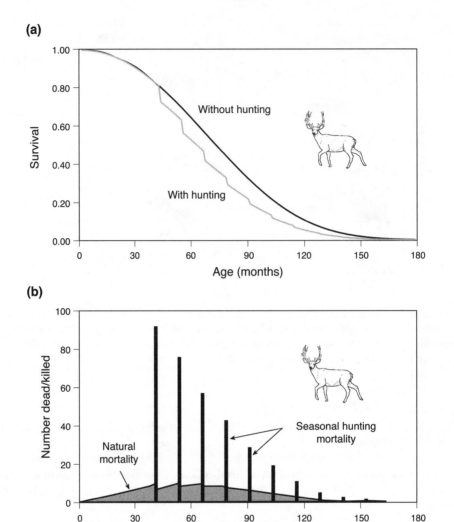

FIGURE 10.14. Example application of the demographic outcome of hunting in a hypothetical cohort of 1,000 white-tailed deer subject to competing risks of Gompertz (a = 0.002; b = 0.01) and hunting (one month seasonal of 0.10): (a) survival rates with and without hunting; (b) death distribution for natural and hunting causes of death.

Trophy hunting: African lions

For trophy hunting, African lions are an especially sought-after species due to their iconic status as a member of Africa's Big Five, a term denoting the five most dangerous species (Whitman et al. 2004; Lindsey et al. 2012, 2013; San Diego Zoo 2016). Lion-hunting safaris are a major source of income for many countries, not only because of the high licensing fees for the trophy animals but also because the safari itself is a major source of revenue with charges for food, lodging, guides, equipment, and travel. With the exception of rhinoceroses, lions generate the highest revenue per hunt of any species in Africa. In light of the conservation status of African lions across

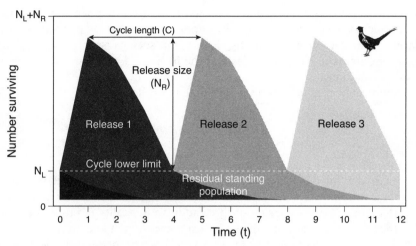

FIGURE 10.15. Release-pulse cycle for the hypothetical game farm (pheasants) release and restock in table 10.10.

Table 10.10. Four-day cycle (C=4) for releases of a hypothetical population of pheasants subject to a fixed survival schedule

	Time	Release number −1	1	2	3	Time sums	Cycle sums
	t	20				20	
Release cycle 1	t+1	10	100			110	286
	t+2	5	90			95	
	t+3	1	60			61	
	t+4	0	20			20	
Release cycle 2	t+5		10	100		110	286
	t+6		5	90		95	
	t+7		1	60		61	
	t+8		0	20		20	
Release cycle 3	t+9			10	100	110	286
	t+10			5	90	95	
	t+11			1	60	61	
	t+12			0	20	20	
	t+13				10		
	t+14				5		
	t+15				1		
	t+16				0		
	Cohort sums		286	286	286		

Note: The number of pheasants released ($N_R = 100$) die off at the rates shown in the release number columns. Note that the population cycles between $20 = N_L$ (lower limit just before next release) and 110 individuals (maximum limit the day of new releases).

the continent, hunting policies must be based on, among other factors, informed demographic strategies. This is especially important because lion populations are particularly sensitive to trophy hunting due to the social disruption caused by the removal of dominant breeding males and the likelihood of infanticide when a new male takes over a pride.

A baseline mortality model can be presented as follows:

$$\mu_x = e^{a_0 - a_1 x} + c + e^{b_0 + b_1 x} \tag{10.19}$$

This Siler model (Siler 1979) yields the age- and sex-specific parameters and trajectories that are presented in fig. 10.16. As Barthold and her coworkers note (Barthold et al. 2016), this model is the sum of three additive mortality hazards: the first term models the decrease in mortality rates over early ages (infant and juvenile), the middle term is the Makeham term (Makeham 1860) and captures age-independent mortality, and the last term captures the exponential increase in mortality with age as essentially the Gompertz model (Gompertz 1825).

Sustainable trophy-hunting policies for lions in particular, and all trophy species in general (e.g., elephants; buffalo; leopards; rhinoceroses), require a deep understanding of the population and behavioral biology of the species. Indeed, no simple model or even sets of demographic models are sufficient for creating such policies. However, insights into different types of hunting-policy trade-offs are possible using simple models. Our purpose here is to use the age- and sex-specific mortality rate model for both sexes of African lions contained in the paper by Barthold and her colleagues (2016), a model that is based on some of the best actuarial data available for this species. We use the mortality rates for males to create two heuristic models of trophy-hunting mortality based on two competing risks: natural mortality, denoted $\mu_x(n)$, and hunting mortality, denoted $\mu_x(h)$.

Total lion age-specific mortality, denoted $\mu_x(total)$, equals the sum of the baseline natural mortality, denoted $\mu_x(n)$, and mortality due to trophy hunting:

$$\mu_x(total) = \mu_x(n) + \mu_x(h) \tag{10.20}$$

Hunting mortality is based on scaling natural mortality by a factor s, thus $\mu_x(h) = s\mu_x(n)$, where s denotes the scalar. For example, hunting mortality scaled by a factor of $s = 2.0$ is equal to twice the baseline natural mortality.

Here we explore two models:

Model 1. *Scaled hunting mortality*, where we assume that all male lions, 6 years old and greater, can be taken as trophies but that the hunting mortality scalar, s, varies.

Model 2. *Minimum age hunting*, where we adjust the minimum age at which a male lion can be killed in yearly increments from 6 to 14 years. Thus male lions are subject to natural mortality up to the minimum trophy age at which time their hunting mortality increases by a factor of 3 over the baseline natural mortality.

Model 1 yields the highest number of trophy lions per newborn cohort because they are killed while still relatively young (i.e., before natural mortality reduces their numbers). Table 10.11 shows that 6.3 male lions per 100 newborns could be killed if hunting policies were based on allowing the equivalent of a fivefold increase of natural mortality (fig. 10.17). At least two problems arise from this strategy. First, the sex ratio at this scalar level is skewed to nearly 4 to 1 mature females to mature males. Second, the dominant pride males are nearly all relatively young because increasing mortality by fivefold starting at early maturity prevents the vast majority of lions from reaching advanced ages.

Model 2 yields the greatest number of trophy males if the minimum legal lion age for killing is newly mature 6-year-olds (table 10.12, fig. 10.17). The number of trophy

Table 10.11. Output of model 1 showing number of harvestable male lions per 100 newborns (lion trophies) and female-to-male sex ratio relative to level of hunting mortality scaled from baseline mortality

Scalar	Lion trophies	Female-to-male ratio
1.0	0.0	1.3
1.5	2.6	1.7
2.0	3.9	2.0
2.5	4.7	2.3
3.0	5.2	2.6
3.5	5.6	2.9
4.0	5.9	3.2
4.5	6.1	3.5
5.0	6.3	3.8

Table 10.12. Output of model 2 showing number of harvestable male lions per 100 newborns (lion trophies) and female-to-male sex ratio relative to minimum kill ages

Minimum kill age (years)	Lion trophies	Female-to-male ratio
No hunting	—	1.3
6	5.2	2.2
7	4.7	1.9
8	4.1	1.8
9	3.5	1.6
10	2.8	1.5
11	2.2	1.4
12	1.5	1.4
13	1.0	1.4
14	0.5	1.3

Note: Hunting mortality is scaled threefold relative to baseline natural mortality.

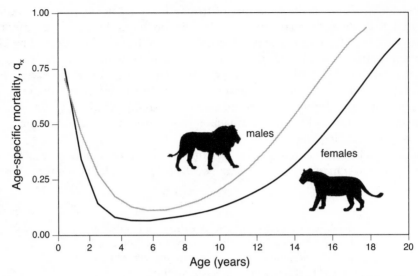

FIGURE 10.16. Age- and sex-specific mortality in free-ranging African lions. African lions in the Serengeti using Siler model and parameters in Barthold et al. (2016). Values for model parameters (from birth) a_0, a_1, c, b_0, and b_1 are 0.150, 0.727, 0.025, −4.388, and 0.297, respectively, for males, and 0.280, 1.277, 0.032, −5.019, and 0.287, respectively, for females. End-stage ages for African lion infants and juveniles, large cubs, subadult, and young adults are 1, 2, 4, and 6 years, respectively. Prime adults are over 6 years with males and females both reproducing up to 15 to 16 years. Only 8% of males and 12% of females survive to full maturity (6 years).

lions decreases with an increase in the minimum killing age, dropping to 1.0 or less at 13 years minimum age or greater. With the policy where only 1 in 100 newborn males lives to 14 years or greater, these older males would be extraordinarily rare. The female-to-male ratio is maximized at a minimum killing age of 6 years and decreases with increasing age.

Poaching: Rhinoceroses

Background

African rhinoceroses, including both the black rhino and the white rhino, are classified as critically endangered and are thus among the most at-risk species of large mammals. At one time there were hundreds of thousands of rhinos of both species distributed throughout much of sub-Saharan Africa, but by 1970 the numbers for black rhinos were reduced to 65,000 and by 2016 there were only 2,500 remaining in small pockets in Zimbabwe, South Africa, Kenya, Namibia, and Tanzania. Habitat loss is one of the main reasons for their population reductions. But currently 90% of all rhino deaths are attributable to poachers killing them for their horns, which can be used to make ornamental handles or traditional Asian medicines. Rhinos are also killed for trophies and meat, and their hides are made into shields and good-luck charms.

The black rhino, or hook-lipped rhino, is a browser found in the transitional zone between grassland and forest. In contrast, the white rhino, or square-lipped rhino, is

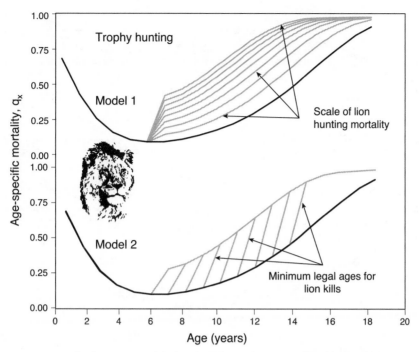

FIGURE 10.17. Trophy-hunting models for the African lion using baseline mortality rates given in fig. 10.16. Model 1 assumes that lions 6 years old and greater can be killed and scales age-specific male mortality above this from the baseline to fivefold greater in increments of 0.5. Model 2 assumes that the minimum age at which male lions can be killed varies from 6 to 14 years in single-year increments at which time the hunting mortality is increased by threefold over the baseline at and beyond that age.

a grazer found in woodland interspersed with grassy areas (Nowak 1991). Females of both species reach sexual maturity after 4 to 6 years and males after 7 to 10 years. After a gestation period of around 450 days, females give birth to a single calf with an interbirth period of 2 to 5 years (duToit 2006). Annual mortality at different ages is estimated as 15% for 0–1 years, 4% at 1 to 4 years, 8% at 5–6 years, and 5% for individuals greater than 6 years (Anderson-Lederer 2013). The life span of rhinos is estimated to be around 40 to 50 years.

Competing risk model

We constructed a multiple-decrement life table (table 10.13) for rhinoceros management based on natural mortality of the age-specific rates, given by Anderson-Lederer (2013), through age 6 years, after which time the mortality rates were based on the Gompertz model ($a = 0.02$; $b = 0.0453$). We used annual loss (mortality) due to poaching at 0.10 for 6 years and older (Ferreira et al. 2015) and loss due to removal at 0.10/year but only for ages 5 through 10 years (Linklater et al. 2011).

The results of this hypothetical, semirealistic, multiple-decrement life table analysis reveal that, for these particular poaching and removal rates, slightly less than half of all mortality is due to natural mortality, much of it occurring in the first decade of

Table 10.13. Rhinoceros multiple-decrement life table in which cohort of 1,000 newborn individuals are subject to sources of competing risk

| Age (x) | Cause i force of mortality | | | Cohort survival | Deaths by cause or removal | | | Total deaths or removals | Age-specific reproduction | Net reproduction |
| | Natural | Poaching | Removal | l_x | Natural | Poaching | Removal | d_x | m_x | $l_x m_x$ |
(1)	(2)	(3)	(4)	(5)	(6)	(7)	(8)	(9)	(10)	(11)
0	0.1500	0.00	0.0000	1000	139.3	0.0	0.0	139.3	0.0	0.0000
1	0.0400	0.00	0.0000	861	33.7	0.0	0.0	33.7	0.0	0.0000
2	0.0400	0.00	0.0000	827	32.4	0.0	0.0	32.4	0.0	0.0000
3	0.0400	0.00	0.0000	795	31.2	0.0	0.0	31.2	0.0	0.0000
4	0.0400	0.00	0.0000	763	29.9	0.0	0.0	29.9	0.0	0.0000
5	0.0800	0.00	0.1000	733	53.7	0.0	67.1	120.8	0.0	0.0000
6	0.0800	0.10	0.1000	613	42.7	53.4	53.4	149.6	0.3	0.1838
7	0.0275	0.10	0.1000	463	11.4	41.4	41.4	94.2	0.3	0.1389
8	0.0287	0.10	0.1000	369	9.5	33.0	33.0	75.4	0.3	0.1106
9	0.0301	0.10	0.1000	293	7.9	26.2	26.2	60.3	0.3	0.0880
10	0.0315	0.10	0.1000	233	6.5	20.8	20.8	48.2	0.3	0.0699
11	0.0329	0.10	0.0000	185	5.7	17.3	0.0	23.0	0.3	0.0555
12	0.0344	0.10	0.0000	162	5.2	15.2	0.0	20.4	0.3	0.0486
13	0.0360	0.10	0.0000	142	4.8	13.2	0.0	18.0	0.3	0.0425
14	0.0377	0.10	0.0000	124	4.4	11.5	0.0	15.9	0.3	0.0371
15	0.0395	0.10	0.0000	108	4.0	10.0	0.0	14.0	0.3	0.0323
16	0.0413	0.10	0.0000	94	3.6	8.7	0.0	12.3	0.3	0.0281
17	0.0432	0.10	0.0000	81	3.3	7.6	0.0	10.8	0.3	0.0244
18	0.0452	0.10	0.0000	70	3.0	6.6	0.0	9.5	0.3	0.0211
19	0.0473	0.10	0.0000	61	2.7	5.7	0.0	8.3	0.3	0.0183
20	0.0495	0.10	0.0000	53	2.4	4.9	0.0	7.3	0.3	0.0158
21	0.0518	0.10	0.0000	45	2.2	4.2	0.0	6.4	0.3	0.0136
22	0.0542	0.10	0.0000	39	2.0	3.6	0.0	5.6	0.3	0.0117
23	0.0567	0.10	0.0000	33	1.7	3.1	0.0	4.8	0.3	0.0100
24	0.0593	0.10	0.0000	29	1.6	2.6	0.0	4.2	0.3	0.0086

25	0.0621	0.10	0.0000	24	1.4	2.2	0.0	3.6	0.3	0.0073
26	0.0649	0.10	0.0000	21	1.2	1.9	0.0	3.1	0.3	0.0062
27	0.0680	0.10	0.0000	18	1.1	1.6	0.0	2.7	0.3	0.0053
28	0.0711	0.10	0.0000	15	1.0	1.4	0.0	2.3	0.3	0.0044
29	0.0744	0.10	0.0000	12	0.9	1.1	0.0	2.0	0.3	0.0037
30	0.0778	0.10	0.0000	10	0.7	1.0	0.0	1.7	0.3	0.0031
31	0.0815	0.10	0.0000	9	0.7	0.8	0.0	1.5	0.3	0.0026
32	0.0852	0.10	0.0000	7	0.6	0.7	0.0	1.2	0.3	0.0022
33	0.0892	0.10	0.0000	6	0.5	0.6	0.0	1.0	0.3	0.0018
34	0.0933	0.10	0.0000	5	0.4	0.5	0.0	0.9	0.3	0.0015
35	0.0976	0.10	0.0000	4	0.4	0.4	0.0	0.7	0.3	0.0012
36	0.1022	0.10	0.0000	3	0.3	0.3	0.0	0.6	0.3	0.0010
37	0.1069	0.10	0.0000	3	0.3	0.3	0.0	0.5	0.3	0.0008
38	0.1118	0.10	0.0000	2	0.2	0.2	0.0	0.4	0.0	0.0000
39	0.1170	0.10	0.0000	2	0.2	0.2	0.0	0.4	0.0	0.0000
40	0.1225	0.10	0.0000	1	0.2	0.1	0.0	0.3	0.0	0.0000
41	0.1281	0.10	0.0000	1	0.1	0.1	0.0	0.2	0.0	0.0000
42	0.1341	0.10	0.0000	1	0.1	0.1	0.0	0.2	0.0	0.0000
43	0.1403	0.10	0.0000	1	0.1	0.1	0.0	0.2	0.0	0.0000
44	0.1468	0.10	0.0000	1	0.1	0.1	0.0	0.1	0.0	0.0000
45	0.1536	0.10	0.0000	0	0.1	0.1	0.0	0.1	0.0	0.0000
46	0.1607	0.10	0.0000	0	0.1	0.0	0.0	0.1	0.0	0.0000
47	0.1681	0.10	0.0000	0	0.0	0.0	0.0	0.1	0.0	0.0000
48	0.1759	0.10	0.0000	0	0.0	0.0	0.0	0.1	0.0	0.0000
49	0.1841	0.10	0.0000	0	0.0	0.0	0.0	0.0	0.0	0.0000
50	0.1926	0.10	0.0000	0	0.1	0.0	0.0	0.1	0.0	0.0000
					455.3	302.7	242.0		9.6000	1.0000

Note: Sources of competing risk include natural mortality (lifetime), poaching mortality (> 5 years), and removal (5–10 years old). Poaching and removal rates shown in the table yield a net reproduction rate of replacement (i.e., NRR = 1.0), and expectation of life at birth is 8.3 years (when removal is considered a "mortality factor"). Net reproduction and expectation of life at birth in the absence of both poaching and removal are 4.16 and 19.2 years, respectively.

rhinoceroses' lives. Slightly less than a third of all deaths occurred with this 10% annual poaching rate, and slightly over a quarter of the loss in the population was due to the 10% removal rate of animals between 5 and 10 years old. More broadly, this analysis illustrates the use of the multiple-decrement life table approach for exploring trade-offs in management options and outcomes of poaching, or hunting more generally, on the survival outcome of an endangered species of an African large game animal.

Captive breeding in conservation biology

Background

In light of recent trends of decreasing biodiversity (Butchart et al. 2010) and increasing numbers of species threatened with extinction (Ricketts et al. 2005), captive breeding strategies have gained increased attention as a means to at least partially mitigate these trends. Nearly 20% of extant vertebrate species were classified as threatened in 2010, ranging from 13% of birds to 41% of amphibians (Hoffmann et al. 2010). It has been suggested that captive breeding of mammals in zoos is the last hope for many of the best-known endangered species (Alroy 2015) despite a number of practical (Kawata 2012; Whitham and Wielebnowski 2013; Fa et al. 2014), behavioral (Slade et al. 2014), and financial (Conway 1986; Conde et al. 2013) limitations. A schematic of captive breeding is given in fig. 10.18. One of the advantages of captive breeding programs is that they can serve as an "insurance policy" against threats like disease or pressure from nonnative species until reintroduction into the wild is possible (Conde et al. 2011; Conde et al. 2019). Moreover, managed *ex situ* populations in zoos can assist biodiversity conservation by serving as genetic and demographic reservoirs of critically endangered species—a "last resort" strategy (Fa et al. 2011). Here we evaluate data from an African lion captive breeding program to explore various demographic trade-offs in such programs.

Demography captive breeding: African lion

Although African lions are not currently threatened by extinction, their numbers are declining rapidly across Africa (Bauer et al. 2015). Indeed, the IUCN (International Union for Conservation of Nature) "Red List of Threatened Species" reported a population reduction by 43% from 1993 to 2014 (Bauer et al. 2016) and thus classified this species as "vulnerable"; in other words, there is the expectation that it will become endangered unless the circumstances that are threatening its survival and reproduction improve.

The demographic parameters (birth and death rates) required for estimating the production capabilities for African lion captive breeding programs were obtained from a database compiled at Species360, an international nonprofit organization that maintains an online database of wild animals under human care (Species360 2018). The birth and death information include data on 153 female African lions, which is illustrated in the event history graphic shown in fig. 10.19. Analysis of these data show that the 153 female lions produced a total of 654 offspring for a net reproductive rate (both sexes) of 4.3 offspring/female, or $R_0 = 2.15$ female offspring/female. The expectation of life for this population was 14.3 years. This was nearly sevenfold greater than the 2.1-year life expectancy of free-ranging African lions that was implied

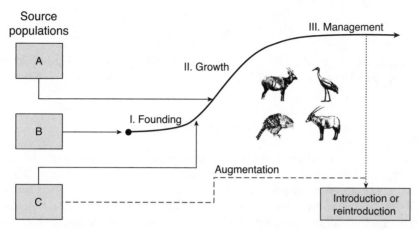

FIGURE 10.18. Schematic diagram of phases of a captive breeding program, including founding, growth, and management. Individuals can be introduced or reintroduced to the wild and/or used to augment populations when the population is sufficiently large and thus enters the management phase Inset images: Eastern bongo (upper left), whooping crane (upper right), kakapo (lower left), and Arabian oryx (lower right). (*Source*: Adapted from Fa et al. 2011)

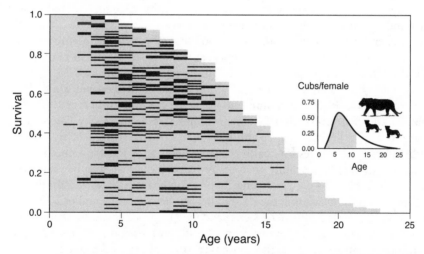

FIGURE 10.19. Event history chart of captive African lion reproduction (n = 153). Each horizontal line represents a single lion, the length of which corresponds to her life span. Within-life-line ticks depict a parturition, with lighter marks denoting the production of 1–2 cubs and darker ones 3–5 cubs. Inset: Smoothed age-specific reproductive schedule. Shaded area from 2 through 10 years indicates the reproductive window during which 85% of all lion offspring are produced. (*Source*: Species360 2018)

by the mortality rate estimates of Barthold et al. (2016). Mean age of reproduction of T = 7.1 years, with around 85% of all offspring produced between ages 2 and 10 (inset in fig. 10.19). Using T as an approximation for generation time, the intrinsic

rate of increase, r, is calculated as $r = \dfrac{\ln(R_0)}{T} = \dfrac{\ln(2.15)}{7.1} = 0.107$, and the finite rate of increase is $\lambda = e^r = 1.11$. Finally, doubling time $DT = \dfrac{\ln(2)}{r} = \dfrac{0.6931}{0.1067} = 6.5$ years. This

rate suggests that a breeding zoo population of 10 lions would increase to 20 in less than 7 years and to over 150 in 25 years.

Consider Further

Like chapter 9, many of the foundational concepts and models on which much of the content in this chapter rests are contained in the earlier chapters (e.g., demography basics; life tables; stable population theory) and a number of the entries in appendixes I–III. Deeper sources for various methods covered in this chapter include the recent review of wildlife demography, including population processes, analytical tools, and management applications (Lebreton and Gaillard 2016); the book on wildlife demography (Skalski et al. 2005); and the manual on wildlife techniques (Silvy 2012). Two important papers involving the demography of bears but which contain general demographic concepts include one on spatially explicit capture-recapture in grizzly bears (Morehouse and Boyce 2016) and one on how regulated hunting reshapes the life history of brown bears in Sweden (Bischof et al. 2018).

Unique perspectives on comparative demography of aging include the little-known monographs on the demography of survival and old age by the late Finnish demographer Väinö Kannisto (1994, 1996). Papers on health demography, from which many of the methods and concepts can be adapted for use in population biology, include the work by prominent health demographers such as Mark Hayward and his colleagues (Hayward et al. 1998, Hayward and Gorman 2004; Hayward and Warner 2005), Eileen Crimmins and her colleagues (Crimmins et al. 1994, 1996; Crimmins and Beltran-Sanchez 2011; Crimmins 2015), and Vicky Freedman, Emily Agree, and their colleagues (Agree and Freedman 2011; Freedman et al. 2011, 2013).

Other important sources for health-related demography include a number of entries in volume 2 of the demography treatise by Graziella Caselli and her European colleagues (Caselli et al. 2006a), including chapters on health, illness, and death (Gourbin and Wunsch 2006), measuring the state of health (Sermet and Camboi 2006), medical causes of death (Mesle 2006), dependence and independence of causes of death (Wunsch 2006), and the relationship between morbidity and mortality by cause (Egidi and Frova 2006). Important literature on harvesting theory and practice includes the papers or books on demographic models of fish, forest, and animal resources by Wayne Getz (Getz 1984; Getz and Haight 1989) and on harvesting spatially distributed populations by Niclas Jonzén and his coauthors (Jonzén et al. 2001). Chapters in the *Handbook of Population* (Poston and Micklin 2005) are relevant to much of the content in this chapter, including ones on ecological demography (Poston and Frisbie 2005), health demography (Kawachi and Subramanian 2005), and the demography of population health (Hayward and Warner 2005). Anderson, Li, and Sharrow (2017) provide new insights into mortality patterns and causes of death through what they refer to as a "process point of view model." Finally, two important recent papers concerning the analysis of wild populations include one by Koons et al. (2014) describing methods for studying cause-specific senescence in the wild and another by Ergon et al. (2018) describing the utility of mortality hazard rates in population analysis.

Biodemography Shorts

Science is impelled by two main factors: technological advances and a
guiding vision. Without the proper technological advances the road ahead
is blocked. Without a guiding vision there is no road ahead.

Carl Woese (2004, 173)

Most of the content in this chapter falls into overlapping categories of what we call
biodemography "shorts" (S). This includes novel material that goes beyond the structured chapters and thus has the potential to inform, enlighten, and, in some cases,
inspire. We also include a few simple concepts and methods that readers are virtually
certain to be aware of but are presented as a reminder of their demographic utility.
Beyond this, we include material that extends the margins of demography.

These shorts also present questions that expand the breadth of demographic tools
to unexplored aspects of society. For example, How can life table analysis be used to
estimate the career length of a professional athlete? Or, if a population of a wild primate were decreasing, how many years would it take for the population to reach
one-half of its present size? We also include material that is mainstream demography
but is seldom included in demography texts, including how fish are aged and how
demographic principles can be applied to forensic analysis. These examples demonstrate the multiple contexts in which demography is relevant.

Group 1: Survival, Longevity, Mortality, and More

Survival and longevity

S 1. A different way to look at survival—life increments

The difference between any two survival curves may be described as the collection
of all horizontal distances between the ages at which a given proportion survive in
one cohort and the ages at which this same proportion survive in another cohort (Feeney 2006). Feeney refers to these distances as the increments to life (fig. 11.1). For
example, if 80% in cohort A survive to 60 years and 80% in cohort B survive to
70 years, the "increment to life" would equal 10 years. In the example, using data
from the 2015 life tables for US males and females (fig. 11.2), the life increment for
cohort survival is equal to 0.928. Whereas the survival to this level in the male
cohort occurs at age 36, survival to this level in the female cohort does not occur
until age 50, a 14-year difference. This horizontal distance is referred to as the 14-year
life increment between males and females. Thus, out of 1,000 individuals of each

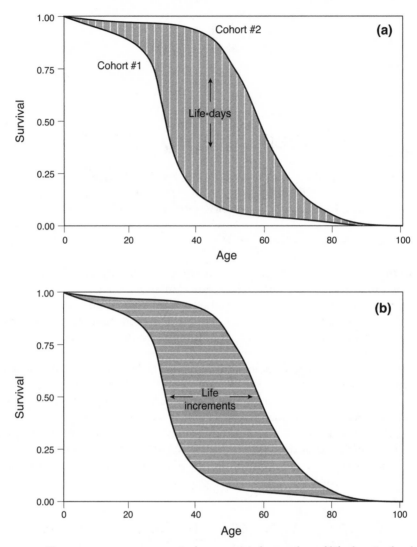

FIGURE 11.1. Two ways to compare survival curves: (a) the number of life-days in the shaded area equals the differences in respective expectations of life; (b) the life increments in the shaded area depict the differences in the time required to attain identical levels of survival.

sex, the first 72 males would die by age 36 but the first 72 females would not be dead until age 50.

S 2. A 1-year survival pill is most effective near e(0)

If a magic pill could be produced to reduce mortality to zero for one single age, which age would yield the maximum increase in life expectancy at birth? The outcome to this question depends on trade-offs among a number of actuarial components, including (a) survival *to* a given age and (b) mortality *at* that age as well as (c) mortality

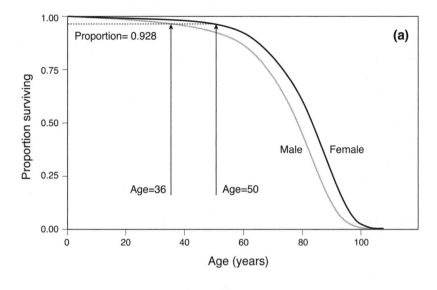

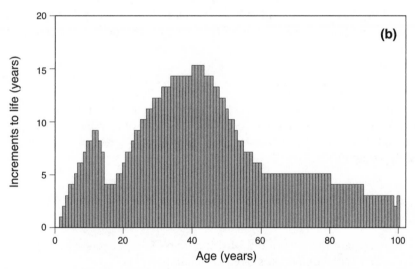

FIGURE 11.2. (a) Sex-specific survival curves for 2015 US life tables illustrating the 14-year "increment to life" for females relative to males, the horizontal distance between the ages when the proportion surviving equals 0.928; (b) Time-discrete increments to life for US men and women, 2015 life tables. Each bar represents the number of years longer a female cohort requires to attain identical levels of survival as a male cohort.

levels and patterns *after* that age and, in turn, (d) life-years *remaining* to the individuals surviving the zero-mortality age who would not have otherwise survived. The answer, for the 2015 data for US females (fig. 11.3), is that eliminating mortality at age 0 *or* at age 77 will have the same impact. The rule of thumb for the effect of mortality on life expectancy is that the optimal period for saving lives by reducing mortality is near the life expectancy of the population (Vaupel 1986).

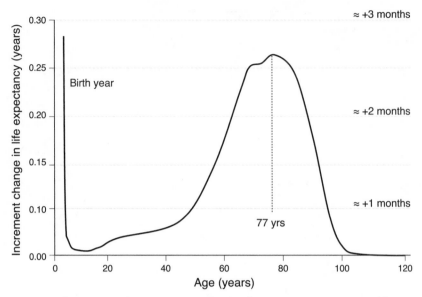

FIGURE 11.3. Reducing mortality to zero at either birth (age 0) or at 77 years adds, respectively, 0.30 years (110 days) and 0.27 years (100 days) to life expectancy (data for 2015 US females).

S 3. Redundancy increases system reliability

The reliability theory, from engineering, can be applied to failure associated with demographic aging in living organisms. The reliability of a system in series, denoted S_s, with independent failure events of the components is the product of the reliabilities of the components (Gavrilov and Gavrilova 2006).

$$S_s = p_1 p_2 p_3 p_4 \cdots p_n$$

where $p_1 \ldots p_n$ are the reliabilities of the system's components. The reliability of a parallel system with components of equal reliability, p, is

$$S_s = 1 - (1-p)^n$$

where n denotes the number of redundant units in each component. Failure models in systems engineering possess three important properties: (1) a system constructed of nonaging elements behaves like an aging object where aging is a direct consequence of the redundancy of the system; (2) at very high ages, the phenomenon of aging apparently disappears; and (3) systems with different initial levels of redundancy have very different failure rates in early life. However, these differences disappear as failure rates approach the upper limits. This means that in a system in series, all of the components must succeed for the system to succeed, but in a simple parallel system only one of the units must succeed for the system to succeed (fig. 11.4).

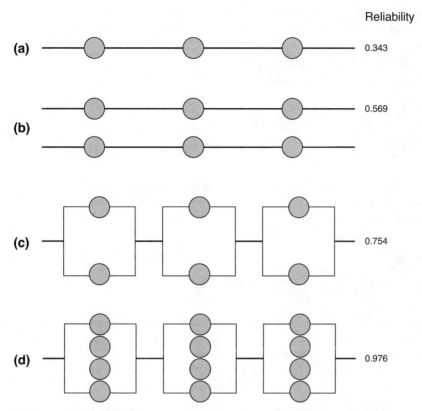

Reliability

(a) ———————————————— 0.343

(b) ———————————————— 0.569

(c) ———————————————— 0.754

(d) ———————————————— 0.976

FIGURE 11.4. Redundancy drastically increases reliability. Each circle represents a component required for system function with a failure probability of 0.3. The logical schematics
for a three-component system are (a) series with reliability probability of 0.343, (b) two
systems each in series with a reliability probability of 0.569 (i.e., at least one of the
two systems will not fail), (c) parallel system with a redundancy of two with a reliability
of 0.754, and (d) parallel system with a redundancy of four with a reliability of 0.976.

S 4. Five important and useful period measures of longevity

Different period measures of longevity, with special emphasis on the tempo effect,
can be summarized as follows (after Bongaarts 2008):

Measure 1: *Life expectancy*

$$e_0(t) = \int_0^\infty \exp\left\{-\int_0^x \mu(a,t)da\right\}dx \qquad (11.1)$$

where $\mu(a, t)$ is the force of mortality at age a and time t.

Measure 2: *Cross-sectional average length of life*

$$CAL(t) = p_c(a, t-a)da \qquad (11.2)$$

where $p_c(a, t-a)$ equals the proportion of survivors at age a and time t for the cohort born at time $t-a$. CAL(t) sums proportions of cohort survivors at time t, and it therefore equals the size of the population at time t in which births have occurred at a constant rate of 1 per year in the past.

Measure 3: *Tempo-adjusted life expectancy*

$$e_0^*(t) = \int_0^\infty \exp\left\{-\int_0^x \frac{\mu(a,t)}{1-r(t)}\,da\right\}dx \qquad (11.3)$$

This is a variant of the conventional period life expectancy, but the tempo effect in the force of mortality is removed by dividing this rate by $1-r(t)$. The variable $r(t)$ denotes the increments to life due to mortality improvements at time t.

Measure 4: *Lagged cohort life expectancy*

$$LCLE(t) = e_0^c(c) = e_0^c(t - e_0^c(c)) \qquad (11.4)$$

LCLE(t) is estimated as the life expectancy of the cohort born at time c with the lag between t and c equal to the life expectancy of the cohort: $c = (t - e_0^c(c))$. LCLE equals the life expectancy of the cohort that reaches its mean age at death at time t. This concept applies primarily to human populations.

Measure 5: *Average weighted cohort life expectancy*

$$ACLE(t) = \int_0^\infty \omega(a,t) e_0^c(t - ac)da \qquad (11.5)$$

where $\omega(a, t)$ are the weights for the life expectancy of cohorts born at time $t-a$.

There are several comparisons between these measures that should be noted (Bongaarts 2008). First, three of these period longevity indicators, CAL(t), $e_0^*(t)$, and LCLE(t), are virtually identical. Second, the conventional life expectancy e_0 is substantially higher than three other period measures. Finally, the weighted average cohort life expectancy, ACLE(t), is much higher than the other four indicators because the weights applied to the life expectancies of cohorts alive at time t are highest for the youngest cohorts, and this measure is heavily influenced by the mortality that young cohorts will experience in the future.

S 5. A simple equation for producing survival curves

Given different values for the parameters a and b, a relatively simple equation (11.6) can be used to model survival over a wide variety of patterns (see fig. 11.5). Note that convex survival patterns are produced when $a > 1$ and $b = 1$, and concave survival patterns are produced when $a = 1$ and $b > 1$.

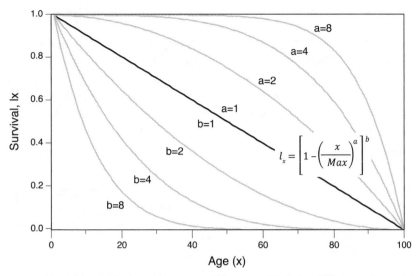

FIGURE 11.5. Example survival curves generated by eqn. (11.6) for different values of the parameters a and b.

$$l_x = \left[1 - \left(\frac{x}{\text{Max}}\right)^a\right]^b \qquad (11.6)$$

S 6. Computation of life overlap of two or more persons

The average number of years that the lives of two persons will overlap, both of whom are subject to the same survival, equals the sum of products of the squared probabilities that each will live to age x from birth. Given that l_x denotes the probability that a newborn individual will survive to age x, the probability that two persons, from the same cohort, will survive to that age is simply $(l_x)^2$.

Let $\bar{l}$ denote the sum of the probability of overlapping,

$$\bar{l} = \sum_{x=0}^{\omega} (l_x)^2 \qquad (11.7)$$

Then this concept can be generalized to determine the average number of years n individuals will all overlap:

$$\bar{l} = \sum_{x=0}^{\omega} (l_x)^n \qquad (11.8)$$

Using the US female 2015 life table for computations, two newborn individuals subject to these life table rates will on average share slightly over 73 years of life

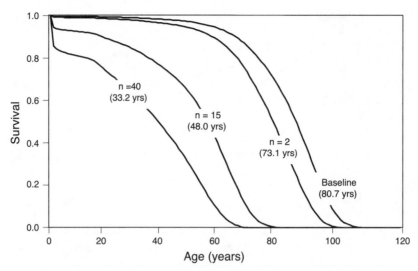

FIGURE 11.6. Average overlap of n = 2, 15, and 40 lives relative to the baseline cohort survival of one individual.

together (fig. 11.6), and the lives of 15 and 40 newborn individuals will share 48 years and 33 years, respectively.

S 7. Mortality of 10-year-old girls is incredibly low

Mortality in 10- and 11-year-old girls (and to a slightly lesser extent for same-aged boys) is extremely low in developed countries. One can ask the question, How many of a newborn cohort of 100,000 individuals would survive to 100 years if they were subject to the age-specific mortality rates of 10-year-old girls for their entire life span? Using the 10-year-old mortality rate for 2015 US females, the annual survival of this cohort would be

$$100,000 \times (1 - 0.00009)^{100}$$

$$= 99,104 \text{ individuals}$$

In other words, over 99% of the original cohort would become centenarians. Additional math reveals that $\approx 90,000$, $\approx 40,000$, and ≈ 10 individuals would survive to 1,000, 10,000, and (a remarkable) 100,000 years, respectively.

Mortality

S 8. Reparameterization of Gompertz using modal age, M

The Gompertz equation is the most commonly used equation to describe mortality. But a change in the parameterization of this equation has been proposed by Missov et al. (2015) that expresses the Gompertz force of mortality in terms of the rate parameter b and the old-age modal age of death, M.

The Gompertz equation is given as

$$\mu(x) = ae^{bx} \tag{11.9}$$

Reparameterized with the mode, M, is given as

$$\mu(x) = be^{b(x-M)} \tag{11.10}$$

The formula for the mode is

$$M = \frac{1}{b}\ln\frac{b}{a} \tag{11.11}$$

Density of death, d(x), is given as

$$d(x) = a\left[\exp\left\{bx - \frac{a}{b}(e^{bx} - 1)\right\}\right] \tag{11.12}$$

Whereas in the Gompertz equation there is a negative association between a and b, with this reparameterization there is a positive association between M and b. Using the old-age mode M instead of the mortality at starting age a has two major advantages. First, statistical estimation is facilitated by the lower correlation between the estimators of model parameters. Second, estimated values of M are more easily comprehended and interpreted than estimated values of a.

S 9. Excess mortality rate

Excess mortality rate can be defined as the death rate in the general population due to the excess risk imposed by a specific disease (Lenner 1990). Let N denote the total population, n denote the subgroup in the population with a disease, a denote the deaths among those who are sick (among n persons), and b denote the deaths among all of the population without the disease (N − n). The excess risk of death within population n, in relation to the corresponding population without the disease, is

$$\frac{a}{n} - \frac{b}{(N-n)} \tag{11.13}$$

The expected number of deaths in population n due to this excess risk is

$$n\left\{\frac{a}{n} - \frac{b}{(N-n)}\right\}$$ (11.14)

and the death rate in the total population, N, due to the excess risk is

$$\frac{n}{N}\left\{\frac{a}{n} - \frac{b}{(N-n)}\right\}$$ (11.15)

S 10. Mortality elimination

In a previous short (S2) we considered the impact on life expectancy of a magic pill that could reduce mortality to zero for one single age class. Here we consider the effect of eliminating all mortality through middle and advanced ages. The effects on the overall death distribution for eliminating mortality through age 70 is shown in fig. 11.7. This mortality elimination has a minimal to moderate impact on life expectancy because mortality is low already through the majority of these ages. Additionally, because more people live to older ages and are thus exposed to the higher risk of dying, the distribution of death scales more upward rather than to the right (i.e., shifting to older ages). Using the US female 2015 period rates, where $e_0 = 81.4$ years, the elimination of mortality through ages 50, 60, 70, 80, 90, and 100 years yields hypothetical increases in life expectancies to 83.6, 85.1, 87.1, 90.4, 95.7, and 103.3 years, respectively.

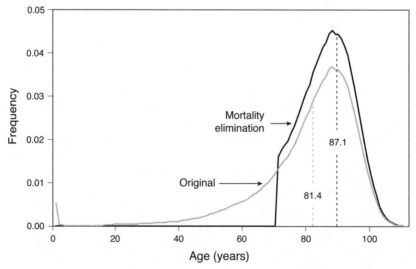

FIGURE 11.7. Effects of mortality elimination before age 70 on the proportion dying at each age. The original baseline mortality (light gray curve) is based on data for 2015 US females.

S 11. Quiescent period emerges in cohort but not period mortality

The quiescent phase (Q-phase) is a unique property of mortality observed in data from Swedish women (fig. 11.8) in 1920, where the hazard trajectory remains relatively low and constant during the prime ages for reproduction and investment in both personal capital and relationships with others (Engelman et al. 2017). This age segment of mortality was originally identified by Finch (2012) and is observed *only* in the cohort (longitudinal) mortality rates but, interestingly, not in the period (cross-sectional) mortality rates (see fig. 11.8).

S 12. Probability of same-year deaths of sisters is extremely small

What is the likelihood that two sisters aged 25 and 30 will die in exactly the same year 60 years later? The qualitative answer to this problem is that this probability is extremely small because it is the product of two sets of probabilities, both of which are small. The first set involves the joint probability that each *survive* to 60 years, which is the probability of surviving from 25 to 85, $_{60}p_{25} = 0.448$, and the probability of surviving from 30 to 90, $_{60}p_{30} = 0.265$, the product of which is 0.119. The product of these probabilities is the likelihood of them both surviving the next 60 years, which is around 12%. The second set of probabilities involves the joint probability of them *both dying* the sixtieth year. The probabilities of dying between 85 and 86 (q_{85}) and between 90 and 91 (q_{90}) are 0.079 and 0.140, respectively, and the product of these two probabilities is 0.011. Therefore, there is slightly over a 1% probability that both would die between the sixtieth and sixty-first year, given they both survive

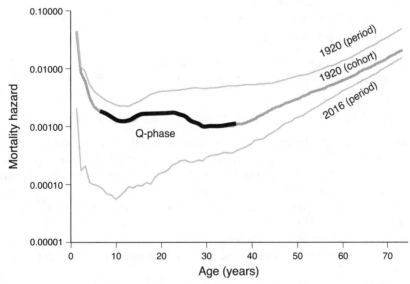

FIGURE 11.8. Q-phase (black segment) in the age-specific mortality hazard curve for the 1920 Swedish female cohort. Note the overall downward trend beginning around age 8 and continuing through age 40. The period mortality rates for the 1920 and the 2016 cohorts are presented for perspective (HumanMortalityDatabase 2018). Adapted from Engelman et al. (2017).

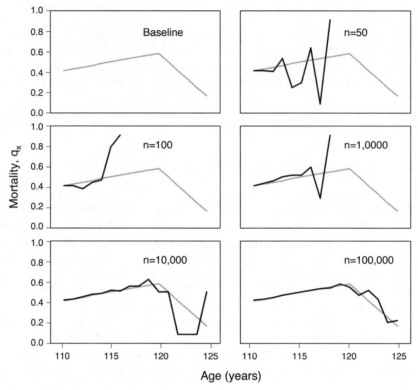

FIGURE 11.9. Mortality, q_x, estimates at advanced ages when mortality, q_x, is extraordinarily high. Light gray curve is baseline. Note that only when n > 10,000 does the estimate (black line) begin to show the slowing and decrease of mortality rates at the oldest ages.

60 more years. Finally, the joint probability that they both survive the next 60 years, and then both die in the next year, is $(0.011) \times (0.119) = 0.0013$, or a 1-in-767 chance.

S 13. Estimating mortality trajectories in supercentenarians

One of the greatest challenges in human actuarial studies is that of estimating mortality at the most advanced ages, where numbers are small and mortality rates are high (Mesle et al. 2010). One of the reasons for this is illustrated in fig. 11.9, which shows that, relative to a hypothetical mortality baseline (shown in light gray), the variance in mortality renders estimates useless with small initial numbers. This is because the die-off itself reduces the numbers further, and thus mortality estimates at older ages become progressively less reliable due to the high variance. This concept is also illustrated in the million Medfly study published by Carey and his colleagues (1992).

S 14. Evidence for the absence of life span limits—105-year-old Italian women

One of the most intense ongoing debates in demography, biodemography, and gerontology is whether there is a fixed upper limit to human life span (Fries 1980;

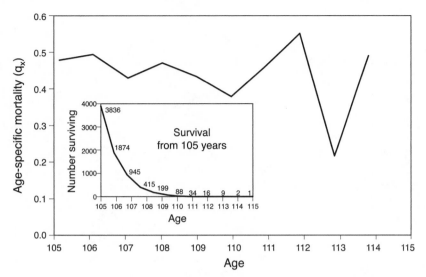

FIGURE 11.10. Mortality trajectory of over 3,800 Italian women from 105 through 114 years. Inset shows survival curve with sample sizes. (*Source*: Barbi et al. 2018)

Olshansky et al. 1990, 2001; Vaupel et al. 1998; Gavrilov and Gavrilova 2012, 2015). Although the hypothesis that mortality is an ever-increasing function of age was rejected in fruit flies with the observation that mortality in two separate species leveled off at advanced ages (Carey et al. 1992; Curtsinger et al. 1992), it is not clear whether this is also true in humans. The concept of *mortality deceleration* at older ages is important because it implies that a fixed limit does not exist.

Italian demographers (Barbi et al. 2018) assembled mortality data on over 3,800 women who had survived 105 years and reported that mortality, although extremely high (i.e., > 0.5/year), was essentially age independent (fig. 11.10). These results provide evidence for extreme-age mortality plateaus in human "longevity pioneers"; in other words, there is no evidence for specific life span limits. Note, however, that as demonstrated in S13, the sample sizes at the most extreme ages are very small, so the exact shape of these latest-age trajectories cannot be discerned. Both Salinari (2018) and Gavrilov and Gavrilova (2019) offer alternative perspectives on the interpretation of human mortality data at the most advanced ages.

S 15. Advancing front of old-age survival—traveling demographic waves

In S14 the analysis of the mortality trajectory suggests that there are no sharply defined limits to human life span because of the age independence of age-specific mortality in extremely old Italian women (Barbi et al. 2018). This concept of the absence of human life span limits was also the conclusion of an analysis that used a completely different approach (Zuo et al. 2018). Using death distributions over time, these researchers showed that for 5 decades in 20 developed countries old-age survival follows an advancing front, like a traveling wave, the concept of which is illustrated with data on US females (fig. 11.11). Note that the endpoints of the solid gray horizontal bars in this figure are consistently advancing with age over the different decades, which suggests that there are no fixed limits to life span.

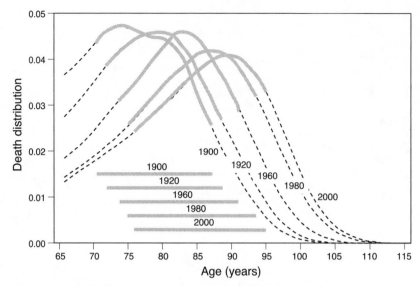

FIGURE 11.11. Example of "traveling waves" of death distributions for US females in selected decades. The solid gray curves indicate the distributions from 25% to 90% mortality, and the solid gray horizontal bars indicate the number of years between these two points in the survival curves (plot concept adopted from Zuo et al. 2018).

S 16. Actuarially speaking, 92 years old is halfway to 100

Danish biodemographer Kaare Christensen asked the question, "At what age is mortality exactly half of total mortality to reach 100?" He discovered that the answer for Danish women was around 92 years. In other words, the force of mortality over the first 92 years of life is identical to the force of mortality over the next 8 (i.e., from 92 to 100). We generalized this question by considering the equivalences between two sets of ages: (1) fixed ages, such as 30, 60, and 100 years old; and (2) the age when mortality to those ages is exactly half the total. These equivalencies are given in fig. 11.12 for US women, which show that mortality from birth to age 12 equals mortality from 12 to 30, mortality from birth to 51 equals mortality over the next 9 years (to age 60), and mortality from birth to 93 years equals mortality over the next 7 years (to age 100).

S 17. Why the oldest person in the world keeps dying

What is the probability that a randomly chosen woman in a hypothetical stationary population of $n = 1,000$ (distributed over a stable age structure) who is subject to the 2015 period mortality rates will become the oldest person alive at some time during a century? What is the average age and length of reign for these oldest women?

This problem was motivated by an article in the *New York Times* titled "Why the Oldest Person in the World Keeps Dying" (Goldenberg 2015) and by a hypothetical problem where the oldest member of a certain tribe becomes king (for problem description and its analytical solution, see Keyfitz 1985, 74–75). To explore these questions, we created a life table population based on the 2015 US female life table,

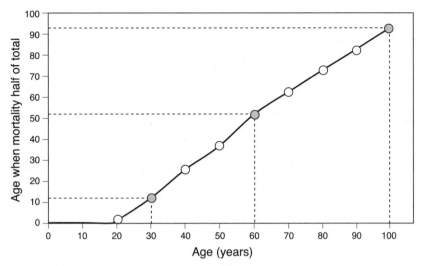

FIGURE 11.12. Mortality equivalencies—age at which mortality from birth is exactly half total mortality to a given age. For example (shaded circles), total mortality from birth to ages 30, 60, and 100 (x-axis) is exactly half at ages 12, 51, and 93, respectively (y-axis). Computed using life table rates of US women (2000) from Berkeley Mortality Database.

simulated individual-level mortality for each of 1,000 individuals in the stationary population from their current age forward through their deaths (and extinction of all members) for 100 years, and recorded the oldest individual alive during each year. The simulations were repeated 100 times.

The results of one of the simulation trials are presented in fig. 11.13. Overall (all 100 simulations) the trials yielded an average of ≈ 22 individuals who became one of the oldest persons living through the 100-year series of simulations (i.e., ≈ 2% chance). Around 60% of the reigns lasted from 1 to 3 years and less than 2% of the reigns lasted for 10 years or more. This concept of "oldest person" could be applied to retirement communities, towns, and cities, and, if death is generalized as "exit from current state," it can also apply to oldest on a team and oldest in a company or a department.

Roland Rau and colleagues (Rau, Ebeling et al. 2018) connected the problem of how often the oldest person alive dies with queuing theory. In their model the population older than 110 years followed a Poisson distribution with parameter λ/μ, where λ depicts the rate at which people turn 110 and μ represents the force of mortality. Their simulation studies also showed that the waiting time between the deaths of the (respective) oldest person alive follows an exponential distribution with parameter μ and with a mean duration of $1/\mu$ if $\lambda > \mu$. However, they did not have a clear-cut result for the case of $\lambda \leq \mu$.

S 18. Life span limits: Challenges in data interpretation

Improvements in survival with age decline after age 100, and the age at death of the world's oldest person has not increased since 1990 (fig. 11.14). These observations led Dong et al. (2016) to suggest that the maximum life span of humans is fixed. Their

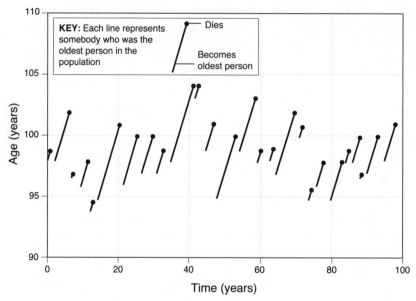

FIGURE 11.13. Example result for the oldest living women in 100-year simulations involving a stable, stationary population of 1,000 individuals. A total of 26 women became the oldest person living for this particular simulation.

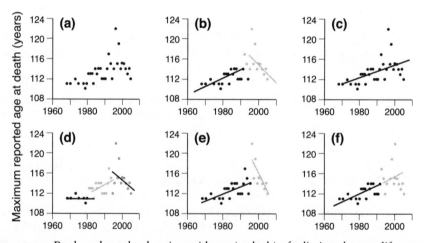

FIGURE 11.14. Replotted graphs showing evidence (or lack) of a limit to human life span based on paper by Dong et al. (2016) and brief communications: (a) original data plot of trends in maximum recorded ages of death; (b) regression partitioning by Dong et al. (2016); (c), (d), and (e) show nonpartitioned linear regression, multiple partitions, and regression with partition shifted two years, respectively (Hughes and Hekimi 2017); and (f) shows linear regressions if most extreme maximum reported age at death (122 years) is shifted 7 years to the right (Brown et al. 2017).

article was followed by several papers criticizing both their approach and their conclusions. Some of these comments included (1) questionable assumptions about how the data set should be partitioned, that there is only one plateau, and that the correct year to partition is 1994 (Hughes and Hekimi 2017); (2) the use of the same data set both to propose a hypothesis (change in trend in human longevity) and to test the hypothesis, and a lack of support for their broader argument from a biological perspective (Rozing et al. 2017); (3) their fitting separate regressions did not ensure that the predicted probability of surviving to some age is higher than the probability of surviving to an older age (Lenart and Vaupel 2017); and (4) that their small sample size is problematic (i.e., 21 individuals for 1968 and 12 individuals for 1995–2006), there was no comparison with alternative models, and the results were largely dependent on a single exceptional case—that of Jeanne Calment who died at age 122 years (Brown et al. 2017). Responses by Dong and his colleagues to all of these concerns followed each of the respective commentaries arising from their original paper (Dong et al. 2016). Fig. 11.14a–f contains graphic perspectives on the different ways in which the same data on trends in maximal ages of death can be visualized and interpreted.

Life course

S 19. Relative age

My late friend Stan Ulam used to remark that his life was sharply divided
into two halves. In the first half, he was always the youngest person in the group;
in the second half, he was always the oldest. There was no transitional period.
MIT mathematician Gian-Carlo Rota (1996)

Let A_C be calendrical age, defined as the number of years since a person is born, and A_R be relative age, defined as the percentage of people in the world whose calendrical age is less than that person, which is also a person's percentile in the world population age structure. Using the data in fig. 11.15, when a woman born in 1950 is age 10 she is older than a quarter of the people on the planet, when this same woman is only age 20 she is older than half of the people in the world, and at age 30 she is older than nearly two-thirds of the world's population (Hayes 2012). After age 60 her relative age is in the 90s, which suggests that, *relatively speaking,* she ages more slowly. Note that this figure shows that the relative age of a person born in the year 2000 advances less rapidly relative to his/her calendrical age than someone born in 1950.

S 20. Retirement age based on remaining life expectancy

Historically, there have been two conventional measures for characterizing the burden of population aging: the old-age dependency ratio (OADR), which is the ratio of people above working age to people of working age, and the potential support ratio at 65 (PSR65), which is the ratio of the number of people age 20 to 64 to the number of persons 65 and over. Because of the ever-increasing burden of old age on societies, many countries are revisiting the original default retirement age of 65 that was put forward nearly 100 years ago in Germany and adopted by many countries, including the US (Costa 1998). A concept that is gaining serious attention in many countries,

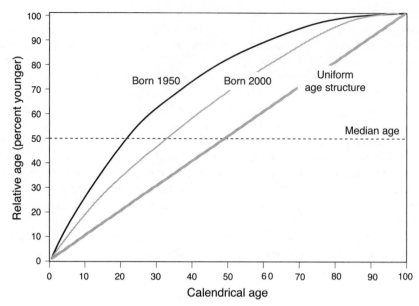

FIGURE 11.15. Relationship of calendar age and the percent of individuals in the world who are younger (redrawn from Hayes 2012).

and one that has been adopted by Denmark, is retirement age based on the age when there are X years of remaining life expectancy (Canudas-Romo et al. 2018).

$$e_y^p(t) = X \tag{11.16}$$

As shown in fig. 11.16, this approach to determining the retirement age results in a shifting target that changes over time and that changes with different values of X.

S 21. Effects of classmate age differences

A continuum of ages exists at school entry due to the use of a single school cutoff date—making the "oldest" children approximately 20 percent older than the "youngest" children (Bedard and Dhuey 2006). Relative age has been associated with higher intelligence, school success, identity formation, peer-perceived competence and leadership, success in sports, and positive self-perception and self-esteem (Jeronimus et al. 2015). This concept is illustrated and explained further in fig. 11.17.

Sports career

S 22. Major League Soccer (MLS)

What is the likelihood of a 21-year-old soccer player drafted into the MLS, the main professional soccer league in the US, playing past his thirtieth birthday? Life table

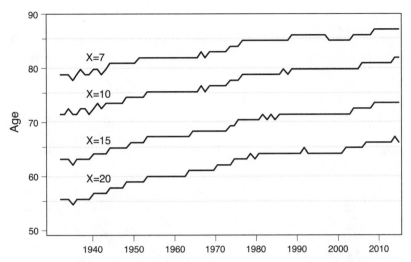

FIGURE 11.16. Age at which expectation of life for US women age 65 equals X. (*Source*: Berkeley Mortality Database, http://www.mortality.org/)

approaches can be used to address this question. Using data from Boyden and Carey (2010), table 11.1, a player's chances would be slightly over 10% that he would still be playing as a 30-year-old. His chance of playing beyond his thirty-second birthday is virtually nil.

S 23. National Basketball Association (NBA)

What are the scoring patterns and career durations of professional basketball players in the NBA? Event history curves can be used to address this question. The patterns of point scoring by year in the NBA and career lengths (fig. 11.18) show that highest scoring for most players occurs between their second and eighth year in the league, that these are the players destined to play for at least a decade, and that the superstars who play from 12 to over 20 years are among the highest scorers in their early to mid careers.

Group 2: Population, Statistical, Epidemiological, and Catastrophic

Population

S 24. Growth rate required for 40-year-old men to have two 20-year-old wives

In a certain hypothetical primitive community, men marry at age 40 and women at age 20. How fast does the community have to be increasing for each man to have two wives? What growth rate is required for all 40-year-old men to have two wives? Assuming a stable population with men and women increasing at the same rates, and

(a)

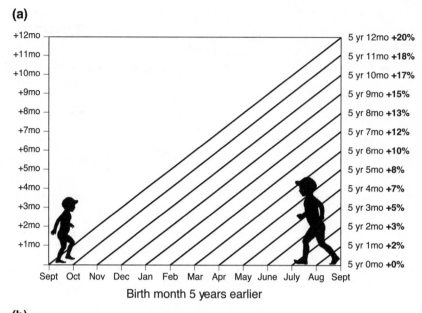

+12mo –		5 yr 12mo **+20%**
+11mo –		5 yr 11mo **+18%**
+10mo –		5 yr 10mo **+17%**
+9mo –		5 yr 9mo **+15%**
+8mo –		5 yr 8mo **+13%**
+7mo –		5 yr 7mo **+12%**
+6mo –		5 yr 6mo **+10%**
+5mo –		5 yr 5mo **+8%**
+4mo –		5 yr 4mo **+7%**
+3mo –		5 yr 3mo **+5%**
+2mo –		5 yr 2mo **+3%**
+1mo –		5 yr 1mo **+2%**
		5 yr 0mo **+0%**

Sept Oct Nov Dec Jan Feb Mar Apr May June July Aug Sept

Birth month 5 years earlier

(b)

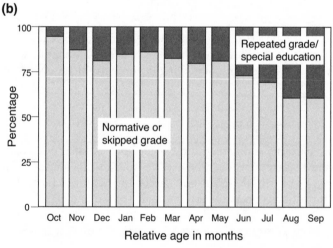

Repeated grade/ special education

Normative or skipped grade

Percentage

0 — 25 — 50 — 75 — 100

Oct Nov Dec Jan Feb Mar Apr May Jun Jul Aug Sep

Relative age in months

FIGURE 11.17. (a) Age differences in 5-year-olds with September cutoff date for entering school; (b) graph from study by Jeronimus and colleagues (redrawn from Fig. 1 in Jeronimus et al. 2015) showing that, for each additional month, relative age was associated with 17% lower odds of grade repetition and 47% increased odds of skipping a grade. The relatively young quartile (Jul to Sep) was almost four times more likely to repeat a grade (29.6% vs. 8.2%) and over twenty times less likely to skip a grade (0.3% vs. 7%) than the relatively old quartile (Oct to Dec).

Table 11.1. Life table analysis of career length of Major League Soccer (MLS) players who participated from 1996 through 2007

Time in MLS (years)	Cohort survival (remaining active in MLS)	Period survival	Period mortality	Fraction exiting	Expectation of remaining career
	l(x)	p(x)	q(x)	d(x)	e(x)
0	1.0000	0.7100	0.2900	0.2900	3.4
1	0.7100	0.7500	0.2500	0.1775	3.6
2	0.5325	0.7600	0.2400	0.1278	3.6
3	0.4047	0.7600	0.2400	0.0971	3.5
4	0.3076	0.7600	0.2400	0.0738	3.5
5	0.2338	0.7800	0.2200	0.0514	3.5
6	0.1823	0.7700	0.2300	0.0419	3.3
7	0.1404	0.8500	0.1500	0.0211	3.1
8	0.1193	0.8800	0.1200	0.0143	2.6
9	0.1050	0.8100	0.1900	0.0200	1.9
10	0.0851	0.6900	0.3100	0.0264	1.2
11	0.0587	0.0000	1.0000	0.0587	0.5
12	0.0000			0.0000	

Source: Boyden and Carey (2010).
Note: Here x denotes the time in the league from draft forward.

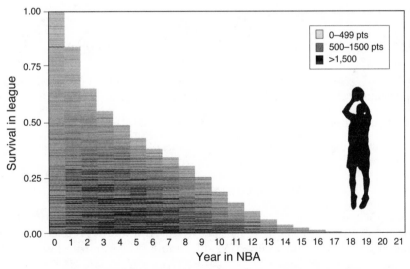

FIGURE 11.18. Event history chart for career scoring each year for each of 2,575 NBA players. Raw data (1940–2005) on NBA players downloaded from SportsData.com in 2007 (https://www.sportsdata.ag/about-us/group-set-up/). The first- and second-year "survival" in the NBA was low primarily because of differences in NBA recruitment strategies in the 1940s through the 1960s and more recent years.

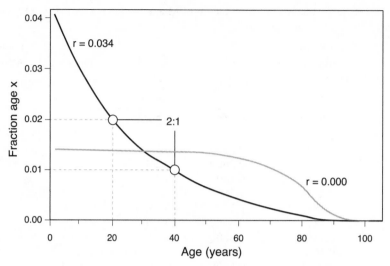

FIGURE 11.19. Stable age distribution in a population growing at 3.4% annually. Note that the fractions age 20 and 40 are 0.02 and 0.01, respectively. This produces a 2-to-1 ratio of 20- to 40-year-olds.

equal numbers of boy and girl babies, Keyfitz and Beekman (1984, 60, problem 41) show the growth rate, r, needed is the solution to the equation

$$r = \frac{1}{20} \ln\left(\frac{2 l_{40}^{*}}{l_{20}} \right) \tag{11.17}$$

where l_{20} and l_{40}^{*} denote survival to ages 20 and 40 for females and males, respectively. Assuming that sex-specific survival differences are negligible, a first approximation of this rate is $r = \frac{\ln(2)}{20} = 0.035$, which is within 0.001 of the population growth rate estimated using the age-specific survival rates of US females and assuming male rates are identical (fig. 11.19).

S 25. Interaction of mortality and fertility on r

How does the interaction between mortality and fertility impact r? Let l^{A} and l^{B} and m^{A} and m^{B} denote the respective age-specific survival and reproductive schedules shown in fig. 11.20a–b for species A (Medfly) and B (Mexican fruit fly). Then let

$$[r(l^{B}, m^{B}) + r(l^{A}, m^{A}) - r(l^{B}, M^{A}) - r(l^{A}, m^{B})] \tag{11.18}$$

The results given in table 11.2 show that substituting the Mexfly's survival and reproduction schedules for the Medfly's respective schedules increased r for the Medfly by 1.08-fold (l_x substitution) and 1.39-fold (m_x substitution), respectively. In contrast, substituting the Medfly's survival and reproduction schedules for the Mexfly's respective schedules *reduced* r in the Mexfly by essentially the same factors as it reduced r for the Medfly. Interestingly, even though survival in the Medfly was higher

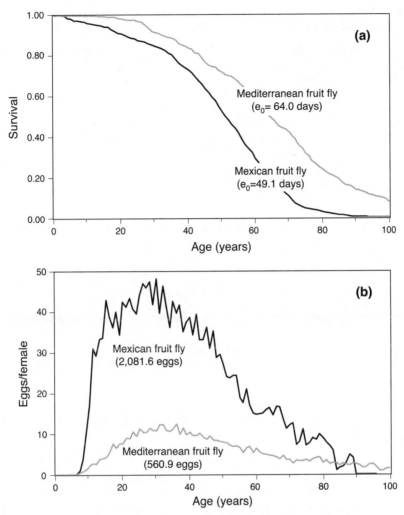

FIGURE 11.20. (a) Survival and (b) reproductive schedules for the Medfly (*Ceratitis capitata*) and the Mexican fruit fly (*Anastrepha ludens*).

than in the Mexfly, when the Medfly's survival was combined with the Mexfly's reproduction, r was reduced rather than increased. This was because generation time increased with the higher survival rate, the net result of which was to decrease growth rate, r. (From Keyfitz and Beekman 1984, 72, problem 35. The impact on r of generation time changes is illustrated in Lewontin 1965.)

S 26. Sex ratio in stable populations is independent of growth rate

The sex ratio in a stable population is given by the ratio

$$SR = \frac{se^{-rx}l_x^M}{e^{-rx}l_x^F} \tag{11.19}$$

Table 11.2. Values of intrinsic rate of increase, r, for the original and switched survival and reproductive schedules for the Medfly (*C. capitata*) and the Mexican fruit fly (*A. ludens*)

Species' origin		Net reproductive rate	
l_x schedule	m_x schedule	NRR	r-value
*C. capitata**	*C. capitata**	412.7	0.160
C. capitata	*A. ludens*	1,694.7	0.223
A. ludens	*C. capitata*	331.2	0.172
*A. ludens**	*A. ludens**	1,424.2	0.238

Note: Starred (*) entry indicates original l_x and m_x values for each species. Using eqn. (11.18) yields an interaction value of 0.0027.

Since both sexes are increasing at the same rate in the population, the exponential terms cancel, yielding

$$SR = \frac{s l_x^M}{l_x^F} \qquad (11.20)$$

where s is the sex ratio at birth and l_x^M and l_x^F denote the male and female survival schedules, respectively. Thus the sex ratio at age x is simply the ratio of the male and female survival rates to this age adjusted by the sex ratio at birth.

S 27. Populations can experience "transient" stationarity

Transient stationarity is when birth and death rates are equal for populations in the process of convergence to a stable, stationary status (Rao and Carey 2018). For the data in fig. 11.21, the initial age structure is based on the US population in 2000, the fertility rates are based on a standard age-specific fertility schedule in humans scaled to a net reproduction of 1.0, and the age-specific survivorship schedule is based on the female rates in 2006. N(t + 1)/N(t) is the ratio of the number in the population at time t + 1 and the number at time t. Frequency, in the inset, refers to the frequency distribution of the population at each age. Point A corresponds to the starting growth rate, point B is when the growth rate first reaches replacement level (i.e., transient stationarity), and point C is when the growth rate is constant at zero (i.e., fixed stationarity). For these data, replacement levels of growth required approximately 40 years from the start (i.e., A to B) and another 60 years to become fixed (i.e., B to C). Note, also, the small oscillations around stationarity after B as the age structure converges to C.

S 28. Effect on growth rate of female migration

How much does the intrinsic rate of growth decrease if 80% of all 0- to 7-day-old (newly emerged) adult Drosophila emigrate to new regions? This can be solved by taking the difference between the following two equations:

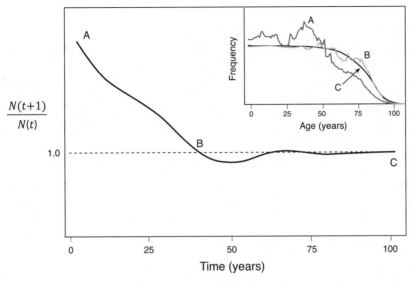

$$\frac{N(t+1)}{N(t)}$$

FIGURE 11.21. Change in growth rate in a hypothetical population that is projected forward 100 years (with age structure inset). See text for description. (*Source*: Rao and Carey 2017)

$$0.2\sum_{\alpha}^{\alpha+7} e^{-rx}l_x m_x + \sum_{\alpha+8}^{\omega} e^{-rx}l_x m_x = 1 \qquad (11.21)$$

$$\sum_{\alpha}^{\beta} e^{-rx}l_x m_x = 1 \qquad (11.22)$$

The difference gives the extent to which r is reduced when the contribution of young females (0 to 7 days postemergence) is reduced by 80%. (Fruit fly version of Keyfitz and Beekman 1984, 88, problem 22.)

S 29. Person-years ever lived

There were 1,000 births in a population in the year 1600 and 50,000 in the year 1900. If births were growing exponentially during the period with a constant rate of increase, how many people lived over the period? To solve,

$$50,000 = 1000 e^{300r} \qquad (11.23)$$

thus

$$r = 0.01304 \qquad (11.24)$$

Total births are given by

$$\text{births} = \int_0^{300} B(t)\,dt \qquad (11.25)$$

$$= \int_0^{300} 1000\,e^{0.01304t}\,dt \qquad (11.26)$$

So the total persons who lived in this 300-year period, P_{300}, is estimated as

$$P_{300} = 3{,}758{,}000 \qquad (11.27)$$

(From Keyfitz and Beekman 1984, 60, problem 43.)

S 30. Tempo effects

A tempo effect is an inflation or deflation of the period incidence of a demographic event (e.g., births, marriages, deaths) resulting from a rise or fall in the mean age at which the event occurs. Bongaarts and Feeney (2008) illustrated that a tempo effect causes a shift in death rates with the following hypothetical case. Consider a stationary population with a life expectancy at birth of 70 years. Suppose the exact age of death of each individual is predetermined until the invention of a "life extension" pill that adds 3 months to the life of any person who consumes it. If everyone in the population takes this pill on January 1 of year T, there will be no deaths during the first 3 months of the year. The number of deaths in year T will fall by 25%, and the mean age at death will rise from 70 to 70.25 years. Because the pill's effect is the same at all ages, the level of the force-of-mortality function is also reduced by 25%, and the age to which each value of the function is attached increases by 0.25 years. This decrease in value of the force-of-mortality function, together with the shift to older ages, causes life expectancy at birth, as conventionally calculated, to increase to ≈ 73 years for year T. A full description of tempo effects, including life table distortion and turbulence on longevity estimation, is given in Barbi et al. (2008).

S 31. Birth rates in stable versus stationary populations

How do the intrinsic rates of birth (b) and death (d) differ between stable and stationary populations? The birth rates for stable and stationary populations are $(\Sigma e^{-rx}l_x)^{-1}$ and $(\Sigma l_x)^{-1}$, respectively. Since $(b-d)=r$, the death rate, d, in the stationary population equals the birth rates. The death rate in the stable population is simply $d=(b-r)$. (Variation on Keyfitz and Beekman 1984, 56, problem 6.)

S 32. Super-exponential growth

Cohen (1995) notes that at certain times in history, humans have experienced super-exponential growth, where super-exponential growth is a form of mathematical titration—a chained power involving n numbers a—where

$$^{n}a = a^{a^{\cdot^{\cdot^{\cdot^{a}}}}} \qquad\qquad (11.28)$$

(from Bromer 1987).

This denotes n copies of a combined by exponentiation, right to left. This mathematical titration can be compared to addition (a + n), multiplication (a × n), and exponentiation (a^n). As an example of titration, first consider the outcome of a simple exponentiation where a = 3 and n = 2; then

$$^{2}3 = 3^3 = 27$$

However, if the first exponent is exponentiated, as in titration (i.e., n = 3), then

$$^{3}3 = 3^{3^3} = 3^{27} = 7{,}625{,}597{,}484{,}987$$

Increasing n by only 3 (i.e., a = 3 and n = 6) yields staggeringly large numbers, that is,

$$^{6}3 = 2.65912 \times 10^{36{,}305}$$

Note that the super-exponential population rates of growth mentioned by Cohen (1995) are extraordinarily small relative to any of the rates shown in the examples above.

S 33. Mean age of death in a growing population

Whereas the mean age in a stationary population is $\dfrac{1}{\mu}$, the mean age in a population at rate r with constant mortality (μ) is $\dfrac{1}{(\mu + r)}$. For example, if $\mu = 0.0125$ and the growth rate is 1% (r = 0.01), then the mean age in this hypothetical population equals 1/0.0225, or 44.4 years. This value is approximately half the mean age of death for individuals, which is 1/0.0125 = 80 years. (From Keyfitz and Beekman 1984, 90, problem 28.)

S 34. Equality in stable populations

Demographers James Vaupel and Francisco Villavicencio (2018) demonstrated that the "life-lived and -left" concepts, originally derived for use in estimating the age structure of stationary populations (see chapters 2 and 9), can also be used to estimate age structure in stable populations. Assume all ages are unknown, but individuals are followed through death. Then one can derive the underlying, although unknown, survival schedule of the population from a cohort perspective as

$$l(x) = \frac{N(x,t)}{N(t)} = \frac{D^+(x,t) + rN^+(x,t)}{D^+(0,t) + rN^+(0,t)} \tag{11.29}$$

as well as the unknown age structure at initial time t,

$$c(x,t) = \frac{N(x,t)}{N(t)} = e^{-rt}\frac{D^+(x,t) + rN^+(x,t)}{N(t)} \tag{11.30}$$

Vaupel and Villavicencio (2018) use their model to estimate the survival curve and growth rate of a historical Swedish population and suggest several applications, including estimating age structure at the time of a census and extending the general concept to multistate populations (also see Carey, Silverman, and Rao 2018).

Statistical

S 35. Anscombe's quartet

Anscombe (1973) showed four fictitious data sets (fig. 11.22), each containing 11 (x, y) pairs, with the same statistical output, including mean of the x's (9.0), mean of the y's (7.5), regression line (y = 3 + 0.5x), sum of squares (110.0), and the correlation coefficient ($r^2 = 0.667$). Plot (a) matches what one would expect. Plot (b) shows a smooth, curved relationship that is not captured if summarized as a linear regression. Plot (c) shows how a single outlier can lead to a regression line that is misleading, because the slope of the line does not match the linear line of all other points. And plot (d) shows a regression line that is almost perpendicular to the true pattern that is apparent in the data. These plots illustrate that statistical metrics alone (e.g., standard deviation; mean; regression coefficient) may be not only inadequate but also misleading.

S 36. Descriptive statistics

Computation of basic descriptive statistics is the starting point for virtually all demographic analyses. These include the mean, standard deviation, and mode, examples of which are given in fig. 11.23a–b, for the death distributions of two species of fruit flies. These death distributions correspond to the d_x schedule in the life table (normalized) and the probability density function (pdf) in formal statistics.

S 37. Comparing lifetime longevity using the t-test

To test whether the mean of a sample (identical to expectations of life at age 0, e_0), $\bar{x}_1$ (= 36.1 days), differs from the mean of another sample, $\bar{x}_2$ (= 48.6 days), requires the calculation of the standard error of the mean (SEM):

$$SEM = \sqrt{\frac{var_1}{N_1} + \frac{var_1}{N_1}} \tag{11.31}$$

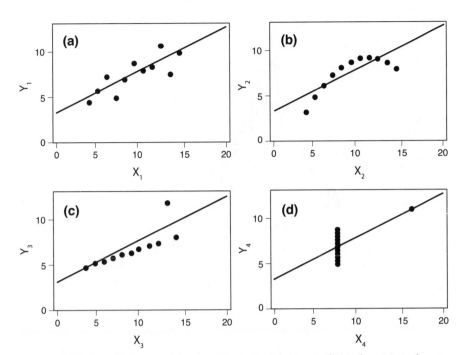

FIGURE 11.22. Anscombe's quartet is four data sets with identical simple statistical properties: mean, variance, correlation, and linear regression (Anscombe 1973). Visual inspection immediately shows how their structures are different.

$$= \sqrt{\frac{(15.1)^2}{1000} + \frac{(18.2)^2}{1152}} \qquad (11.32)$$

$$= 0.718 \qquad (11.33)$$

The t statistic is then computed as

$$t = \frac{\bar{x}_1 - \bar{x}_2}{SEM} \qquad (11.34)$$

$$= \frac{12.5}{0.718} \qquad (11.35)$$

$$= 17.4 \qquad (11.36)$$

(Lee 1992).

This test statistic gives a 2-tailed test p-value of $p < 10^{-5}$; in other words, the differences in the mean ages of death for these two cohorts is highly significant.

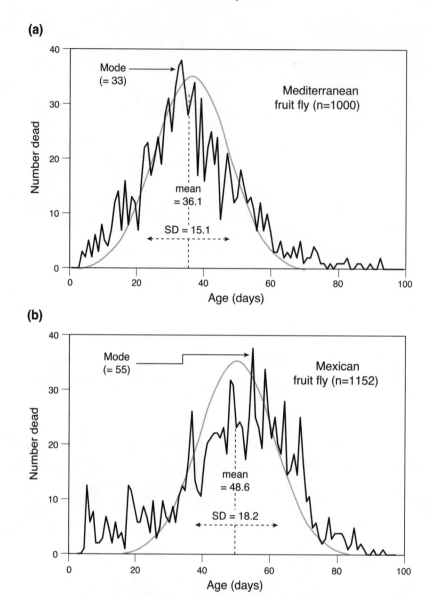

FIGURE 11.23. Death distributions and descriptive statistics for (a) the Mediterranean fruit fly (Carey et al. 1998) and (b) the Mexican fruit fly (Carey et al. 2005). Light gray curve shows the normal distribution for perspective.

S 38. Conditional probability of Alzheimer's disease (AD)

What is the probability that, given a person has a particular disease, the person is female? What is the probability that, given a person is female, she has the disease? These questions are conditional probabilities that can be understood by creating a 2×2 table of all possible combinations. For example, suppose data show that out of 237 males (M) and 425 females (F), 33 of the men and 97 of the women have Alzheimer's disease (A) while the remaining 532 are healthy (H).

	Male M	Female F	
Healthy H	n(MH) 204	n(FH) 328	n(H) 532
Alzheimer's A	n(MA) 33	n(FA) 97	n(A) 130
	n(M) 237	n(F) 425	N 662

The conditional probability that, given a person has the disease, the person is female, is given by

$$P(A|F) = \frac{n(FA)}{n(A)} = \frac{97}{130} = 0.746 \qquad (11.37)$$

The conditional probability that, given a person is female, she has Alzheimer's disease, is given by

$$P(F|A) = \frac{n(FA)}{n(F)} = \frac{97}{425} = 0.228 \qquad (11.38)$$

Restated, these results reveal that in the sample population there is around a 3-in-4 probability a person with AD is a female and nearly a 1-in-4 probability that a female has AD.

S 39. Peak-aligned averaging

Peaks occur in a variety of demographic contexts, including both age-specific reproduction and age-specific mortality. Averaged peaks tend to "flatten out" because of the heterogeneity (variability) in the timing of the peaks for individual cohorts. What technique can be used to preserve the peaks? Müller and his colleagues (2001) used the following technique to avoid biases associated with averaging peaks across cohorts. Assume that the hazard rate of the jth cohort has a peak at location θ_j. To estimate θ_j, the first step is to estimate the hazard rate for the jth cohort, $j = 1, \ldots, N$ (see methods in chapter 3). Next, the location $\hat{\theta}_j$ of the peak of the estimated hazard rate is obtained. This provides an estimate of the cohort peak location θ_j. The estimated average peak location for all cohorts is obtained by averaging all individual cohort peak locations:

$$\hat{\theta} = \frac{1}{N} \Sigma_{j=1}^{N} \hat{\theta}_j \tag{11.39}$$

Scaled alignment of the hazard rates of individual cohorts to the average peak location $\hat{\theta}$ is then achieved by transforming the time coordinate t for each of the N cohorts as follows:

$$t_j' = \frac{t\hat{\theta}}{\hat{\theta}_j} \tag{11.40}$$

for the data of the jth cohort. This time-scale transformation maps all individual peak locations θ_j. After transforming the time scale (originally in days) separately for each cohort, the methods outlined in chapter 3 (Mortality) are used again to obtain the final peak-aligned hazard rate estimates (see fig. 11.24a–b).

S 40. Correlation of ages of death for twins

The conventional wisdom in society in general and in gerontology in particular is that long-lived persons tend to have long-lived relatives (Christensen et al. 2006). Although being the child of long-lived parents will increase a person's odds for achieving older ages, the correlation is not particularly strong. McGue and his colleagues (1993) concluded from the analysis of age of death in twins that longevity is "moderately heritable" (see fig. 11.25). Although not detectable from the scatter plots shown in fig. 11.25a and 11.25b, Hjelmborg and his colleagues (2006) found that the genetic influences on life span, for both monozygotic and dizygotic twins, are minimal prior to age 60 but increase thereafter.

S 41. Two persons having same birthday in a group of 80 is near 100%

What are the chances that at least two students in a class of 80 students will have the same birthday? First, we give the answer, which is a 99.99% chance. This nonintuitive result (the "birthday paradox") can be understood by noting that

- The first person has a 100% chance of a unique number (of course)
- The second has a $(1 - 1/365)$ chance (all but 1 number from the 365)
- The third has a $(1 - 2/365)$ chance (all but 2 numbers)
- The seventy-ninth has a $(1 - 78/365)$ chance (all but 78 numbers)
- The eightieth has a $(1 - 79/365)$ chance (all but 79 numbers)

The probability of at least two persons *not* having the same birthday, p(different), is then

$$p(\text{different}) = 1 \times \left(1 - \frac{1}{365}\right) \times \left(1 - \frac{2}{365}\right) \cdots \times \left(1 - \frac{79}{365}\right) \tag{11.41}$$

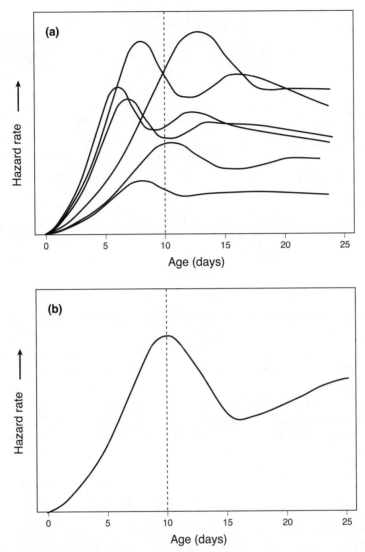

FIGURE 11.24. Illustration of peak-aligned averaging in Medfly cohorts reconstructed from Figs. 1–3 in Müller et al. (2001): (a) hazard rates in multiple cohorts; (b) peak-aligned estimated hazard rates from 0 to 25 days.

Given that

$$e^x \approx 1 + x$$

then

$$1 - \frac{1}{365} \approx e^{-1/365}$$

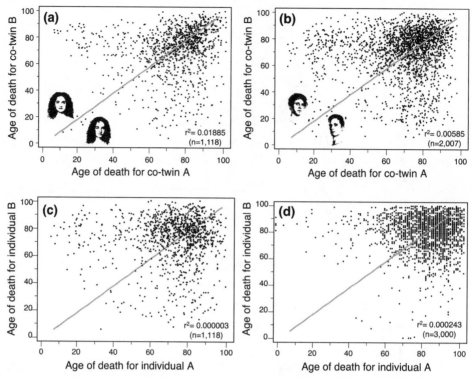

FIGURE 11.25. Scatter plots of paired life spans in (a) Danish monozygotic (identical) twins (both sexes); (b) Danish dizygotic (fraternal) twins (both sexes); (c) randomly paired Danish monozygotic twins (i.e. unrelated individuals); and (d) computer-generated deaths of paired hypothetical individuals subject to death rates of 2015 US females. (*Source*: Kaare Christenson, with permission)

$$p(\text{different}) = e^{-(1+2+\ldots 79)/365} \tag{11.42}$$

$$= 0.0001 \tag{11.43}$$

Thus the probability of at least two persons in a group of 80 having the same birthday is $(1 - 0.0001)$, or 0.9999 ($= 99.99\%$). A general formula (close approximation) for the number (n) needed in a sample with m probability that two or more will match out of T possibilities is given as

$$n \approx \sqrt{-2\ln(1-m)} \times \sqrt{T}$$

(for example, see Mathis 1991; Brink 2012).

Epidemiological

S 42. Prenatal famine exposure

The timing of exposure to a stress such as famine in relation to the stage of pregnancy may be of critical importance for later health outcomes independent of intermediary life conditions. The circumstances of the Dutch famine (Hunger Winter) of 1944–1945 at the end of World War II, which resulted in civilian starvation, have been used to examine the relationship between nutrition during pregnancy and birth outcomes and morbidity later in life (Ekamper et al. 2014). The Dutch famine resulted from an embargo on the transport of food supplies imposed by the German occupying forces, and the demographic consequences experienced by pregnancy cohorts depended on the timing of the famine exposure (fig. 11.26). For those with prenatal exposure, there was an increase in mortality in a national birth cohort of men. For those with a famine exposure in early gestation, there was an increase in mortality; but there was no increase in mortality for those exposed only in the late gestation period. These results suggest that the timing of exposure in relation to the stage of pregnancy may be of critical importance for determining later health outcomes and that intermediary life conditions do not modify or mediate the relationship between famine exposure and mortality.

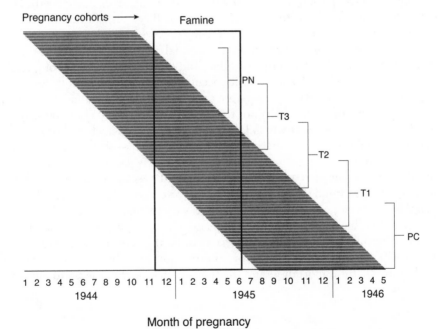

FIGURE 11.26. Classification of famine exposure categories by pregnancy cohorts. PN: exposed in the immediate postnatal period; T3: exposed in the third trimester of pregnancy; T2: in the second trimester; T1: in the first trimester; PC: exposed prior to the estimated date of conception (Ekamper et al. 2014). Daily calories per day averaged around 1,500 prefamine, 600 during the famine, and 2,000 postfamine.

S 43. Direct evidence of demographic selection

One of the few papers in the literature that presents strong evidence of demographic selection in human populations is the study by Ho and his colleagues (2017). This study examined mortality in a population-representative sample of residents of Aceh, Indonesia, for five years after the Indian Ocean earthquake and tsunami in 2004. The concept and one of the key results are shown in fig. 11.27. Among males mortality at all ages is higher for those who, at the time of the tsunami, were living in communities that were unaffected by the disaster. The gap between communities increases steadily with age, just before age 50, and reaches nearly 10 percentage points by age 75. The authors of this study suggest that this is evidence that males who were living in tsunami-affected areas and survived the disaster were positively selected and that the heterogeneity of the population changed. As a result, males who were living in tsunami-affected areas had lower subsequent mortality rates.

S 44. Quantal analysis in bioassays

Two models used in the analysis concerned with quantal response in bioassays have demographic components. The first is probit analysis used in toxicology, developed by Finney (1964). As Carey (1993) notes, the statistics of the life table are identical with those of probit analysis if age is viewed as a dose of time in that (1) life table and probit analysis are both concerned with quantal outcomes (alive/dead); (2) the d_x schedule in the life table is the analog of the normalized distribution of deaths per increment of doses in probit analysis; (3) the log dose vs. mortality curve in a bioassay is typically presented as the cumulative normal distribution, which is the complement of the survival curve in the life table; and (4) it is assumed in probit analysis that the log dose at which 50% mortality occurs (L_{50}) represents both the mean dose and the median dose. The expectation of life at birth in the life table can be redefined as the exact mean age of death at birth of the cohort and is usually close to the median age of death (i.e., the age x at which $l_x = 0$). Thus the age at which half of the cohort is dead in a life table can be reconceived in terms of the LD_{50}—that is, people die of an "overdose" of time.

Another model considered here in the context of bioassays is Abbott's correction (Abbott 1925)—a formula derived for adjusting the outcome of insecticide bioassays due to the effects of natural mortality. Abbott's correction is essentially a double-decrement life table (competing risk) concept over a single age interval. The formula for determining what Abbott referred to as the effective kill rate (EKR) is $EKR = (A - B)/A$, where A and B denote the fraction of individuals that died in a control group (natural mortality) and the fraction that died in the treatment group, respectively.

Catastrophic

S 45. Napoleon's Grand Armée—March to Moscow

This scenario illustrates creative use of data and both design and graphic concepts for visualizing survival in both space and time (Tufte 2001). The French invasion of Russia led by Napoleon began on June 24, 1812, when Napoleon's Grand Armée

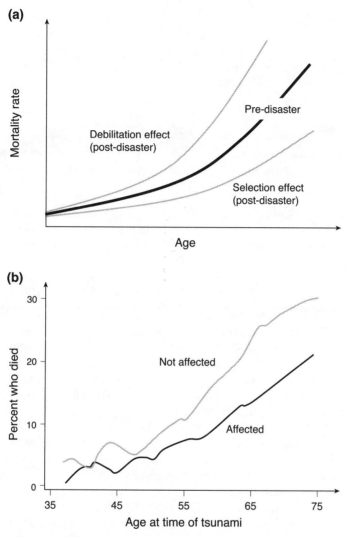

FIGURE 11.27. (a) Hypothetical mortality curves pre- and postdisaster showing debilitation and selection effects postdisaster. (b) Mortality in the five years after the 2004 tsunami, comparing the percentage of males who died in communities affected and unaffected by the tsunami (redrawn from Fig. 2 in Ho et al. 2017).

crossed the Neman River in Belarus. He returned with only 27,000 troops when he crossed back into Belarus at the Berezina River in November. The campaign ended on December 14, 1812. A total of 380,000 soldiers in the Grande Armée were killed and another 80,000–100,000 were taken prisoner (Zamoyski 2005; Mikaberidze 2007). Fig. 11.28 shows the survival curve for troops over this period; in the inset, the gray-shaded stream is the advance from Belarus to Moscow and the black-shaded stream depicts the retreat. The width of the streams is proportional to the number of troops remaining.

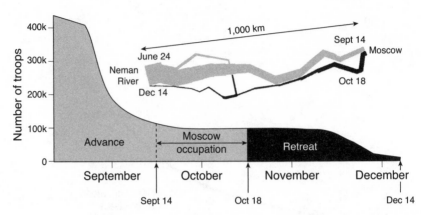

FIGURE 11.28. Troop attrition during Napoleon's march to Moscow in the French invasion of Russia of 1812. Inset: Charles Minard's famous cartographic depiction of numerical data on a map of Napoleon's disastrous losses suffered during the Russian campaign of 1812.

S 46. The Donner Party disaster—Mediation by age, gender, and kinship

In this scenario demographic data of a historical disaster uncovers several trends. The Donner Party disaster is one of the best-known stories in American lore (Wallis 2017). A total of 87 men, women, and children set out for California in wagon trains from Illinois in the summer of 1846, but because of unfortunate mishaps, poor decisions, and bad luck, they became trapped by early heavy snowfall in the eastern Sierra Nevada Mountains in November. Four months later only 48 of the original 87 members of the party arrived in California alive.

Grayson (1990) notes that the fate of the Donner Party members was mediated almost entirely by three factors that are evident in the data presented in table 11.3. First, there were age differences in survival such that the young and old died at higher rates than the middle-aged. Indeed, with one exception, no individuals of either sex beyond 49 years survived. Second, there were sex differences in mortality. Males not only died at roughly twice the overall rate as females, but many males also died earlier in the disaster period. Third, kinship size and association were important to survival; individuals not associated with a kinship group died at much higher rates than those who were associated with a group. There were some kinship groups in which no members died (Wallis 2017). Despite the conventional wisdom that women are the more frail sex, the Donner Party statistics and a wealth of other evidence (Widdowson and McAnce 1963; Widdowson 1976) shows that this is not the case—males typically are the first to succumb to extreme conditions. This is also consistent with the so-called male-female health-survival paradox, which is that men are physically stronger and have fewer disabilities, but they have substantially higher mortality at all ages compared with women (Wingard 1984; Oksuzyan et al. 2008, 2009).

S 47. Actuarial analysis of the Titanic disaster: Hypothesis testing

The passenger ship Titanic collided with an iceberg on her maiden voyage during the night of April 14, 1912. Less than three hours later the ship sank, resulting in the loss of 1,517 lives, or more than two-thirds of the 2,207 passengers and crew. In order

Table 11.3. Demographic characteristics and mortality of the Donner Party, 1846–1847

Characteristic	Number	Deaths per number at risk	Mortality rate	Mortality ratio
Number in party	90	42/90	47%	—
Age				
Unknown	2	2/2	100%	-
< 5	19	11/19	58%	6.6
6–14	21	2/21	10%	1.0
15–34	34	16/34	47%	3.3
35 >	14	11/14	79%	8.4
Sex				
Male	55	32/55	58%	2.0
Female	35	10/35	29%	1.0
With kin group				
Yes	72	27/72	38%	1.0
No	18	15/18	83%	2.0

Source: Table 1 in McCurdy (1994).
Note: Mortality ratio computed relative to the lowest mortality in the group.

to test hypotheses regarding who survived the disaster, Frey and his colleagues (2009) modeled surviving the Titanic disaster as a tournament with-risk averse contestants divided into three categories: economic, natural, and social. Their analysis supported their hypotheses that there was higher survival for first-class passengers, for crew members, for deck crew versus engine crew, for people in their prime, for women of reproductive age, for women with children, and for children versus men. There was no evidence of higher survival for passengers traveling alone versus in groups, nor for British subjects versus other nationalities. Interesting perspectives on victimization in natural disasters, including those on the Titanic disaster, include the papers by Dudasik (1980) and Rivers (1982).

Group 3: Familial, Actuarial, and Organizational

Family

S 48. Strategy for choosing the "best" spouse

What is the best strategy for maximizing the chances of finding the richest man to marry? Suppose a princess wants to marry the richest man in the land. Although she can interview each man to ask questions about his wealth status, he is out of the running once she decides to interview the next man. This scenario, applied to a hypothetical marriage strategy, is known in statistics as the "secretary problem," which is in a set of generic problems referred to as "optimal stopping problems." In the case of the secretary problem, the question is how to devise a strategy for assessing a pool of applicants but having to decide on-the-spot whether to hire or move to the next interview, where all dismissed secretaries interviewed will be hired by rival companies. It turns out that the optimal stopping rule can be explained for a wide range of similar questions,in this case for maximizing the chances of marrying the richest man, as follows. First, interview $\left(\frac{1}{e} \times n\right)$ men $= (0.37 \times n)$ men, where e is the exponential number 2.718 . . . and n is the number in the pool. Second, out of the remaining $0.63 \times n$ men, pick the next man who exceeds the richest in the first set of $(0.37 \times n)$ men. This strategy gives the princess a 37% chance of marrying the richest man in the pool. The derivation and proof of this result are given in Bruss (2000).

S 49. Stopping rules for number of girls

A couple decides to have four children. What is the probability that there will be 0, 1, 2, 3, and 4 girls? Using the binomial theorem yields the formula

$$P[x = k] = \binom{n}{k} p^k (1-p)^{n-k} \tag{11.44}$$

$$= \left(\frac{n!}{k!(n-k)!}\right)[p^k(1-p)^{n-k}] \tag{11.45}$$

where p, n, and k denote the probability of having a girl, the number of children, and the number of girls, respectively. The complete distribution is given in table 11.4. Slightly over a third (37.5%) of all families with four children will have two girls (and thus two boys) and half will have either one girl or three girls. One of eight families will have either no girls (all boys) or all girls.

S 50. Children already born and left to be born

Does the identity "life lived and left" apply to the parity progression life table and, if so, how? Known as Carey's equality, in which the fraction of a cohort age x equals

Table 11.4. Probability distribution of k girls in a family of n = 4 children

Number of girls	Probability of k girls
k	P
0	0.0625
1	0.2500
2	0.3750
3	0.2500
4	0.0625
	1.0000

Note: The probability of having a girl is denoted p = 0.5 and of having a boy is (1 − p) = 0.5.

(a)

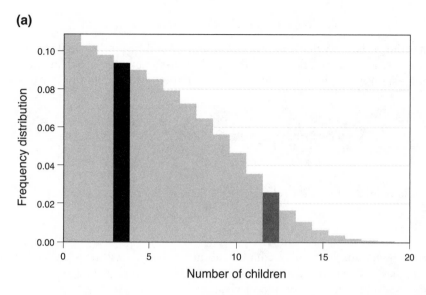

(b)

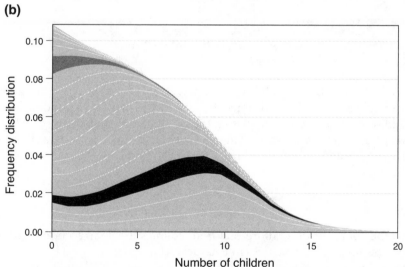

FIGURE 11.29. Illustration of the Carey equality applied to the parity progression table: (a) fraction of individuals in the parity progression table for French-Canadian women who are parity 3 and parity 12; (b) parity-specific fraction of individuals who will bear either 3 or 12 more children. (*Data source*: LeBourg et al. 1993)

the fraction of a cohort with x years to live (Vaupel 2009), the concept can also be applied to parity progression life tables (Feeney 1983) as shown in fig. 11.29a–b. These graphs show that the fraction of the cohort of French-Canadian women who bore 12 children equals the fraction of women who still have 12 children to bear, given their current parity.

S 51. Family size with boy and girl target numbers

What is the probability that a family will have $n = 7$ children if the couple has children until they obtain exactly 2 boys (b) and 2 girls (g)? Assuming the probabilities of having either a son or a daughter are equal at $p = q = 0$, then the formula to solve this is

$$P(N = n) = \frac{(n-1)!}{(g-1)!(n-g)!} p^{g-1} q^{n-g} p + \frac{(n-1)!}{(b-1)!(n-b)!} q^{b-1} q^{n-b} q \quad (11.46)$$

(Formula from Keyfitz and Beekman 1984, 126, problem 16.)

S 52. Marriage duration

What is the number of years a couple who never divorce will be together until one of them dies if they married at 25 (bride) and 30 (groom) years of age? This solution requires summing the age-specific survival probabilities, given by

$$_1p_{25}^f \, _1p_{30}^m + _2p_{25}^f \, _2p_{30}^m + _3p_{25}^f \, _3p_{30}^m + \cdots$$

where $_np_{25}^f$ denotes the probability of a 25-year-old female surviving n additional years beyond age 25 and $_np_{30}^m$ denotes the probability of a 30-year-old male surviving n additional years beyond 30. Using the 2015 US sex- and age-specific life table rates yields a value of 42.9 years. Thus the average couple marrying at these specified ages (i.e., 25 and 30 years) will remain married for nearly 43 years before one or the other dies.

S 53. Contraceptive effectiveness

What is the contraception failure rate if all couples use contraception with equal effectiveness (e) and all have the same natural fecundability (f_n)? The monthly risk of conception is $(1-e) f_n$ (Bongaarts and Potter 1983). The probability of not failing in one month is $1 - [(1-e)f_n]$ and the one-year failure rate is $F = 1 - (1 - [(1-e)f_n]^{12}$.

S 54. Likelihood of golden wedding anniversary

What is the probability of celebrating a golden wedding anniversary and how frequently does the husband die before the wife? Consider 500 couples in which 25-year-old women married 40-year-old men. The results of simulations (fig. 11.30) show that only 37 of 500 couples survive to celebrate their fiftieth wedding anniversary, and the husband dies first in 31 of these 37 marriages (i.e., 84%). The average marriage lasted 30.4 years and the husband died first 80% of the time. Men were widowers for 13.6 years before they died and women were widows for 26.7 years before they died.

FIGURE 11.30. Distribution of deaths from time of marriage for 500 paired couples where the wife and husband were 25 and 40 years old, respectively. Mortality schedules are from the Berkeley Mortality Database (US, 2015).

S 55. What is a life worth? Working life estimation in children

The idea of assigning a dollar value to a person's life seems odd to some people, distasteful to some, and repugnant to still others (Peeples and Harris 2015). Although demographers study life course concepts (Herzman 1999; Billari 2003), assigning a dollar value to a person's life is not part of the pedagogy. However, for tort (civil wrongs) attorneys, this question is routinely addressed in wrongful death or injury cases, such as motor vehicle accidents or medical malpractice. Estimating the earning capability of a minor child requires information about the likely educational attainment of the child (Gill and Foley 1996) and a determination of the wage-earning profile associated with each level of education (Tamborini et al. 2015).

For example, in a case described by Spizman (2016), a child, Michael Boone, age 7, was born with a traumatic brain injury on September 11, 2012, and his parents are seeking compensation for wrongful injury based on what Michael would have earned had he been normal. Information used to estimate lifetime earning potential includes socioeconomic and educational information about the family, which is then used to compute the probabilities of a person's level of education. These probabilities, in turn, are used to estimate the probability of lifetime earnings for an individual in the case of childhood impairment or death or in wrongful death suits.

$$\text{Predicted Lifetime Earnings} = \sum_{i=1}^{n} S_i P_i$$

Table 11.5. Probability of earnings of male (Michael Boone) in hypothetical case study

Education level	Probability of attainment	Lifetime earnings at the education level (million $)	Earnings adjusted for probability
	P_i	S_i	$P_i \times S_i$
< High school	10.2%	$1.13	$115,000
High school diploma	48.1%	1.54	741,000
Some college	9.5%	1.76	167,000
Bachelor's degree	27.1%	2.43	659,000
Advanced degree	5.1%	3.05	156,000
		TOTAL	$1,838,000

Source: Adapted from Spizman (2016).

where S_i and P_i denote the probability of attaining education level i and lifetime earnings at education level i, respectively. In this particular case Michael's lifetime earnings were estimated to be slightly over $1.8 million (table 11.5).

Actuarial science

Most of the entries in this subsection are based on information contained in *Basic Life Insurance Mathematics* by Norbert (2002), including the storytelling progression from simple interest computation to the concepts of a mutual fund. We include this content as our modest attempt at introducing the actuarial sciences, an area primarily concerned with life insurance, pensions, and health care, into biodemography. Professional actuaries typically publish in journals separate from demography, such as the *British Actuarial Journal*, the *North American Actuarial Journal*, the *Journal of Insurance Issues*, and the *Journal of Risk & Uncertainty*.

S 56. Yield on deposit, P, at interest rate, r

What will be the 10-year yield of a $100 investment, P, at 5% interest rate (r = 0.05)? This yield, A, can be computed as

$$A = 100(1 + 0.05)^{10}$$

$$= \$163$$

(11.47)

S 57. Payments for financial goal

Mary, age 25, would like to have $200,000 when she is 50 to invest for her retirement. At 5% interest, how much does she need to deposit monthly to reach this goal? Let P denote principal (original amount invested or borrowed), A the account balance, r the interest rate (written as a decimal), Y the number of years, PMT the regular payment amount, and n the number of payments per year (n is

12 for monthly payments). If P is invested at interest rate r for Y years, the balance is given by

$$A = P(1+r)^Y \tag{11.48}$$

If a person has a savings account earning r interest and deposits the PMT amount n times a year and continues this for Y years, the amount of money in the account, A, is given by the formula

$$A = PMT \left[\frac{\left(1+\dfrac{r}{n}\right)^{nY} - 1}{\dfrac{r}{n}} \right] \tag{11.49}$$

If a person borrows P at an interest rate of r and pays it back by making n equal payments a year for Y years, the size of the payment is given by the formula

$$PMT = \frac{P\left(\dfrac{r}{n}\right)}{\left[1 - \left(1+\dfrac{r}{n}\right)^{-nY} \right]} \tag{11.50}$$

Solving for PMT with respect to A in eqn. (11.49) gives the monthly payment amount as PMT = \$1,169.18.

S 58. Monthly mortgage payments for 30-year mortgage

At age 35, Mary decides to borrow \$250,000 to buy a house. With a 30-year mortgage at 5% interest, what are her monthly payments to retire this loan when she is 65? Solving for PMT in eqn. (11.50) yields monthly payments of \$1,342.05.

S 59. Payments to retire credit card debt

Mary runs up credit card debt of \$5,000 at an interest rate of 18%. How much does she need to pay per month to retire this debt in 4 years if she makes no more charges? Solving for PMT in eqn. 11.50) reveals that she needs to pay \$146.87 monthly to retire this debt in 4 years. She will have paid \$7,049.76, or over 40% in interest payments.

S 60. Bank savings for retirement

Upon celebrating her fiftieth birthday, Mary decides to invest money for her retirement at age 70. Her initial strategy is to deposit a capital amount of $S_0 = \$100,000$

into a savings account and withdraw the entire amount, with earned compound interest, in 20 years. The account bears interest rate $i = 5\%$ per year. Assuming a 15% probability that Mary will die before 70, what is the expected amount at her disposal after 20 years? In one year her investment will increase to

$$S_1 = S_0 + S_0 i = S_0 (1 + i) \tag{11.51}$$

In two years it will increase to

$$S_2 = S_0 (1 + i)^2 \tag{11.52}$$

and in 20 years it will have accumulated to

$$S_{20} = S_0 (1 + i)^{20} \tag{11.53}$$

In 20 years her $100,000 initial investment will have grown to

$$S_{20} = \$100,000 (1 + 0.05)^{20}$$

$$S_{20} = \$265,330$$

In light of the 15% probability that Mary will die before 70, the expected amount at her disposal after 20 years is the product of her likelihood of survival to age 70 and her accumulated cash:

$$0.85 \times S_{20} \tag{11.54}$$

or

$$0.85 \times 265,330$$

$$= \$225,530$$

S 61. A small-scale mutual fund concept

Having rethought her retirement strategy, Mary decides to make arrangements with her friends, Emily and Olivia, also both 50 years old. Each of the three deposit $100,000 in the savings account, and those who survive to 70 will then share the total accumulated capital $(3 \times S_{20})$ equally. What is Mary's expected amount at her

Table 11.6. Possible outcomes of saving scheme with three participants

	Name			L_{70}	$3S_{20}/L_{70}$	Probability
	Mary	Emily	Olivia			
1	+	+	+	3	S_{20}	$(0.8)(0.8)(0.8) = 0.512$
2	+	+	−	2	$1.5S_{20}$	$(0.8)(0.8)(0.2) = 0.128$
3	+	−	+	2	$1.5S_{20}$	$(0.8)(0.2)(0.8) = 0.128$
4	+	−	−	1	$3S_{20}$	$(0.8)(0.2)(0.2) = 0.032$
5	−	+	+	2	$1.5S_{20}$	$(0.2)(0.8)(0.8) = 0.128$
6	−	+	−	1	$3S_{20}$	$(0.2)(0.8)(0.2) = 0.032$
7	−	−	+	1	$3S_{20}$	$(0.2)(0.2)(0.8) = 0.032$
8	−	−	−	0	—	$(0.2)(0.2)(0.2) = 0.008$

disposal after 20 years? The possible outcomes of a saving scheme with three participants (where + and − refer to survival to or death before age 70, respectively) is shown in table 11.6.

The per survivor amount at age 70 is $3S_{20}/L_{70}$. Mary now has the following possibilities:

- With probability 0.512 she and both of her friends survive to age 70 and they will each possess $3S_{20}/3$ (= \$265,330).
- With probability $2 \times 0.128 = 0.256$ she and one other friend survives to age 70 and she will split ($3S_{20}/2$) with the surviving friend (= \$397,995 each).
- With probability 0.032 she is the sole survivor and she acquires the total savings of $3S_{20}$ (= \$795,989).
- With probability 0.200 she dies and gets nothing.

This strategy is superior to the earlier one with separate savings because, at the very least, she will receive \$265,530 if she survives to age 70 as in the sole investor case. Mary's expected amount at her disposal after 20 years is

$$(0.512 \times S_{20}) + (0.256 \times 1.5 \times S_{20}) + (0.008 \times 3 \times S_{20})$$

$$= 0.92S_{20} = 0.92 * \$265,330 \qquad (11.55)$$

$$= \$244,103$$

This is nearly \$20,000 more than if she were the sole investor.

S 62. A large-scale mutual fund concept

Having thought through the advantages of the small-scale mutual fund concept over the sole investor concept, Mary now decides to extend this concept to a large number of participants. We assume that a total of L_{50} persons, all 50 years old, agree to join an investor's consortium similar to the one described for the three. Then the total savings after 20 years is $L_{50}S_{20}$. For example, if $L_{50} = 1{,}000$ persons, then the total savings will be \$265.33 million. What is the expected amount at Mary's disposal after 20 years? By the law of large numbers, the proportion of survivors tends to the individual survival probability of 0.80. Therefore, as the number of participants increases, the individual share per survivor tends to

$$\frac{1}{0.8}S_{20} \tag{11.56}$$

Mary now is faced with the following situation:

- With probability 0.8 she survives to 70 and gets $\dfrac{1}{0.8}S_{20}$ (= \$331,662).
- With probability 0.2 she dies before 70 and gets nothing.

The expected amount at Mary's disposal after 20 years is

$$0.8 \times \frac{1}{0.8}S_{20} = S_{20} = \$265{,}330 \tag{11.57}$$

Thus, the *bequest* mechanism of the mutual fund scheme has raised Mary's expectations of a future pension to what it would be with the individual savings contract if she were immortal. This is what is meant by the concept that "insurance risk is diversifiable"; that is, the risk can be eliminated by increasing the size of the portfolio.

S 63. Burden in old-age pension

What is the old-age pension burden considering the ratio of individuals over 65 relative to individuals of working age from 20 to 65 years (Keyfitz and Beekman 1984)? How does this burden differ between a stationary population (growth rate, $r = 0$) and a population increasing at an annual rate of 2% ($r = 0.02$)? The old-age burden can be expressed as

$$\text{old-age burden} = \frac{\sum_{66}^{\omega} e^{-rx} l_x}{\sum_{20}^{65} e^{-rx} l_x}$$

where r and l_x denote the intrinsic rate of increase and survival to age x, respectively. Using the 2015 US female life table data yields old-age burdens of 0.1845 and 0.3808 for annual growth rates of 2% and 0%, respectively. This corresponds to 5.4 and 2.6 working-age individuals per pension individual over age 65.

Organization

S 64. Rejuvenating organizations

What is the best recruitment strategy to keep an organization young? Organizations including firms, universities, societies, political bodies, teams, or national academies seek ways to rejuvenate. Dawid and his colleagues (2009) demonstrated that the best way to keep an organization young is through a mixed strategy of recruiting both young and old (fig. 11.31), and contrary to intuition, recruiting those of middle age is the least effective strategy for maintaining a younger age structure.

S 65. Gender parity in university hiring

How much time would be required for the University of California (UC) system to reach gender parity in the faculty (50% women) if the replacement strategy for attrition (deaths; movement; retirement) was to only hire 30-year-old assistant professors? The answer to this question is dependent upon the current age structure (fig. 11.32a), which is a 7:3 ratio of men to women faculty, and the fraction of new hires that are women (fig. 11.32b). Unpublished simulations by J. R. Carey based on the 2013 age distribution of UC faculty revealed that half a century would be required to attain true gender parity if 50% of the new hires were women, and it would take approximately 15 years if 100% of new hires were women.

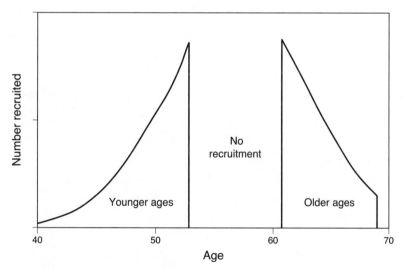

FIGURE 11.31. Optimal recruitment density for keeping an elite society young (redrawn from Dawid et al. 2009). Also see Leridon (2004).

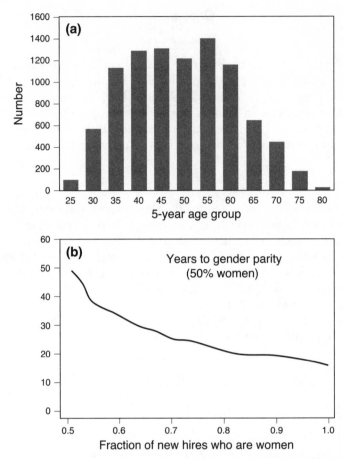

FIGURE 11.32. (a) Age structure in 2013 of the approximately 9,500-member faculty (70% men; 30% women) in the 10-campus University of California system; (b) time to parity if all new hires are 30 years old (unpublished simulation by J. R. Carey).

Group 4: Biomedical and Biological

Biomedical

S 66. Some persons are cured

One of the main assumptions in classical survival analysis is that all subjects eventually experience the event of interest. But there are many situations where a fraction of the subjects will never experience the event and thus can be said to be cured (Amico and Keilegom 2018). There are many situations, aside from health and disease recurrence, where life table methods are used: for example, in economics, where finding a new job is the event of interest but where some persons may never find a new job; in engineering, where time to failure is of interest but failure never occurs; the time when someone marries (some never marry); or the time to rearrest prisoners (some are never rearrested). A simple formula that accounts for a subset of individuals who do not succumb (i.e., transition) is given as

$$P = pl_0 + (1-p)l_0 e^{-\beta t}$$

where P denotes the probability of survival up to time t, p denotes the probability of being cured, and B is the hazard function for the uncured population (Berkson and Gage 1952).

S 67. Chances of false positives in cancer diagnoses

A woman is told that her mammogram results were positive for breast cancer. Given that 1% of women develop breast cancer in any given year, out of 10,000 women there will be 100 with cancer and 9,900 without cancer. Given also that 80% of these cancers are found through mammograms, but 9.6% of the 10,000 screenings are false positives, what is the likelihood the woman actually has breast cancer? At issue is the reliability of the test, given the information above.

Using the number out of 10,000 as the probabilities (to avoid using decimals), we provide a 2×2 table of true and false, positive and negative outcomes.

	True	False	
Positive (X)	80	950	1,030
Negative (not X)	8,950	20	8,970
	9,030	970	10,000

The probability that this woman has breast cancer given that her test was positive, denoted Pr(A|X), is computed as

$$Pr(A|X) = \frac{\text{True positives}}{\left(\text{True positives} + \text{False positives}\right)} \qquad (11.58)$$

$$= \frac{80}{(80+950)} = 0.0776 \qquad (11.59)$$

The woman's probability of actually having breast cancer, given that her mammogram results were positive, is only around 8%.

The formal theory underlying this result is Bayesian theory (Bayes 1763), based on the model

$$\Pr(A|X) = \frac{\mathrm{pr}(X|A)\,\mathrm{pr}(A)}{\Pr(X|A)\,\Pr(A) + \Pr(X|\text{not }A)\,\Pr(\text{not }A)} \qquad (11.60)$$

where

- $\Pr(A|X)$ is what we want to know, which is the chance of having cancer (A) given that the test is positive (X). In other words, how likely is the woman to have cancer with a positive test result?
- $\Pr(X|A)$ is the true positive, in other words, the chance that the test is positive (X) given that she has cancer (A).
- Pr (A) is the cancer rate, which in this example is a 1% chance of having cancer.
- Pr (not A) is the noncancer rate, and in this example the chance of not having cancer is 99%.
- $\Pr(X|\text{not }A)$ is the false positive, in other words, the chance of a positive test given that the woman does not have cancer (9.6%).

A simplified version of Bayes' theorem is given as

$$\Pr(A|X) = \frac{\mathrm{pr}(X|A)\,\mathrm{pr}(A)}{\Pr(X)} \qquad (11.61)$$

Bijak and Bryant (2016) note that Bayesian theory in demography is especially suited for both sparse data and data that are unreliable or incomplete. They provide a detailed overview of the applications of Bayesian theory in demographic forecasts, limited data, and complex models.

S 68. Visualizing health—hypertension

The metrics concerned with what constitutes healthy and unhealthy levels for individuals, such as body mass index (BMI), cholesterol levels, and hypertension, are typically presented in categories. For example, the American Heart Association (2018) published guidelines for cholesterol levels within three categories: HDL (good) cholesterol), LDL (bad) cholesterol, and triglycerides. Similarly, the American College of Cardiology (2017) has guidelines that include categories for hypertension of normal, elevated, stage 1, stage 2, and hypertensive crisis. Important aspects lacking in nearly all health-related information, such as these types of guidelines, are perspectives on both intra- and inter-individual variation through the life course. For example, do persons with high cholesterol and/or high blood pressure tend to live shorter lives? A perspective on this question with respect to systolic blood pressure is presented in fig. 11.33 using longitudinal data from the Baltimore Longitudinal Study (https://www .blsa.nih.gov/).

This event history chart and the insets shed light on the relationship of blood pressure and health, including the large variation in blood pressure both within and

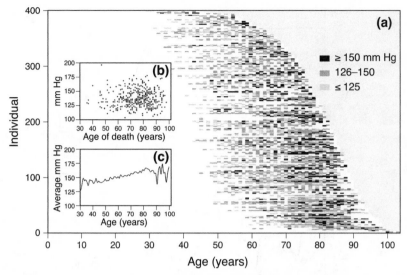

FIGURE 11.33. Systolic blood pressure readings from 400 deceased individuals who were participants in the Baltimore Longitudinal Study. Each horizontal line depicts an individual, with segments shaded according to the age at which they were examined and their blood pressure. (a) Main graphic—event history chart showing age of entry into the study and the periodic (2- to 3-year) readings of systolic blood pressure rank-ordered top to bottom from shortest- to longest-lived. The irregularities on the left are due to different ages of study entry. (b) Relationship of blood pressure over duration of individual's participation and age of death. (c) Average of all individuals in the study by age. (*Data source*: Angelo J. G. Bos, MD, PhD (contractor); Medstar Research Institute (contractor); National Institute on Aging, Clinical Research Branch, Longitudinal Studies Section)

between individuals (fig. 11.33a), the near absence of a correlation between lifetime blood pressure levels and longevity (fig. 11.33b), and the steady increase in average levels with age (fig. 11.33c).

S 69. Trajectories of chronic illness

Elderly persons who are sick enough to die generally follow trajectories of decline over time that are characteristic of three main groups illustrated in fig. 11.34. Lynn and Adamson (2003) note that US Medicare claims show that 20%, 20%, and 40% of those who die have courses consistent with the first, second, and third groups, respectively. The remaining 20% of decedents are split between those who die suddenly and those whose pattern of decline has not yet been classified.

S 70. The Will Rogers phenomenon: Migrating frailty groups

The American humorist Will Rogers noted, "When the Okies left Oklahoma and moved to California, they raised the average intelligence in both states." This is a heterogeneity concept that is relevant to demography, where the migration of one subgroup of frail individuals to another can decrease mortality in both groups. Consider

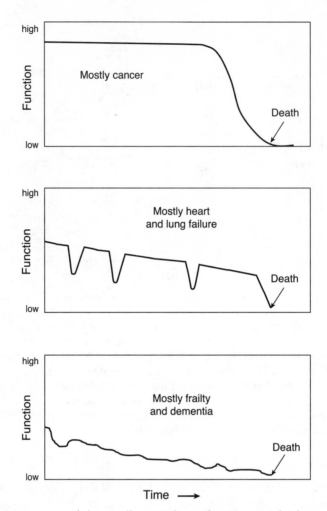

FIGURE 11.34. Trajectories of chronic illness (redrawn from Lynn and Adamson 2003).

the following two groups where the numbers indicate the percentage of survival of their respective subgroups, each comprised of equal numbers:

> Group A [10, 20, 30] with arithmetic mean of 20%
> Group B [40, 50, 60] with arithmetic mean of 50%

If the subgroup experiencing 40% survival in group B is moved to group A, then

> Group A [10, 20, 30, 40] with arithmetic mean of 25%
> Group B [50, 60] with arithmetic mean of 55%

The Will Rogers phenomenon was first used in medicine by Feinstein and colleagues in describing the outcome of *stage migration*, where earlier diagnosis of lung cancer metastases results in patients who previously would have been classified in a "good" stage being assigned to a "bad" stage. They state, "Because the prognoses of those who

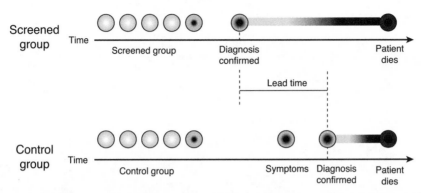

FIGURE 11.35. Lead time bias schematic. If earlier treatment has no effect on the natural history of a cancer, earlier detection still gives the appearance of an increase in survival. That increase is simply the "lead time" that comes from detecting the tumor earlier (hence the term "lead time bias"), before it becomes clinically apparent on its own (redrawn from *Science Based Medicine* blog, https://sciencebasedmedicine.org/).

migrated, although worse than that for other members of the good-stage group, was better than that for other members of the bad-stage group, survival rates rose in each group without any change in individual outcomes" (Feinstein et al. 1985, 1604).

The same stage migration effect was noted by Sormani et al. (2008) in their work on multiple sclerosis. In these cases a group of people were moved from the list of healthy to the list of sick individuals because they had a disease and were sicker than the average healthy person. Taking them off the "healthy individual" list increases the life expectancy (or average health) of the healthy list. The schematic presented in fig. 11.35 illustrates this concept, known as *lead time bias*.

Animal biodemography: Selected studies

S 71. Whooping crane population

Unregulated hunting and loss of habitat reduced the whooping crane population to 21 individuals in 1941 (Cannon 1996). Conservation efforts then led to a limited recovery of 603 birds in 2015, which is a 28.7-fold increase after 74 years. There are several questions that can be addressed: (1) How many population doublings occurred? (2) What was the average time per doubling? (3) What is the average rate of exponential increase?

For the first question we have

$$28.7 = 2^n$$

$$\ln(28.7) = n\ln(2)$$

$$n = \frac{\ln(28.7)}{\ln(2)}$$

$$= 4.8 \text{ doublings}$$

(11.62)

The answer to the second question is

$$\frac{74}{4.8} = 15.4 \text{ years per doubling} \qquad (11.63)$$

And the answer to the third question is

$$28.7 = e^{74r} \qquad (11.64)$$

$$r = \frac{\ln(28.7)}{74}$$

$$= 0.0454 \ (= \text{exponential rate of increase})$$

S 72. Gorilla population decrease

The worldwide mountain gorilla population numbers around 700 individuals. If it decreases by 1% per year, how many years will it take to reach half its present size?

$$P_t = P_t(1 - 0.01)^t \qquad (11.65)$$

$$0.5 = 0.99^t$$

$$t = \frac{\ln(0.5)}{\ln(0.99)}$$

$$= 69.0 \text{ years}$$

or about 7 decades. This is also the same number of years that it would take for the population to double if it grew at 1% annually.

S 73. Longevity minimalists: Force of mortality in mayflies

Understanding the life history strategy of short-lived species, such as mayflies, will provide insights into the more general aspects of selection for life span. That is, the results may shed light on the question, What factors favor the evolution of abbreviated life spans? Some insights into this question are possible from examination of mayfly life histories. Requirements for species with extremely short life spans may include (1) synchronous emergence to concentrate adults so that the likelihood of finding a mate is maximized; (2) the inability to feed due to their vestigial mouthparts, which preempts the need for individuals to spend time foraging for food; and (3) a "general" nymphal habitat (i.e., nearby lake or stream), which requires that

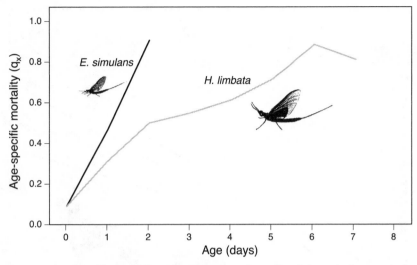

FIGURE 11.36. Age-specific mortality of two species of mayfly adults—the brown drake (*E. simulans*) and the giant mayfly (*H. limbata*). (*Source*: Carey 2002)

females fly only a short distance to deposit their eggs (Carey 2002). As an illustration, the life expectancy for the mayfly commonly known as the brown drake (*Ephemera simulans*) and the species known as the giant mayfly (*Hexagenia limbata*) is 2.0 and 2.6 days, respectively. The mortality data in fig. 11.36 reveal essentially the equivalent of "walls of death" built into the mayfly life history—a trait that is not found in species (including humans) that reproduce repeatedly over longer life spans.

S 74. Dog-to-human year conversions depend on dog breed

Is one year of a dog's life equal to seven years of human life? Unlike many other animals in which the larger the body size the greater the longevity, dogs have the inverse relationship; that is, smaller dogs generally live longer than large dogs (Patronek et al. 1997; Cooley et al. 2003). Thus, as illustrated in fig. 11.37, the translation of dog years to human years depends on the dog breed.

S 75. Impact of parasitoid insect depends on all-cause mortality

How much does a pupal parasitization rate of 95% impact egg-to-adult mortality in a pest holometabolous insect species? The impact depends on existing prepupal mortality. For example, if prepupal mortality is 0.9 (i.e., survival=0.1), then survival will be decreased from 0.1 to 0.005, thus mortality increases from 0.90 to 0.995, or by 0.095. However, if preadult mortality is only 10% (survival=0.90), then a 95% pupal mortality will result in a survival reduction from 0.90 to 0.045. This equals a mortality increase from 10% to 99.5%—an increase of nearly 90%. This example illustrates that understanding the impact of a biological control agent (or more generally any mortality factor) cannot be understood independent of all-cause mortality.

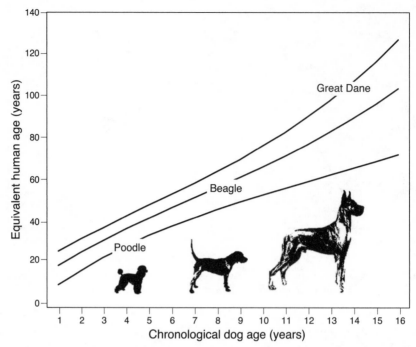

FIGURE 11.37. Relationship of dog chronological age and the equivalent human age normalized according to remaining expectation of life (Patronek et al. 1997).

S 76. Pace and shape of aging

Baudisch (2011) noted that in order to compare patterns of aging across species it is important to de-confound two dimensions of age-specific change, which are the pace (rate) and the shape (pattern). She introduced a formal analytical framework where the standardized age, x_s, can be calculated as $x_s = x/L$, where x denotes age and L denotes life expectancy. Thus standardized age equals "one life expectancy" and standardized mortality, $\mu_s(x_s)$, equals nonstandardized mortality, $\mu(x)$, times life expectancy (i.e., $\mu_s(x_s) = \mu(x)L$).

Baudisch used various ratios among the following five life table metrics to characterize the shape of aging: (1) $x_{0.01}$ denotes the age at which 1% of a cohort remains alive (=96 and 104 years for 1935 and 2015 US female cohorts, respectively); (2) $\mu(x_{0.01})$ denotes mortality at age $x_{0.01}$ (=0.2853 and 0.3842 for 1935 and 2015 US female cohorts, respectively); (3) $\mu(e_0)$ denotes mortality at e_0 (=0.0232 and 0.0245 for 1935 and 2015 US female cohorts, respectively); (4) $\mu(0)$ denotes mortality at birth (=0.008297 and 0.000352 for 1935 and 2015 US female cohorts, respectively); and (5) $\bar{\mu}$ denotes average mortality (=0.0162 and 0.0133 for 1935 and 2015 US female cohorts, respectively).

Descriptions and formulas for the ratios are given in table 11.7 along with examples comparing the shape values for each using the period life tables for US women in 1935 and 2015. With the exception of the first ratio involving the age at which 1% of a cohort remain alive relative to the life expectancy at birth, all ratios were

Table 11.7. Various shape parameters of aging

#	Ratio	Formula	Value by year 1935	2015
1	Age at which 1% remain alive relative to the expectation of life	$\dfrac{x_{0.01}}{e_0}$	1.55	1.39
2	Mortality at the age when 1% remain alive relative to mortality at birth	$\dfrac{\mu(x_{0.01})}{\mu(0)}$	34.4	1091.8
3	Mortality at expectation of life relative to mortality at birth	$\dfrac{\mu(e_0)}{\mu(0)}$	2.8	69.6
4	Mortality at the age when 1% remain alive relative to average lifetime mortality	$\dfrac{\mu(x_{0.01})}{\bar{\mu}}$	17.6	28.9
5	Mortality at expectation of life relative to average lifetime mortality	$\dfrac{\mu(e_0)}{\bar{\mu}}$	1.4	1.8

Source: Baudisch (2011).

Note: The first parameter (#1) shows the relationship between longevity measures, and the remaining parameters show the relationship between mortality measures. The values, by year, are for US female period life tables for 1935 and 2015 (HumanMortalityDatabase 2018).

higher in 2015 than in 1935. Closer examination of changes in each ratio provides different types of insights into the rate and pace of aging as outlined in Baudisch (2011).

S 77. Population extinction

Pielou (1979) derived an expression relating the probability of population extinction at time t, denoted $p_0(t)$, to birth rate (b), death rate (d), and population size (N), given as

$$p_0(t) = \left[\frac{d \langle \exp[(b-d)t] \rangle}{b \langle \exp[(b-d)t] \rangle} \right]^N \tag{11.66}$$

The birth and death rates in this expression are related to $p_0(t)$ in two ways, both as a difference (growth rate) and as a ratio (number of births per number of deaths). It follows that two populations can have the same growth rate but different ratios, with each relationship implying different extinction outcomes.

If $b > d$, then as time approaches infinity,

$$p_0(t) = \left[\frac{b}{d} \right]^N \tag{11.67}$$

Continued existence is not assured since the probability of extinction remains finite. However, this probability becomes smaller the more the birth rate exceeds the death rate and the larger the size of the initial population.

S 78. *Spatial mark-recapture*

A major deficiency of classical capture-recapture (CR) methods used to estimate abundance is that they do not consider the spatial structure of the ecological processes that give rise to the encounter data (Royle et al. 2014). Spatial capture-recapture (SCR) methods resolve a number of the technical problems that arise from conventional methods by making ecological processes explicit in the model—density, organization, movement, and spatial usage (Parmenter et al. 2003; Efford 2004; Royle et al. 2014). Highly mobile animals are bound to use areas beyond the immediate area covered by sampling devices or surveys, and the difficulty in determining this effective sampled area has long been recognized (Bondrup-Nielsen 1983). How much individual home ranges overlap with sampling efforts also influences how likely individuals are to be detected, a source of heterogeneity in detection probability that cannot be addressed mechanistically in traditional CR approaches. In essence SCR models involve (1) the location of *spatial encounter histories* of individuals; (2) the number N (population size) in a circumscribed area, A; (3) the locations of s_i, I = 1, 2 . . . , N of all individuals within the area or state-space, S; (4) the collection of *activity centers* (≈ home ranges) for all individuals; and (5) the density D of the population expressed as number per unit area, D = N/A(S). The broad concept of SCR is illustrated in fig. 11.38.

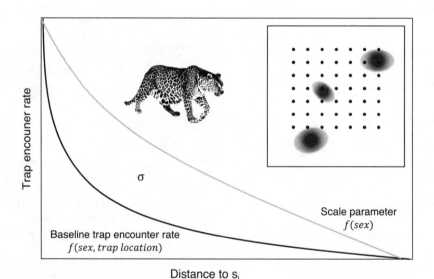

FIGURE 11.38. Schematic graphic of the trap encounter rate versus distance for individual i's center of activity, S_i. Inset: Schematic of study area with traps (grid of dots) and centers of activity (shaded areas), one of which is completely within the study area and the other two at the margins. Note that part of the activity of these two hypothetical individuals falls outside the study area. SCR methods correct for the reduced encounters for these individuals. (*Source*: Royle et al. 2014)

S 79. Immortality assumption in vectorial capacity model

Since the 1950s the assumption that mosquitoes do not senesce (i.e., that mosquito mortality remains constant with age) had been used to assess the role of mosquitoes in pathogen transmission and predict public health consequences of vector control strategies. This assumption was first articulated by MacDonald (1952, 1957), who reasoned that environmental insults, disease, and predation would kill mosquitoes before they had an opportunity to die of old age. Garrett-Jones (1964) developed a model of this assumption that is now ubiquitous in vector biology (e.g., mosquito-vectored diseases such as malaria and dengue fever) that he defined as the *vectorial capacity* (VC) model,

$$VC = \frac{A}{-\ln(p)} \qquad (11.68)$$

where p = daily survival rate and the numerator A is

$$A = ma^2bp^n \qquad (11.69)$$

In this formula m = number of female mosquitoes per host, a = daily blood feeding rate, b = transmission rate among exposed mosquitoes, and n = extrinsic incubation period.

Since −ln(p) equals the average daily death rate, its inverse equals the expectation of remaining life. Therefore, the vectorial capacity (VC) is based on an assumption of immortality; in other words, a constant mortality at all ages where the expectation of remaining life is identical at all ages. A newborn adult mosquito has the same remaining expectation of life as a 100-day-old mosquito under this assumption. Styer and her colleagues (2007) demonstrated that mosquitoes do indeed senesce and thus they do experience age-specific mortality. This model has subsequently been modified to include an age-specific version of vectorial capacity that incorporates age-specific mortality (Novoseltsev et al. 2012).

Evolutionary demography

Evolutionary demography brings together concepts and methods from evolution, population ecology, human demography, anthropology, genetics and genomics, statistics, epidemiology, and public health (Kaplan 2002; Sear 2015). Its focus is on how demographic processes influence evolution and how evolution shapes the demographic properties of organisms across the Tree of Life. Demography and evolutionary biology have a long history, starting when Darwin was famously influenced by Malthus when developing his ideas on natural selection (Carey and Vaupel 2005; Sear 2015). The conceptual interlinkages of the two fields are captured in a passage taken from Carey and Vaupel (2005, 84):

Nothing in biology, Dobzhansky has asserted, makes sense except in the light of evolution (Dobzhansky 1973). An equally valid overstatement is that nothing in evolution can be understood except in the light of demography. Evolution is driven by population dynamics governed by age-schedules of fertility and survival. Lotka emphasized this in his pathbreaking research. Since the work of Lotka, models of the evolution of fertility, mortality, and other life-history patterns have been based on stable population theory.

In this subsection we introduce what we consider to be four of the most important demographic concepts related to evolutionary demography: the intrinsic rate of increase, r, as a fitness measure; life history trade-offs; the r-K continuum; and Fisher's reproductive value.

S 80. Intrinsic rate of increase as a fitness measure

As originally defined in chapter 5, the intrinsic rate of increase, r, is the first real root of the recursive equation

$$1 = \int_0^\infty e^{-rx}\, l(x)\, m(x)\, dx \qquad (11.70)$$

where $l(x)$ and $m(x)$ are age-specific survival and reproduction, respectively. This parameter, r, as a measure of per capita growth rate was first introduced to mathematics by Euler (1760), to demography by Lotka (1907, 1922, 1928) and Sharpe and Lotka (1911), and to ecology by Leslie and Ransom (1940), Birch (1948), and Cole (1954). However, it was Fisher (1930, 1958) who first argued that r was a measure of fitness. He suggested that r is associated with genotypes that follow particular life histories and that selection favors genotypes with the highest values of r (Charlesworth 1994; Roff 2002). Also see the papers by McGraw and Caswell (1996) and Brommer (2000) for important perspectives on r as a fitness parameter.

S 81. Life history trade-offs

One of the most influential papers on life history theory was one by Cornell University biologist Lamont Cole (1954) titled "The Population Consequences of Life History Phenomena." This paper was impactful in two respects; first, because it reaffirmed the use of r as a measure of fitness and, second, because it introduced the concept of life history trade-offs. For example, Cole revealed the relative unimportance of late-life reproduction: "It is strikingly brought to one's attention that the final terms representing reproduction in later life are relatively unimportant in influencing the value of r" (Cole 1954, 133). One of the most transparent approaches for illustrating differences in the impact on r (fitness) of different life history trade-offs is contained in the paper by Lewontin (1965)—see fig. 11.39. This simple analysis reveals that shifts in the age of first reproduction have the largest impact on r.

S 82. The r- and K- continuum

Originally proposed by Robert MacArthur and E. O. Wilson (1967) the concept of r and K selection refers to individuals selected for either rapid reproduction and growth

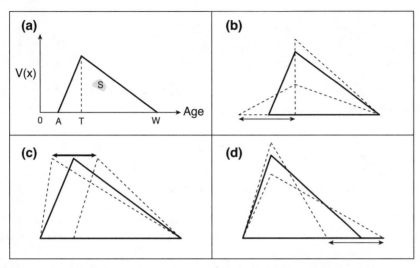

FIGURE 11.39. Schematic of reproductive function, V(x), introduced by Lewontin (1965): (a) generalized V(x), where A is the age of first offspring, T is the turnover point, W is the age of last offspring, and S is the total number of offspring (area under the curve); (b) shifts in age of first offspring, A; (c) shifts in the age of T, the turnover point; (d) shifts in age of last reproduction, W. Not shown are shifts in total reproduction.

(r-selected) or contributions to the population at carrying capacity (K-selected), where r and K denote the intrinsic rate of population growth and carrying capacity, respectively. The letters are based on the parameters of the logistic equation

$$\frac{1}{N}\frac{dN}{dt} = r\left(1 - \frac{N}{K}\right) \tag{11.71}$$

Traits of K-selected species include slow development, reduced resource requirements, delayed reproduction, large body size, and repeated reproduction. In contrast the traits of r-selected species include rapid development, early reproduction, small body size, and single reproduction (Pianka 1978). Insects are typically placed at the r-selected end of the continuum and birds and mammals at the K-selected end.

S 83. Fisher's reproductive value

Reproductive value, first introduced by Fisher (1930, 1958), is the number of offspring that an individual is expected to produce over its remaining life span, given that an individual has survived to age s, after adjusting for the growth rate of the population. It is basically the sum of ratios in which the sum is a relative index of the importance of a female's contribution to future generations. Reproductive value does not possess biological units. The analytical expression for reproductive value is

$$V(x) = \frac{e^{rx}}{l(x)} \int_x^\infty e^{-ry} \, l(y) \, m(y) \, dy \qquad (11.72)$$

This value gives the extent to which a female age x contributes to the ancestry of future generations.

S 84. Survival curve classification

Pearl (1928) classified survival curves as convex, straight, and concave and labeled them as Type I, Type II, and Type III, respectively.

> Type I: Humans in developed countries, many nonhuman primates, and large mammals. Mortality is very low through most of the life course and then increases at older ages.
> Type II: Mortality more or less stays constant or gradually increases with age.
> Type III: Extensive early mortality, such as fish or many plant seedlings (Begon et al. 1996).

The shapes of these three types of survival curves and their corresponding mortality schedules are shown in fig. 11.40. Mortality decreases after the earliest ages and then increases at older ages.

Chronodemography

S 85. Fish otoliths as black box recorders

Otoliths are calcified structures that reside in the inner ear canal and form part of the hearing and balance system in modern fish (Limburg et al. 2013; Starrs et al. 2016). They accrete daily by precipitating aragonite on a protein matrix, laying down growth bands similar to tree rings. This makes what fisheries biologists refer to as biochronological "black box recorders" inasmuch as they allow an individual fish to be retrospectively positioned in space and time throughout its life—that is, the otolith encodes the age, growth, and environmental conditions experienced by the fish (Campana and Thorrold 2001). The growth bands reveal age, the distance between the bands indicates growth rate, and trace elemental and isotopic analyses of microscale banding exposes environmental history. Information on age and growth is important because the data can be used to determine fish growth rates during early life history, estimate pelagic (open sea) larval durations of reef species, and shed light on the effects of physical processes on larval survival from hatch date distributions. Otoliths provide annual sequences of over a century in adult fish (Boehlert et al. 1989) with daily chronologies (microbands) of up to a year during the larval and juvenile stages. Otolith chemistry is an accurate proxy for concentrations of some trace elements in the ambient environment and serves as a proxy for ambient salinity and thus can be used to reconstruct a history of anadromous (freshwater-to-ocean) migra-

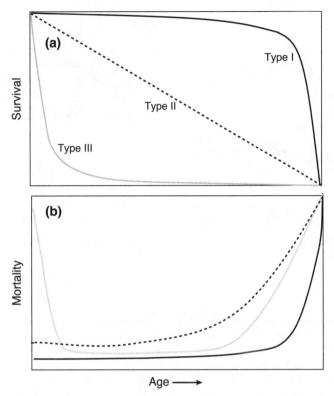

FIGURE 11.40. (a) Classification of survival curves into three types (after Pearl 1928; Deevey 1947); and (b) mortality curves associated with each survival curve type.

tions in species such as salmon (Campana and Thorrold 2001). A schematic of concepts related to the use of otoliths is shown in fig. 11.41.

S 86. Tree rings: Records of age and injury

Chronodendrology uses the age of trees to reconstruct environmental history. The cross-dating principle states that matching patterns in ring widths or other ring characteristics (such as ring density patterns) among several tree-ring series allows the identification of the exact year in which each tree ring was formed (Baillie 1999, 2015). For example, it is possible to date the construction of a building, such as a barn, by matching the tree-ring patterns of wood taken from the buildings with tree-ring patterns from living trees. Cross dating is considered the fundamental principle of chronodendrology—without the precision given by cross dating, the dating of tree rings would be nothing more than simple ring counting. A schematic of concepts related to the use of tree life-lines is shown in fig. 11.42.

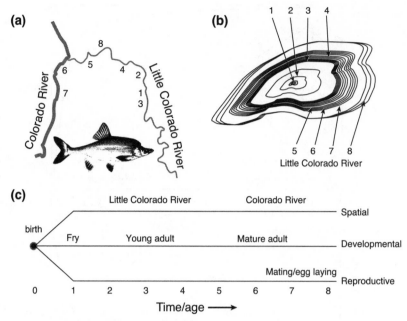

FIGURE 11.41. Schematic of the life course for a hypothetical fish: (a) river habitats, including Little Colorado River that feeds into the main Colorado River; (b) the fish's otolith with annuli (black box) that is used for estimating its age and for reconstructing its environmental spatial history; and (c) fish's life course disaggregated into spatial, developmental, and reproductive components. (*Source*: Information from Limburg et al. (2013) for drawing (a) and (b))

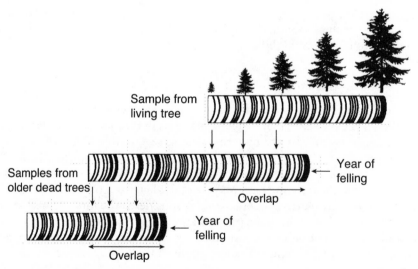

FIGURE 11.42. Cross-dating tree rings showing overlap and juxtaposition of good and bad years for trees of different ages but with overlapping growth periods.

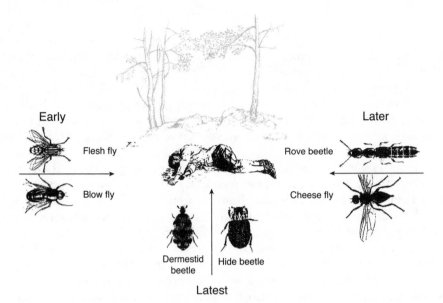

FIGURE 11.43. Schematic of the postmortem succession of insect species colonizing a corpse.

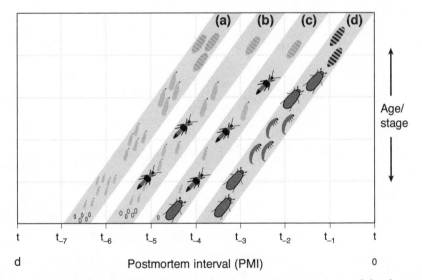

FIGURE 11.44. Lexis diagram–like schematic showing species succession and developmental age and stage concept for estimating postmortem interval: (a) early stage—blow flies, flesh flies, and staphylinid beetles; (b) midstage—blow flies, flesh flies, and staphylinids in addition to cheese skipper flies; (c) late-middle stage—hide beetles and clown beetles; (d) late stages—hide beetles and dermestid beetles.

S 87. Forensic entomology: Insects as crime tools

Forensic entomology is concerned with the application of basic ecological, entomo-logical, and demographic principles to forensic questions, particularly those concerned with estimating the postmortem interval (PMI)—the time from death of an individual to discovery of its remains (Nenecke 2001; Gennard 2012; McDermid 2014). PMI estimates are based on an evaluation of the insect species present on the body or at the scene, the stage or stages of the species, and the succession patterns of the insects in the environment in which the body was found (Catts and Goff 1992; Tomberlin et al. 2011).

There are essentially two concepts that are used to estimate the PMI (Wells and Lamotte 2010; Gennard 2012): (1) the *development model*, based on the concept that the minimum PMI can be estimated from the time required for insects to develop through the different stages; and (2) the *succession model*. The concept for this model is that faunal visitation and colonization of carrion (corpses) is a process of ecologi-cal succession—the orderly and predictable changes in structure of an ecological com-munity through time (Kreitlow 2010; VanLaerhoven 2010). The attractiveness of the corpse to different fauna depends on the state of decay, and therefore the faunal assemblages at a given stage of decay are predictable. The fundamental concepts of forensic entomology are illustrated in figs. 11.43 and 11.44.

Appendixes: Visualization, Description, and Management of Demographic Data

‖‖‖

As the cathedral is to its foundation so is an effective presentation
of facts to the data.

> Willard C. Brinton (quoted in Few 2013)

I rarely think in words at all. My visual images have to be translated . . .

> Albert Einstein

Simplicity is about subtracting the obvious and adding the meaningful.

> John Maeda

Appendix I

Visualization of Demographic Data

Biodemography is a data-rich field involving information on individuals, cohorts, and populations, all of which is assembled across a wide range of historical, geographical, economic, political, biological, and biomedical contexts. For effective communication, it is essential that this information be presented accurately, eloquently, and efficiently in the language of graphics. The strength of graphs is their ability to visualize complex relationships for comparison and pattern identification. Inasmuch as quantitative relationships differ, it follows that the appropriate graph will be dictated by the nature of the demographic data and the messages that the data contain and need to be communicated. Clear, powerful, and strategic techniques of information graphics can enrich the understanding, improve the communication, and aid the interpretation of data. In this appendix we first present the event history chart that has been used in several chapters in this book, and we then give an overview of best practices for other major graph types.

Event History Chart

Individual-level data

Longitudinal data on individuals are often preferred over data that are grouped or cross-sectional, thus graphical techniques that help visualize individual-level data are important. For demographic data there are specific situations where an analysis of individual data is particularly critical; for example, selective changes over a life cycle, compositional changes in a cohort, intra-individual variation, and lifetime comparisons. As a study population ages only a *selected portion* of individuals survive, and at later ages only a small portion of the original cohort will still be alive. Measurements made on young individuals at early ages are thus based on observations of some individuals who do not live to old ages, while data from older individuals represent a select population that has survived. With this selective mortality, longitudinal data of individuals are clearly needed to understand age changes in traits. Individual-level data can also provide insight, for example, into the between-fly variation in egg laying and thus reveal *compositional influences* on cohort characteristics. In a cohort of flies, individual-level data can show whether a decrease in cohort reproduction with age is due to an increase in the fraction of females that lay zero eggs or to an overall decrease in the level of egg laying by each individual. *Intra-individual variation* also requires longitudinal data on individuals. If the periods of more intensive egg laying vary from fly to fly, this intra-individual variation can be lost if the analysis only averages across individuals. For instance, the shape of a peak of egg laying in the averaged or cross-sectional egg-laying graph may not resemble any of the peaks observed for an individual's egg-laying behavior. Finally, individual-level data allow between-fly comparisons to be made in lifetime levels of reproduction and, in turn, on the long-term trajectories of reproduction in each individual over a specified period. In particular, they provide important insights into the reproductive age patterns of flies by comparing high versus low lifetime reproductive rates, early versus late ages of first reproduction, or short versus long lifetimes.

Construction methods

The graphics that integrate individual reproductive data with cohort survival are based on three concepts: (1) the life course of a single individual female is depicted as a horizontal line, the length of which is proportional to her longevity; (2) the age segments of the lines are color coded or shaded according to the number of eggs laid; and (3) the individuals are rank ordered from shortest- to longest-lived so that when the lines are plotted they create a cohort survival (l_x) schedule depicting the number of individuals alive at each age. A schematic showing the structure and organization of individual-level reproductive data and longevity is given in fig. AI.1 and a data example is fully described in chapter 4 and fig. 4.6. The general concept that underlies this technique is that longitudinal data on demographic events, such as reproduction, can be portrayed by color-coding the data on individuals and ordering them according to any number of life history criteria. The cohort survival schedule emerges when the data are plotted for individuals rank ordered from shortest- to longest-lived. Other important relationships can also be visually displayed using this technique, including the rank ordering of the individual-level data by lifetime reproduction or by age of maturity.

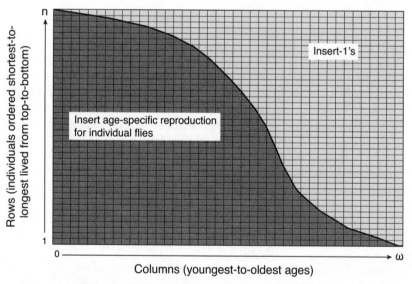

FIGURE AI.I. Illustration of spreadsheet structure for visualizing age-specific reproductive and cohort survival using color-coding feature (e.g., conditional formatting) in software based on numerical values in each cell. Cells to the right of the survival boundary are coded -1's to distinguish (and thus color-code) this area from zero egg-laying days in the cells containing egg-laying information on each female (based on concepts described in Carey et al. 1998).

Overview of Other Major Graph Types

The optimal graphical strategy for communicating the main results of a demographic study depends on what is to be emphasized (e.g., trends; correlations; outliers) and the purpose of the study (e.g., comparisons; distributions; relationships; composition). Indeed, any set of numbers, and thus any numerical data set, can be presented in multiple ways. For example, three hypothetical measurements of 1, 2, and 7 can be presented as time series (1, 2, 7), as rank ordered (7, 2, 1 or 1, 2, 7), as percentages (70%, 20%, 10%), as relative ratios (e.g., 3.5-fold and ½ relative to 2), as a series of differences (7–1; 7–2; 2–1), as a standard deviation (= 3.2), as a range (= 6), or as a mean (= 3 1/3). What needs to be emphasized (e.g., mean) and what additional information needs to be included (e.g., standard deviation and range) depends entirely on the message to be communicated. As a rule of thumb, bar and line charts are useful for comparisons, scatter charts and line and bar histograms for distributions, scatter and bubble charts for relationships, and stacked, 100% stacked, area, and pie charts for composition (Wolfe 2014). In this section we provide overviews and best practice concepts for the most common graphical displays used in biodemography, noting that no graph can be considered good or bad, only more or less effective.

Bar charts

Bar charts are used to show change over time and thus should run left to right (fig. AI.2a–b). Three variations of the basic bar chart include (1) stacked bar charts that are used when you need to compare multiple points to the whole relationship; (2) 100% stacked bar charts that are used when the total value is unimportant and the percent distribution of subcategories is the primary message; and (3) multiple categories side by side, used when comparing values in different categories, for example, trends in male (bar 1) and female (bar 2) marital rates across decades. Bar charts are common and thus familiar to most demographers and biologists. They are also popular because they are easy to read by comparing the endpoints of the bars.

Design best practice guidelines include the following: (1) use horizontal labels and avoid using slanted or vertical pipes that are difficult to read; (2) space the bars appropriately to ensure that the spacing is one-half the width of the bar; (3) start the y-axis at zero to minimize distortions of the comparative values; (4) use consistent colors or shadings unless highlighting a specific data point; and (5) order appropriately by bar value (top to bottom in descending order for horizontal bars).

Dot plot

The dot plot (fig. AI.2c) may be thought of as an adaptation of the scatter plot for use when the vertical axis variable is categorical (Feeney 2014). The originator of this graphical design, William S. Cleveland (1984), argued that dot charts should be replacements for bar charts because they allow for more effective visual decoding of quantitative information and can be used for a wider variety of data sets. Several variations on the dot plot include two non-overlapping variables and a full-scale break where values are forced into small regions of the scale where they may lose resolution.

A scale break should not be used with a bar chart because the break would make the bar sizes meaningless.

Histogram

A histogram is a graph that uses bars to display a distribution (fig. AI.2d). It differs from a bar chart in two respects. First, a histogram is used for continuous data in which the entire range of values is divided into a series of non-overlapping intervals, whereas a bar chart depicts values for categorical data. Second, unlike a bar chart where the bars are separated, the bars in a histogram are contiguous; in other words, there is no white space between the bars. The best practices for constructing histograms are the same as those for constructing bar charts with the exception that no space is inserted between the bars. Typically, histograms use vertical bars though horizontal bars are effective in certain situations. An alternative to a histogram for displaying frequency data is the frequency polygon, where a line is used to depict the shape of the distribution. An advantage of using a single line rather than a series of bars is that it draws the viewer's attention to the shape of the distribution by eliminating any visual component that would draw the eye to values of the individual intervals (Few 2013). Another advantage of a single line in a frequency polygon is that it works well for cumulative distributions.

Scatter plot

Scatter plots show the relationship between items based on two sets of variables (fig. AI.2e). They are best used to visually show correlations within large data sets when the data are encoded simultaneously on the x- and y-axes. Design best practices include the following: (1) start the y-axis values at zero if at all possible, and if multiple variables need to be included, use size and color shading to encode the additional data; (2) use lines to draw attention to a correlation between the variables to show trends; and (3) do not compare more than two trend lines because too many lines make data difficult to interpret.

Bubble chart

Bubble charts are good for displaying nominal comparisons or for ranking relationships (fig. AI.2f). This type of chart is essentially a scatter plot with bubbles that are used to display additional variables. Bubble charts can be used instead of scatter charts if the data have three data series that each contains a set of values. The sizes of the bubbles are determined by the values in the third data series. Best practices include the following: (1) make sure all labels are visible, unobstructed, and easily identifiable with the corresponding bubbles; (2) size bubbles appropriately so they are scaled according to area and diameter; and (3) do not use odd shapes (e.g., triangles; squares) since using shapes that are not entirely circular can lead to inaccuracies.

Area charts

Area charts depict a time series relationship (fig. AI.2g), and they are different from line charts because they can represent volume. There are three different variations of the area chart: (1) standard area charts that are used to show or compare quantitative progression over time; (2) stacked area charts that are best used to visualize part-to-whole relationships inasmuch as they also help show how each category contributes to the cumulative total; and (3) 100% stacked area charts that are used to show distribution of categories as part of a whole where the cumulative total is unimportant.

Best practices for area charts include the following: (1) make the charts easy to read; (2) stacked area charts should be arranged to position categories with highly variable data on the top of the chart and data with low variability on the bottom; (3) start the y-axis value at zero because starting the axis above zero truncates the visualization of values; (4) do not display more than four data categories because too many categories results in a cluttered visual that is difficult to decipher; (5) use transparent colors to clearly distinguish the data from the background; and (6) do not use area charts to display discrete data because the connected lines imply intermediate values that only exist with continuous data.

Line charts

Line charts are used to show series of continuous data, for example, time relationships including trends, acceleration/deceleration, spikes/troughs, and overall variability (fig. AI.2h). Although line charts can be used for demographic data that is either categorical or discrete, they are especially appropriate for visualizing continuous demographic data, such as population rate trends like births, deaths, and growth.

Best practice guidelines for line charts are as follows: (1) Include a zero baseline, if possible, because this provides an important frame of reference. (2) Use a maximum of four or five lines because more lines (i.e., so-called spaghetti charts) are difficult to interpret. If more lines are needed, create two charts or a panel of charts each with a single line. (3) Use only solid lines because dashed lines are distracting. (4) Label lines directly to reduce or eliminate the need for the viewer to visually dart back and forth between the legend and the lines. And (5) use the correct y-axis scale, with the general guideline that the vertical span of the lines should be roughly two-thirds the height of the y-axis.

Box plots

A box plot is just a bar that encodes a range or distribution of values from one end of the bar to the other (fig. AI.2i). The middle of a box plot is a horizontal line that divides the box in two, to mark the center of the distribution, which is usually the median. There are two lines, called whiskers, that encode additional information about the shape of the distribution; one line extends upward from the top of the box and one extends downward from the bottom of the box. The full set of information that is presented in a box plot includes the highest value; the lowest value; the range, or spread, of the values from highest to lowest; the median of the distribution; the range of the middle 50% of the values, or midspread, that is, the value at or above which the highest 25% of the values reside; and the 75th percentile, the value at or below which the lowest 25% of the values reside.

Heat maps

Heat maps display categorical data using intensity of color or shades of gray to represent values or geographic areas (fig. AI.2j). Design best principles for heat maps include the following (Gehlenborg and Wong 2012a; Few 2013): (1) use a simple map outline; (2) use a single color with varying shades or a spectrum between two analogous colors to show intensity, and exercise caution with colors such as red that can stand out and thus give unnecessary weight to the data; (3) intuitively color-code intensity according to relative values; (4) use patterns sparingly because the pattern overlay that indicates a second variable is acceptable but using more than two patterns is overwhelming and distracting; and (5) choose appropriate data ranges (e.g., a numerical range of 3 to 5 that facilitates an even distribution of data), and use plus and minus signs to extend high and low ranges (e.g., 60, 70, 85+). A variation of the heat map is the shaded contour plot, introduced by Vaupel and his coworkers (1997).

Strip plots

Strip plots can be used to display and compare multiple distributions (fig. AI.2k). Strip plots display each value in the data set rather than aggregating the number or percentage of values into intervals. Although strip plots also clearly show the shape of the distribution, they are especially useful when you have a small set of values and wish to show precisely where each value falls along a quantitative scale. There are two ways to make points visible if they are located on the exact same space: (1) use jittering so they are repositioned (e.g., vertically stacked) or (2) make the data point transparent (e.g., open rather than closed circles).

Pie charts

Pie charts are best used to make part-to-whole comparisons (fig. AI.2l). One argument against the use of pie charts is that it is difficult to gauge and thus compare the sizes of the different slices unless they are familiar percentages such as 25%, 50%, and 75%. Design best practice guidelines include the following: (1) use no more than five categories; (2) avoid tiny slices since they become lost; (3) do not use multiple pies side by side because it is difficult to compare slice size between pies; (4) make sure that the slices add up to 100%; and (5) order the slices correctly by either placing the largest segment at 12 o'clock and ordering the remaining slices in descending order counterclockwise, or starting the largest segment at 12 o'clock and continuing in descending order clockwise.

Helpful resources for visualizing data

A good starting point for learning more about best practices in visualizing demographic information graphics is the pioneering book by Edward Tufte titled *The Visual Display of Quantitative Information* (2001). Other key sources for best practices in both graphics and tabulation include Stephen Few's book *Show Me the Numbers* (2013), Matt Carter's *Designing Science Presentations* (2013), and Dona M. Wong's *The Wall Street Journal Guide to Information Graphics* (2010). Additionally, there is the series of nearly three dozen papers in *Nature Methods* on nearly every

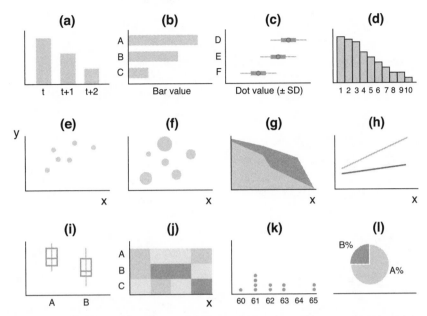

FIGURE AI.2. Common graph types used in demographic research to visualize data: (a) vertical bar graph; (b) horizontal bar graph; (c) dot plot; (d) histogram; (e) scatter plot; (f) bubble plot; (g) area graph; (h) line graph; (i) box plots; (j) heat map; (k) strip plot; (l) pie chart.

topic of information design and information graphs relevant to science and scientists (B. Wong 2010a–f, 2011a–i, 2012; Gehlenborg and Wong 2012a–d; Shoresh and Wong 2012; Wong and Kjaergaard 2012; Krzywinski 2013a–c; Krzywinski and Cairo 2013; Krzywinski and Savig 2013; Krzywinski and Wong 2013; Lex and Gehlenborg 2014; McInerny and Krzywinski 2015; Streit and Gehlenborg 2015; Hunnicutt and Krzywinski 2016a–b).

More advanced and/or specialized topics in information design and graphics include papers on population graphs and landscape genetics (Dyer 2015), data visualization and statistical graphics in big data analysis (Cook et al. 2016), data visualization in sociology (Healy and Moody 2014), multistate analysis of life histories (Willekens 2014), and visualizing mortality dynamics in the Lexis diagram and demographic surfaces (Rau, Bohk-Ewald, et al. 2018; Vaupel et al. 1997). For an overview of typography, see the book (and website) *Butterick's Practical Typography* (Butterick 2015). Sources for best practices in slide preparation include Duarte (2008), and a useful online tool designed to help scientists and cartographers select good color schemes for maps and other graphics is ColorBrewer (2018) hosted by Penn State University.

‖‖‖‖‖ Appendix II ‖‖‖‖‖
Demographic Storytelling

The structural principle of storytelling is used to integrate data and other information into a cohesive whole. The concept of demographic storytelling that we present
here is about the use of narrative structure as a tool for tying together different parts
of demographic concepts and observations. Stories can also be considered thought
experiments that are designed to engage with readers and encourage heuristic and
exploratory thought. As Revkin (2012) notes, science is "full of vexing questions, con-
flict, dead ends, insights and occasional thrilling leaps." Thus a compelling narrative
with a story arc and an episodic presentation in which the story unfolds is far more
compelling than one in which the information is "dumped." Storytelling is emerging
as a powerful and efficient communication concept in a variety of scientific disciplines
ranging from medicine (Krzywinski and Cairo 2013), biology (Knaflic 2015), and
geology (Phillips 2012; Lidal et al. 2013) to computer science (Gershon and Page 2001;
Kosara and Mackinlay 2013), chemistry (Hoffman 2014), and business (Roam 2009,
2014; Knaflic 2015). A well-told scientific story conveys great quantities of information in relatively few words in a format that is easily assimilated by the listener or
viewer (Gershon and Page 2001). People usually find information easier to understand if it is integrated into a clearly told story than if the information is laid out as
bullet points or numbered lists (Munroe 2015). The structural principles of storytelling enable the presenter to integrate multiple panels of information into a more cohesive whole (Krzywinski and Cairo 2013). Demographic stories and demographic
visuals are mutually complementary; whereas the former serves as the connective tissue that aids in linking different parts of complex concepts or data, the later aids the
story by providing optical structure to the narrative (Ma et al. 2012; Borkin et al.
2013; Kosara and Mackinlay 2013).

Demographic Storytelling: Selected Examples

The basics of storytelling are that the story must have structure (i.e., beginning, middle,
end), must have a voice (i.e., yours), and must have character development (i.e., main
theme). The stories themselves are sequences of causally related events that have three
things in common (Ma et al. 2012): (1) they take time to unfold, (2) they hold a person's attention, and (3) they leave a lasting impression. Here we show, in abbreviated form, how four famous persons framed and explained their basic ideas about
demographic concepts using storytelling approaches.

Leonardo Fibonacci: Rabbit breeding

Leonardo Fibonacci was an Italian mathematician who in 1202 introduced a hypothetical model involving population growth over generations that yielded a solution that was a sequence of numbers, which came to be known as *Fibonacci numbers*. His original published story was based on rabbits:

> Suppose a newborn pair of rabbits, one male and one female, is put in the wild. The rabbits mate at the age of one month. At the end of the second month, the female can produce another pair of rabbits. Suppose that the rabbits never die and that each female always produces one new pair, with one male and one female, every month from the second month on. How many pairs will there be in one year?
>
> At two months, the rabbits have mated but not yet given birth, resulting in only one pair of rabbits. After three months, the first pair will give birth to another pair, resulting in two pairs. At the fourth month mark, the original pair gives birth again, and the second pair mates but does not yet give birth, leaving the total at three pair. This continues until a year has passed, in which there will be 233 pairs of rabbits. (Posamentier and Lehmann 2007, 173)

The sequence Fibonacci produced from this thought experiment was 1, 2, 3, 5, 8, 13, 21, 34, 55, 89, 144, and 233 for months 1 through 12, respectively, with the general formula given as

$$X_{n+1} = X_n + X_{n-1}$$

Although this mathematical sequence is an unrealistic model for rabbit population growth, it turned out to apply to a range of relationships found in nature. For example, the number of petals in flowers (e.g., lily $= 3$, buttercup $= 5$; daisies $= 34$) are all Fibonacci numbers as are the number of parents that male honeybees have due to the haplo-diploid mating system of honeybees, where males develop from unfertilized eggs (i.e., 1 parent, 2 grandparents, 3 great-grandparents, 5 great-great-grandparents, and so forth).

The ratio of this Fibonacci sequence is given as

$$\varphi = \frac{X_{n+1}}{X_n}$$

which, in turn, creates another sequence of numbers that converge to what is known as the *golden ratio* (φ), where

$$\varphi = \frac{1+\sqrt{5}}{2} = 1.618$$

This ratio is found in the ratio of spiral diameters of nautilus shells, of hurricanes, and of galaxies (Posamentier and Lehmann 2007). These fundamental mathematical relationships—the Fibonacci numbers and the golden ratio—both flowed from Fibonacci's original story concerned with rabbit breeding. It is unlikely that these numbers would have had the same interdisciplinary resonance had Fibonacci simply used mathematical arguments rather than the rabbit example to introduce his remarkable mathematics.

Thomas Malthus: Population growth and food supply

Thomas Malthus was an English cleric, economist, and demographer who is best known for his theory that populations grow geometrically but food supplies can only grow arithmetically. As a consequence, population growth will always tend to outrun the food supply, a situation that is commonly referred to as Malthusianism. He illustrated this concept using the following story.

> Let us now take any spot of earth, this Island for instance, and see in what ratio the subsistence it affords can be supposed to increase. . . . If I allow that by the best possible policy, by breaking up more land and by great encouragements to agriculture, the produce of this Island may be doubled in the first twenty-five years . . . the very utmost that we can conceive, is, that the increase in the second twenty-five years might equal the present produce. . . . It may be fairly said, therefore, that the means of subsistence increase in an arithmetical ratio. The population of the Island is computed to be about seven million, and we will suppose the present produce equal to the support of such a number. . . . At the conclusion of the first century the population would be one hundred and twelve million and the means of subsistence only equal to the support of thirty-five million, which would leave a population of seventy-seven million totally unprovided for. (Malthus 1798, 8)

In his foreword to the 1959 reprint of Malthus (1798), demographer Kenneth Boulding noted that only two things, or a combination of two things, can bring the population growth of Malthus to an end—declining fertility or increasing mortality. Malthus was not hopeful about declining fertility (i.e., "no progress whatever has hitherto been made [extinction of the passion of the sexes]"), thus the only method of reaching an equilibrium population is increasing mortality, and mortality is increased mainly through misery and starvation (Malthus 1798). Malthus introduced what he called his "Dismal Theorem": if the only ultimate check on the growth of population is misery, then the population will grow until it is miserable enough to stop its growth. He extended this to the "Utterly Dismal Theorem," where he said that any technical improvement can only relieve misery for a while, for as long as misery is the only check on population the improvement will enable the population to grow and will soon merely enable more people to live in misery than before (Malthus 1798).

Charles Darwin: The struggle for existence

Lennox (1991) suggested that Charles Darwin used a thought experiment (story) to provide evidence for his theory's explanatory potential. Darwin's imaginary illustration of predation by wolves, for example, was powerful because the object (the wolf) and process (selection) were concrete and gave the illustration a feeling of experimentation. This illustration was plausible because wolf packs attacking deer herds do not require a stretch of anyone's imagination. Finally, the relationship between the concrete and the abstract terms of the theory are clear; in other words, the crucial element of Darwin's theory of natural selection is linked to a concrete illustration.

> In order to make it clear how, as I believe, natural selection acts, I must beg permission to give one or two imaginary illustrations. Let us take the case of a wolf, which preys on various animals, securing some by craft, some by strength, some by fleetness; and let us suppose that the fleetest prey, a deer for instance, had decreased in numbers, during the season of the year when the wolf is hardest pressed for food. I can under such circumstances see no reason to doubt that the swiftest and slimmest wolves would have the best chance of surviving, and so be preserved or selected. (Darwin 1859, 90)

P. B. Medawar: Actuarial immortality

Sir Peter Brian Medawar was a British biologist and is regarded as the "father of transplantation" because of his seminal research on graft rejection and the discovery of acquired immune tolerance for which he was awarded the Nobel Prize in 1960. In an essay on the evolution of senescence published in 1952 titled "An Unsolved Problem in Biology," he devised a simple story about test tubes to make his case regarding the decreasing importance of older individuals to population fitness. He wrote:

> Imagine now a chemical laboratory equipped on its foundation with a stock of 1000 test-tubes, and that these are accidentally and in random manner broken at the rate of 10 per cent per month. We suppose . . . that the laboratory steward replaces the broken test-tubes monthly. . . . Now imagine that this regimen of mortality and fertility, breakage and replacement, has been in progress for a number of years. What will then be the age-distribution of the test-tube population? The population will have reached the stable-age-distribution in which there are 100 test-tubes aged 0–1 month, 90 aged 1–2 months, 81 aged 2–3 months and so on. This pattern of age-distribution is characteristic of a "potentially" immortal population, i.e., one in which the chances of dying do not change with age. (Medawar 1981, 43–44)

This simple mathematical scenario used test tubes to set the conceptual stage for Medawar to present a story line. He introduced an assumption that each individual test tube was subject to a constant 10% monthly mortality rate and (unrealistically) that each test tube could replace itself at a 10% monthly "reproductive" rate. He showed that the relative contribution of test tubes to the overall test tube population became vanishingly small at advanced ages even though the remaining life expectancy of the oldest individuals was exactly the same as the younger ones. He stated, "This model shows . . . how it must be that the force of natural selection weakens with increasing age—even in a theoretically immortal population, provided only that it is exposed to real hazards of mortality. If a genetical disaster that amounts to breakage happens late enough . . . its consequences may be completely unimportant" (Medawar 1981, 46). This simple story established the foundation of our current theories on the evolution of aging.

Demography Stories through Graphics or Schematics

In addition to narrative stories, a story can also be told through simple graphics. Three examples are highlighted here.

Demographic transition

Demographic transition refers to the decline in mortality and fertility from the high rates characteristic of premodern and low-income societies to the low rates characteristic of modern and high-income societies (Casterline 2003). A schematic of the changes in birth and death rates is presented in fig. AII.1, and the five classic stages of transition with the population rates of birth, death, and growth are described in table AII.1.

The reason for the high birth rates in stages I and II is that many children are needed for farming, but there is also a high death rate of the young. Additionally, religious and social pressures and the lack of family planning result in high births. Falling birth rates in stage III is the outcome of improved medicine and better diet. The falling and low birth rates in stages IV and V are due to increased family planning and good health, improved status of women, and later marriages (Casterline 2003). High death rates in stage I are due to disease, famine, and poor medical knowledge. In stages II and III death rates are falling due to improvements in medical care,

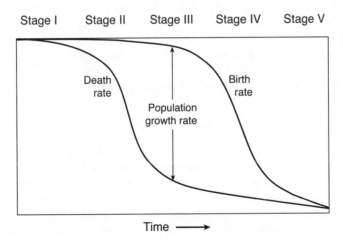

FIGURE AII.1. Schematic of the demographic transition (redrawn from http://www.slideshare .net/jakeroyles/population-34509341).

Table AII.1. Stages of the demographic transition

Population rates	Stages				
	High stationary	Early expanding	Late expanding	Low stationary	Stationary
	I	II	III	IV	V
Birth rate	High	High	Falling	Low	Very low
Death rate	High	Falls rapidly	Falls more slowly	Low	Low
Natural increase	Stable or slow increase	Very rapid increase	Increase slows	Stable or slow increase	Slow increase

water supply, and sanitation with fewer children dying. In stages IV and V the low death rates are the outcome of good health care and reliable food supplies.

Age pyramids

The age pyramid depicting the German population in 1996 (fig AII.2) illustrates the importance of visualizing population age structure in two respects. First, age pyramids reveal the demographic history of a population as the humps and hollows in age structure show the traces of past events, such as wars, epidemics, and economic depressions (Keyfitz 1985). For example, the constrictions in age structure in fig. AII.2 in the few years before, during, and after the birth years of 1916, 1931, and 1945 reflect the effects on birth of World War I, the Great Depression, and World War II, respectively. The deficit of men age 55 and older reflects the impact of the two world wars on male attrition when they were in their late teens and early to mid 20s. The postwar baby boom is seen as bulges in the 20- to 45-year-old age classes. Visualizing age structure with a pyramid can provide visual information on the projection of future populations. In fig. AII.2, the baby boom generation bulge that is apparent in the 1996 projection will be the retiring population in 2016 when the first members of this generation reach age 65. This large bulge must be considered in light of the much smaller population of youth who will be supporting the retirees. An age pyramid makes it clear that today's children are tomorrow's mothers and workers and today's workers and mothers are tomorrow's retirees.

The greatest walk

In a blink of an eye in evolutionary time, humans reached the last continental corner on the earth—Tierra del Fuego in South America (fig. AII.3). It was a journey of over 35,000 kilometers, from Africa through the Levant and Asia and on to the Americas (Djibouti 2013). The "greatest walk" jumps out of the schematic as a transworld "superhighway" with important side streams into Europe and Asia, downward through Southeast Asia and Indonesia, and on to Australia and Tasmania as well as into the easterly interiors of the Americas. The peopling of the world involves the changes in space and time that shaped humans both biologically and culturally. This

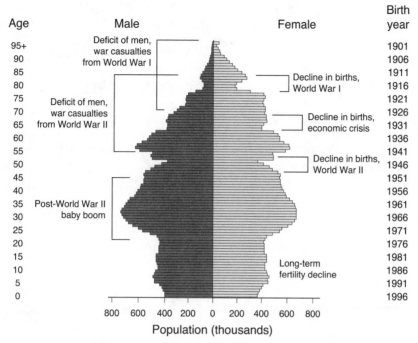

FIGURE AII.2. Age structure of Germany in 1996 (redrawn from http://healthandrights
.ccnmtl.columbia.edu/demography/the_causes_and_effects_of_populations_structures.html).

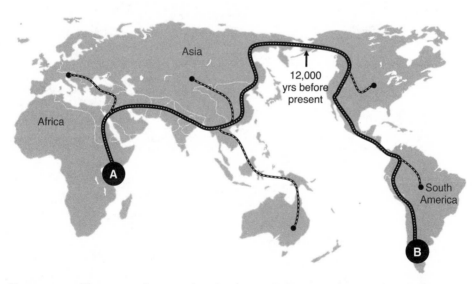

FIGURE AII.3. The greatest human migration began 70,000 years ago in Africa (redrawn
from the Out of Eden Walk created by Jeff Blossom, Center for Geographic Analysis,
Harvard University, 2013).

schematic visualizes a story of a global sweep, intriguing because of the mysteries and puzzles about the mechanisms and rates of spread and profound because of its outcomes.

Helpful Resources for Demographic Storytelling

Some of the best sources for scientific storytelling include the papers by Gerson and Page (2001), Lidal et al. (2013), and Ma et al. (2012). An excellent source on visual storytelling is Duarte (2010). Also see Alley (2013) and Healy (2019) for important perspectives on visualization in storytelling within the context of science writing and presentations.

Appendix III
Ten Visualization Rules of Thumb

The following rules are based on best practices summarized from a wide variety of books, papers, and articles (Tufte 2001; Maeda 2006; Duarte 2008, 2010; Mazza 2009; Few 2013; Wolfe 2014; Butterick 2015).

Rule 1. *Use serif fonts for manuscripts and san serif for slides.* Whereas serif fonts (i.e., letters with a slight projection finishing off the stroke) guide the reader to follow one line at a time and are thus best for manuscripts, sans serif fonts (without these projections) are perceived as simple and pure and are thus easier to read across the room.

Rule 2. *Order tabular data according to main emphasis.* For example, a table comparing population sizes by country should be arranged so that the column containing these sizes is ordered from highest to lowest population rather than, say, alphabetically by country.

Rule 3. *Present comparative tabular data vertically.* The more natural ordering for comparing similar metrics is vertical rather than horizontal. For example, a table comparing population size and mean age should arrange these respective data in columns rather than in rows.

Rule 4. *Use horizontally labeled callouts in figures.* Unnecessary variation in callout lines and labels creates a disorganized figure. If angles are necessary, use fixed angles such as 30° or 45°.

Rule 5. *Use hollow circles for robust symbols.* Hollow circles are flexible and robust plotting symbols, because, unlike the effects of clustering of other symbols, the intersection of a circle with another circle does not form an image of itself.

Rule 6. *Apply numbers and bullet points in different contexts.* It is best to use bullets with lists when order is arbitrary (e.g., ingredients) but numbers when presenting a sequence of steps (e.g., steps in preparing diet). Never use subbullets.

Rule 7. *Present chart titles in oral presentations as conclusions.* Titles should include results in order to immediately provide the viewer with a carry-away message. Example: "Sex-specific mortality trajectories" (suboptimal) versus "Male-female crossover at older ages" (better).

Rule 8. *Use 16:9 presentation aspect ratio for slides.* Aspect ratio is the ratio of width to height of an image. The 16:9 aspect ratio is a product of movie producers, which allows viewers to see a larger picture (relative to a 4:3 aspect ratio).

Rule 9. *Make text easy to read.* Short headings or titles in all uppercase is acceptable, but not for entire sentences since uppercase is difficult to read.

Title case (first letter of all keywords capitalized) can be used for used for most headings. Sentence case (only first word capitalized) is the most natural for reading. All lowercase may be appropriate in selected situations.

Rule 10. *Remember that less is more.* In other words, the opportunity lost from including less is gained in greater emphasis on what is shown (Wong 2011h).

ⅢⅢⅢⅢ Appendix IV ⅢⅢⅢⅢ

Management of Demographic Data

Data coding, management, labeling, archiving, and curating are all essential steps in the life cycle of demographic data. Data management includes planning as well as data documentation, naming, organization, storage, sharing, and retrieval. As in all areas of scientific research, data management in biodemography is something done before a project is started, during its execution, and after its completion. Collectively, data management, also referred to as data curation, encompasses all the processes needed for principled and controlled data creation, maintenance, and management, together with the capacity to add value to data (Miller 2014).

Analysis and visualization of data and its use in modeling, hypothesis testing, and trend identification is central to biodemography. It thus follows that a basic understanding of the principles and methods of data management increases project efficiency by providing clear, systematic guidelines for collecting, analyzing, visualizing, and curating data. This reduces the likelihood that very expensive and/or irreplaceable data will be lost or mislabeled. It also facilitates sharing and reuse and satisfies funding agency requirements for open access. Much of the following is drawn from the book *Data Management for Researchers* (Briney 2015), which is a helpful resource for information on data management.

Data Management Plan and Data Life Cycle

Management of demographic data can be conceptualized around the data life cycle—all steps of the research process begin with collection and move through the various stages of tabulation, visualization, analyses, and summarization, and eventually to publication (fig. AIV.1). Historically, this process typically ended with publication, but now, with the expectation that data can and often are shared and/or reanalyzed for secondary purposes, this process now requires that data be properly labeled, cataloged, and archived (Briney 2015).

Data and Data Documentation

Data, data types, and data categories

Demographer Griffith Feeney notes that dictionary definitions of data tend to be too broad to be useful and specialist definitions tend to be tied to particular applications. He thus defines data as "systematically organized information about the entities

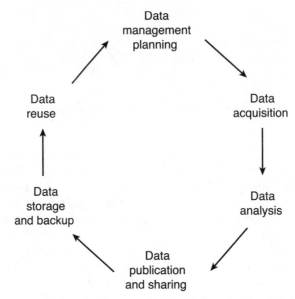

FIGURE AIV.1. Main components and steps in the data life cycle. (*Source*: Briney 2015)

comprising some statistical aggregate" (Feeney 2013). The entities may be persons, households, births, deaths, and most anything else, the only requirement being that individual entries must be clearly defined and identifiable. His reference to "systematically organized" means four specific things: (1) each entity in the aggregate must have a record containing information about it in the form of values of variables (e.g., sex is a variable with values of either male or female); (2) information on record is encoded (e.g., "1" denotes male and "2" denotes female); (3) the code representing the value of a variable occupies the same position on every record and these position assignments refer to the record layout; and (4) the layout is organized as a matrix with the cells in each row giving the values of the variables and the cells in each column giving the values of the variables represented by the column for each individual for which (or for whom) this variable is defined.

There are four categories of data, all of which apply to biodemographic data (Briney 2015). First there are *observational data*, which result from monitoring events in time and at specific locations, such as survey and census data, vital rates from registrations, weather or environmental chamber measurements, and population field counts. Next there are *experimental data* that are generated by researchers under controlled conditions, such as the life table properties of flies subject to different diets and maintained in a programmed environmental chamber, or of rates of seed set in replicated plots of plants. There can also be *simulation data* from computer models of scientific systems, for example, data on projections of future populations or mortality rates or from Monte Carlo simulations of rare species abundance. Additionally, there can be *compiled data*, which refers to data collated from other sources for secondary use or derived from a database containing a variety of data on one topic.

Another category of data is *metadata*. These "data about data" are beginning to replace laboratory and field notes. Whereas field notes are typically informal and quasi-systematic, metadata are highly structured, digital forms of documentation. An example of metadata is given as follows:

> Creator: Stephen C. Thompson
> Date: October 9, 2017
> Title: Dietary restriction in *Drosophila melanogaster*
> Description: Data files on life table properties and reproduction in 400 individually held females of each sex (4 replicates of 100/rep) with access to three levels of yeast and an ad lib control
> Notes: Censored data in replicates 2 and 3
> Subject: *Drosophila melanogaster*
> Identifier: 2017-10-9_DietRestrict.xlsx
> Format: MS Excel

The advantages of a metadata record such as this is that it is not only protocoled so that all the major fields of information are recorded, but the structure enables it to be easily searched and communicated to other users. Field or laboratory notes may or may not possess these traits. Metadata files are especially useful when projects generate large amounts of data.

Methods documentation, data dictionaries, and file codes

Scientific research requires consistency in approach and reproducibility of results. It thus follows that the methods component of any demographic research project must be carefully documented. Methods describe how data are acquired and the particular conditions, locations, and time. Methods tie together different aspects of the data set but are not themselves data. Methods information needs to be saved and *kept with the data* (very important) and it needs to include necessary units (e.g., gm; kg) as well as identification of what the method was for. This is all information that would be included in the methods section of a publication.

Data dictionaries are useful because they make it possible for anyone to access a spreadsheet and quickly understand what each score, or even hundreds of acronyms or abbreviations, stands for. They also provide a context for the data. Data dictionaries include the following types of information (Briney 2015): variable name, variable definition, how the variable was measured, data units, data format, min-max values, coded values and their meaning, representation of null values, precision of measurement, known issues with the data (e.g., missing values), relationship to other variables, and other important notes about the data.

Organization

Folder and file naming conventions

Best practices in naming conventions combine the benefits of good folder organization with the systematic ordering of files within a digital folder. Developing and adopting a good naming convention enables researchers to easily and quickly find a file and avoid duplication of information. Good names convey context about what the file contains by stating information such as experiment type (or number), researcher's name (or initials), sample type (or number), date, site, and version. Names should be descriptive, short, and consistent, use dashes rather than spaces, and follow date conventions such as YYYY-MM-DD or YYYYMMDD. Beginning or ending file names with dates is useful because files will sort chronologically.

The following is an example of best practices in labeling a main folder (Drosophila_CalorieRestriction), subfolders (second and third levels with dates), and files (.xlsx suffix).

```
Drosophila_CalorieRestriction [main folder]
    2016 [subfolder]
        2016-10-30 [subfolder]
            2016-10-30_FlyExpt1.xlsx
            2016-10-30_FlyExpt2.xlsx
            2016-10-30_FlyExpt3_v01.xlsx
            2016-10-30_FlyExpt3_v02.xlsx
        2016-11-14 [subfolder]
        2016-12-2 [subfolder]
    2017 [subfolder]
        2017-03-24 [subfolder]
        2017-06-9 [subfolder]
```

Note that this set of examples of folder and file naming includes what computer scientists refer to as "camel case" and "pothole case," the former referring to names that contain a mix of upper- and lowercase (e.g., FlyExpt1) and the latter to names that separate parts of a name with an underscore (e.g., 2016-10-30_Fly . . .). Best practices in file naming also include provisions for version control; one such strategy illustrated in the example above shows two versions of 2016-10-30_FlyExpt3, the first with v01 and the second with v02, both separated from the previous text with an underscore. Another useful strategy for version control is the use of date as a suffix at the end. Thus the last version is always the latest version. For example, the files

```
FlyCalorieExpt_20181003
FlyCalorieExpt_20181113
FlyCalorieExpt_20181222
```

would refer to a database created on October 3, 2018, and updated on November 13 and December 22. The numbers at the end for year-month-day are interpreted by the computer as the sequence of numbers 20,181,003; 20,181,113; and 20,181,222, respectively.

Storage

The issue of data storage, as part of a larger data management strategy, has two levels, the first of which is short-term storage for backup of working files during active use. For most research data the rule of thumb is to follow the 3-2-1 backup rule. This rule recommends maintaining three copies of data on at least two different types of storage media (e.g., hard drive versus CD), with one off-site copy. The off-site copy protects against fire or natural disaster. The third copy of data is a "just in case" backup.

The second level of concern is long-term storage and preservation—that is, what happens to data after they are no longer in active use. Preservation includes storage and backup as well as careful documentation (e.g., data dictionaries; coding). A rule of thumb is to retain data that support publications and reports and data that are irreproducible; in other words, they are tied to a time and a place.

The main challenge for the long-term preservation of digital information is the rapid evolution of digital technology, including both software and hardware. For example, it is unlikely that data stored in spreadsheet programs that have been discontinued, or stored on floppy discs, will be accessible. Thus rules of thumb for long-term preservation of demographic data include the following: (1) at the end of a project, convert important files into readable formats (e.g., .txt; .rtf; .pdf); and (2) update storage hardware every three to five years to minimize corruption and to ensure that the data are maintained on the more current but ever-evolving hardware technology.

REFERENCES CITED

Abbott, W. S. 1925. A method of computing the effectiveness of an insecticide. *Journal of Economic Entomology* **18**: 265–267.

Agree, E. M., and V. A. Freedman. 2011. A quality-of-life scale for assistive technology: Results of a pilot study of aging and technology. *Physical Therapy* **91**: 1780–1788.

Al-Khafaji, K., S. Tuljapurkar, J. R. Carey, and R. E. Page Jr. 2008. Life in the colonies: Hierarchical demography of social insects. *Ecology* **90**: 556–566.

Alley, M. 2013. *The Craft of Scientific Presentations*. 2nd ed. Springer, New York.

Alroy, J. 2015. Limits to captive breeding of mammals in zoos. *Conservation Biology* **29**: 926–931.

American College of Cardiology. 2017. New ACC/AHA high blood pressure guidelines lower definition of hypertension. Accessed October 21, 2018. https://www.acc.org/latest-in -cardiology/articles/2017/11/08/11/47/mon-5pm-bp-guideline-aha-2017.

American Heart Association. 2018. What your cholesterol levels mean. Accessed October 21, 2018. http://www.heart.org/HEARTORG/Conditions/Cholesterol/AboutCholesterol/What -Your-Cholesterol-Levels-Mean_UCM_305562_Article.jsp#.W1NRkLgnaUm.

Amico, M., and I. V. Keilegom. 2018. Cure models in survival analysis. *Annual Review of Statistics and Its Application* **5**: 311–342.

Amstrup, S. C., T. L. McDonald, and B. F. J. Manly, editors. 2005. *Handbook of Capture-Recapture Analysis*. Princeton University Press, Princeton, NJ.

Anderson-Lederer, R. M. 2013. Genetic management of wild and translocated black rhinoceros in South Africa's KwaZulu-Natal region. PhD diss., Victoria University, Wellington, New Zealand.

Anderson, J. J., T. Li, and D. J. Sharrow. 2017. Insights into mortality patterns and causes of death through a process point of view model. *Biogerontology* **18**:149–170.

Anscombe, F. J. 1973. Graphs in statistical analysis. *American Statistician* **27**: 17–21.

Arias, E., M. Heron, and B. Tejada-Vera. 2013. National Vital Statistics Reports: United States life tables eliminating certain causes of death, 1999–2001. Vol. 61. Division of Vital Statistics. Washington, DC.

Arlet, M. E., J. R. Carey, and F. Molleman. 2009. Species, age and sex differences in type and frequencies of injuries and impairment among four arboreal primate species in Kibale National Park, Uganda. *Primates* **50**: 65–73.

Arthur, W. B. 1981. Why a population converges to stability. *American Mathematic Monthly* **88**: 557–563.

Arthur, W. B. 1982. The ergodic theorems of demography: A simple proof. *Demography* **19**: 439–445.

Avens, L., J. C. Taylor, L. R. Goshe, T. T. Jones, and M. Hastings. 2009. Use of skeletochronological analysis to estimate the age of leatherback sea turtles *Dermochelys coracea* in the western North Atlantic. *Endangered Species Research* **8**: 165–177.

Baillie, M. 1999. *Exodus to Arthur: Catastrophic Encounters with Comets*. B. B. Batsford, London.

Baillie, M. G. L. 2015. *Tree-Ring Dating and Archaeology*. Routledge Library Editions, London.

Barbi, D., J. Bongaarts, and J. W. Vaupel. 2008. *How Long Do We Live? Demographic Models and Reflections on Tempos Effects*. Springer, Rostock, Germany.

Barbi, E., F. Lagona, M. Marsili, J. W. Vaupel, and K. W. Wachter. 2018. The plateau of human mortality: Demography of longevity pioneers. *Science* **360**: 1459–1461.

Barbour, A. B., J. M. Ponciano, and K. Lorenzen. 2013. Apparent survival estimation from continuous mark-recapture/resighting data. *Methods in Ecology and Evolution* 4:846–853.

Barker, D. P. J. 1994. *Mothers, Babies and Diseases*. 1st ed. BMJ Publishing Group, London.

Barot, S., J. Gignoux, and S. Legendre. 2002. Stage-classified matrix models and age estimates. *Oikos* **96**: 56–61.

Barthold, J. A., A. J. Loveridge, D. W. Macdonald, C. Packer, and J. Colchero. 2016. Bayesian estimates of male and female African lion mortality for future use in population management. *Journal of Applied Ecology* **53**: 295–304.

Baskin, C. C., and J. M. Baskin. 1998. *Seeds: Ecology, Biogeography and Evolution of Dormancy and Germination*. Academic Press, San Diego.

Batten, R. W. 1978. *Mortality Table Construction*. Prentice-Hall, Englewood Cliffs, NJ.

Baudisch, A. 2011. The pace and shape of ageing. *Methods in Ecology and Evolution* **2**: 375–383. doi: 10.1111/j.2041–210X.2010.00087.x.

Bauer, H., G. Chapron, K. Nowell, P. Henschel, P. Funston, L. T. B. Hunter, D. W. Macdonald, and C. Packer. 2015. Lion (*Panthera leo*) populations are declining rapidly across Africa, except in intensively managed areas. *Proceedings of the National Academy of Sciences* **112**: 14894–14899.

Bauer, H., C. Packer, P. F. Funston, P. Henschel, and K. Nowell. 2016. *African lion, Panthera leo*. IUCN Red List of Threatened Species 2016. http://dx.doi.org/10.2305/IUCN.UK .2016–3.RLTS.T15951A107265605.en.

Baxter, J. E., S. Vey, E. H. McGuire, S. Conway, and D. E. Blom. 2017. Reflections on interdisciplinarity in the study of childhood in the past. *Childhood in the Past* **10**: 57–71.

Bayes, T. 1763. An essay towards solving a problem in the doctrine of chances. *Philosophical Transactions of the Royal Society of London* **53**: 379–418.

Beamish, E. K., and M. J. O'Riain. 2014. The effects of permanent injury on the behavior and diet of commensal chacma baboons (*Papio ursinus*) in the Cape Peninsula, South Africa. *International Journal of Primatology* **35**: 1004–1020.

Bedard, K., and E. Dhuey. 2006. The persistence of early childhood maturity: International evidence of long-run age effects. *Quarterly Journal of Economics* **121**: 1437–1472.

Beddington, J. R., and D. B. Taylor. 1973. Optimum age specific harvesting of a population. *Biometrics*:801–809.

Begon, M., J. L. Harper, and C. R. Townsend. 1996. *Ecology: Individuals, Populations and Communities*. 3rd ed. Blackwell Science, Oxford, UK.

Bell, R. H. V. 1983. Decision-making in wildlife management with reference to overpopulation. Pages 145–172 in R. N. Owen-Smith, editor, *Management of Large Mammals in African Conservation Areas*. Haum Educational Publishers, Pretoria.

Berkson, J., and R. P. Gage. 1952. Survival curve for cancer patients following treatment. *Journal of the American Statistical Association* **47**: 501–515.

Berrigan, D., J. R. Carey, J. G. Aguilar, and H. C. Hurtado. 1988. Age and host effects on clutch size in *Anastrepha ludens*. *Entomologia Experimentalis et Applicata* **47**: 73–80.

Bijak, J., and J. Bryant. 2016. Bayesian demography 250 years after Bayes. *Population Studies* **70**: 1–19.

Billari, F. C. 2003. Life course analysis. In P. Demeny and G. McNicoll, editors, *Encyclopedia of Population*, Vol. 2. Gale Group, New York.

Birch, L. C. 1948. The intrinsic rate of natural increase of an insect population. *Journal of Animal Ecology* **17**: 15–26.

Bischof, R., C. Bonenfant, I. M. Rivrud, A. Zedrosser, A. Friebe, T. Coulson, A. Mysterud, and J. E. Swenson. 2018. Regulated hunting re-shapes the life history of brown bears. *Nature Ecology & Evolution* **2**: 116–123.

Bock, J. 2002. Learning, life history, and productivity. *Human Nature* 13: 161–197.

Boehlert, G. W., M. M. Yoklavich, and D. B. Chelton. 1989. Time series of growth in the genus *Sebastes* from the northeast Pacific Ocian. *Fisheries Bulletin of the United States* 87: 791–806.

Bogin, B. 1999. Evolutionary perspective on human growth. *Annual Review of Anthropology* 28: 109–153.

Bogin, B., and B. H. Smith. 2000. Evolution of the human life cycle. Pages 377–424 *in* S. Stinson, B. Bogin, R. Huss-ashmore, and D. O. Rourke, editors, *Human Biology: An Evolutionary and Biocultural Perspective*. Wiley-Liss, New York.

Bondrup-Nielsen, S. 1983. Density estimation as a function of live trapping grid and home range size. *Canadian Journal of Zoology* 61: 2361–2365.

Bongaarts, J. 2008. Five period measures of longevity. Pages 237–245 *in* D. Barbi, J. Bongaarts, and J. W. Vaupel, editors, *How Long Do We Live? Demographic Models and Reflections on Tempos Effects*. Springer, Rostock, Germany.

Bongaarts, J., and G. Feeney. 2008. Estimating mean lifetime. Pages 11–27 *in* D. Barbi, J. Bongaarts, and J. W. Vaupel, editors, *How Long Do We Live? Demographic Models and Reflections on Tempos Effects*. Springer, Rostock, Germany.

Bongaarts, J., and R. G. Potter. 1983. *Fertility, Biology and Behavior: An Analysis of the Proximate Determinants*. Academic Press, New York.

Borkin, M. A., A. V. Z. Bylinski, P. Isola, S. Sunkavalli, A. Oliva, and H. Pfister. 2013. What makes a visulaization memorable? *IEEE Transactions on Visualization and Computer Graphics* 19: 2306–2315.

Bowers, N. L., H. U. Gerber, J. C. Hickman, D. A. Jones, and C. J. Nesbitt. 1986. *Actuarial Mathematics*. Society of Actuaries, Itasca, IL.

Boyden, N. B., and J. R. Carey. 2010. From one-and-done to seasoned veterans: A demographic analysis of individual career length in Major League Soccer. *Journal of Quantitative Analysis in Sports* 6. doi: 10.2202/1559–0410.1261.

Brault, S., and H. Caswell. 1993. Pod-specific demography of killer whales (*Orcinus orca*). *Ecology* 74: 1444–1454.

Briney, K. 2015. *Data Management for Researchers*. Pelagic Publishing, Exeter, UK.

Brink, D. 2012. A (probably) exact solution to the birthday problem. *Ramanujan Journal* 28: 223–238.

Bromer, N. 1987. Superexponentiation. *Mathematics Magazine* 60: 169–174.

Brommer, J. E. 2000. The evolution of fitness in life-history theory. *Biological Reviews* 75: 377–404.

Brouard, N. 1986. Structure et dynamique des populations la pyramide des annees a vivre, aspects nationaux et examples regionaux [Structure and dynamics of population pyramids of the years to live, national aspects and regional examples]. *Espace, populations, sociétés: Visages de la population de la France* 4: 157–168.

Brouard, N. 1989. Mouvements et modeles de population [Population movements and models]. Institut de Formation et de Recherche Démographiques, Yaoundé, Cameroon.

Brown, N. J. L., C. J. Albers, and S. J. Richie. 2017. Contesting the evidence for limited human lifespan. *Nature* 546: E6.

Bruss, F. T. 2000. Sum the odds to one and stop. *Annals of Probability* 28: 1384–1391.

Burch, T. 2018. *Model-Based Demography: Essays on Integrating Data, Technique and Theory*. Springer, Cham, Switzerland.

Burnham, K. P., D. R. Anderson, and J. L. Laake. 1980. Estimation of density from line transect sampling of biological populations. *Wildlife Monographs* 72: 1–202.

Butchart, S. H. M., M. Walpole, B. Collen, et al. 2010. Global biodiversity: Indicators of recent declines. *Science* 10: 1164–1168. doi: 10.1126/science.1187512.

Butterick, M. 2015. *Typography for Lawyers*. 2nd ed. O'Connor's, Houston.

Campana, S. E., and S. R. Thorrold. 2001. Otoliths, increments, and elements: Keys to a comprehensive understanding of fish populations? *Canadian Journal of Fisheries and Aquatic Science* **58**: 30–38.

Cannon, J. R. 1996. Whooping crane recovery: A case study in public and private cooperation in the conservation of endangered species. *Conservation Biology* **10**: 813–821.

Canudas-Romo, V., S. Mazzuco, and L. Zanotto. 2018. Measures and models of mortality. In A. S. R. Srinivasa Rao and C. R. Rao, editors, *Handbook of Statistics 39: 405–442.* North-Holland, Amsterdam.

Carey, J. R. 1989. The multiple decrement life table: A unifying framework for cause-of-death analysis in ecology. *Oecologia* **78**: 131–137.

Carey, J. R. 1993. *Applied Demography for Biologists with Special Emphasis on Insects.* Oxford University Press, New York.

Carey, J. R. 1995. Insect demography. Pages 289–303 *in* W. A. Nierenberg, editor, *Encyclopedia of Environmental Biology.* Academic Press, San Diego.

Carey, J. R. 2001. Insect biodemography. *Annual Review of Entomology* **46**: 79–110.

Carey, J. R. 2002. Longevity minimalists: Life table studies of two northern Michigan adult mayflies. *Experimental Gerontology* **37**: 567–570.

Carey, J. R. 2011. Biodemography of the Medfly: Aging, longevity, and adaptation in the wild. *Experimental Gerontology* **46**: 404–411.

Carey, J. R. 2019. Aging in the wild, residual demography and discovery of a stationary population identity. In R. Sears, R. Lee, and O. Burger, editors, *Human Evolutionary Demography.* Open Book Publishers, Cambridge, UK (in press).

Carey, J. R., and J. W. Bradley. 1982. Developmental rates, vital schedules, sex ratios and life tables of *Tetranychus urticae, T. turkestani* and *T. pacificus* (Acarina: Tetranychidae) on cotton. *Acarologia* **23**:333–345.

Carey, J. R., L. Harshman, P. Liedo, H.-G. Müller, J.-L. Wang, and Z. Zhang. 2008. Longevity-fertility trade-offs in the tephritid fruit fly, *Anastrepha ludens,* across dietary-restriction gradients *Aging Cell* **7**: 470–477.

Carey, J. R., and D. S. Judge. 2000a. *Longevity Records: Life Spans of Mammals, Birds, Reptiles, Amphibians and Fishes.* Odense University Press, Odense, Denmark.

Carey, J. R., and D. S. Judge. 2000b. The mortality dynamics of aging. *Generations* **24**:19–24.

Carey, J. R., and D. Krainacker. 1988. Demographic analysis of tetranychid spider mite populations: Extensions of stable theory. *Experimental and Applied Acarology* **4**:191–210.

Carey, J. R., P. Liedo, H.-G. Müller, J.-L. Wang, and J.-M. Chiou. 1998a. Relationship of age patterns of fecundity to mortality, longevity, and lifetime reproduction in a large cohort of Mediterranean fruit fly females. *Journal of Gerontology: Biological Sciences* **53A**: B245–B251.

Carey, J. R., P. Liedo, H.-G. Müller, J.-L. Wang, and J. W. Vaupel. 1998b. A simple graphical technique for displaying individual fertility data and cohort survival: Case study of 1000 Mediterranean fruit fly females. *Functional Ecology* **12**: 359–363.

Carey, J. R., P. Liedo, H.-G. Müller, J.-L. Wang, D. Senturk, and L. Harshman. 2005. Biodemography of a long-lived tephritid: Reproduction and longevity in a large cohort of Mexican fruit flies, *Anastrepha ludens. Experimental Gerontology* **40**: 793–800.

Carey, J. R., P. Liedo, D. Orozco, M. Tatar, and J. W. Vaupel. 1995. A male-female longevity paradox in Medfly cohorts. *Journal of Animal Ecology* **64**: 107–116.

Carey, J. R., P. Liedo, D. Orozco, and J. W. Vaupel. 1992. Slowing of mortality rates at older ages in large Medfly cohorts. *Science* **258**: 457–461.

Carey, J. R., H.-G. Müller, J.-L. Wang, N. T. Papadopoulos, A. Diamantidis, and N. A. Kouloussis. 2012. Graphical and demographic synopsis of the captive cohort method for estimating population age structure in the wild. *Experimental Gerontology* **47**: 787–791.

Carey, J. R., N. Papadopoulos, H.-G. Müller, B. Katsoyannos, N. Kouloussis, J.-L. Wang, K. Wachter, W. Yu, and P. Liedo. 2008. Age structure changes and extraordinary life span in wild Medfly populations. *Aging Cell* **7**: 426–437.

Carey, J. R., N. T. Papadopoulos, S. Papanastasiou, A. Diamanditis, and C. T. Nakas. 2012. Estimating changes in mean population age using the death distributions of live-captured Medflies. *Ecological Entomology* **37**: 359–369.

Carey, J. R., S. Silverman, and A. S. R. S. Rao. 2018. Chapter 5: The life table population identity: Discovery, formulations, proof, extensions and applications. Pages 155–186 *in* A. S. R. Srinivasa Rao and C. R. Rao, editors, *Handbook of Statistics 39*. North-Holland, Amsterdam.

Carey, J. R., and S. Tuljapurkar, editors. 2003. *Life Span: Evolutionary, Ecological and Demographic Perspectives*. Population and Development Review, New York.

Carey, J. R., and R. Vargas. 1985. Demographic analysis of insect mass rearing: Case study of three tephritids. *Journal of Economic Entomology* **78**: 523–527.

Carey, J. R., and J. W. Vaupel. 2005. Biodemography. Pages 625–658 *in* D. Poston and M. Micklin, editors, *Handbook of Population*. Kluwer Academic/Plenum Publishers, New York.

Carey, J. R., and J. W. Vaupel. 2019. Biodemography. In D. Poston and M. Micklin, editors, *Handbook of Population*. Kluwer Academic/Plenum Publishers, New York (forthcoming).

Carey, J. R., P. Yang, and D. Foote. 1988. Demographic analysis of insect reproductive levels, patterns and heterogeneity: Case study of laboratory strains of three Hawaiian tephritids. *Entomologia Experimentalis et Applicata* **46**: 85–91.

Carmichael, G. A. 2016. *Fundamentals of Demographic Analysis: Concepts, Measures and Methods*. Springer, Cham, Switzerland.

Carson, R. 1962. *Silent Spring*. Houghton Mifflin, Boston.

Carter, M. 2013. *Designing Science Presentations*. Academic Press, London.

Caselli, G., and J. Vallin. 2006. Chapter 4: Population dynamics: Movement and structure. Pages 23–48 *in* G. Caselli, J. Vallin, and G. Wunsch, editors, *Demography: Analysis and Synthesis*. Academic Press, Amsterdam.

Caselli, G., J. Vallin, and G. Wunsch, editors. 2006a. *Demography: Analysis and Synthesis—A Treatise in Population*. Four-volume set. Academic Press, Amsterdam.

Caselli, G., J. Vallin, and G. Wunsch. 2006b. Chapter 20: Population models. Pages 249–267 *in* G. Caselli, J. Vallin, and G. Wunsch, editors, *Demography: Analysis and Synthesis*. Academic Press, Amsterdam.

Casterline, J. B. 2003. Demographic transition. In P. Demeny and G. McNicoll, editors, *Encyclopedia of Population*. Gale Group, New York.

Caswell, H. 1978. A general formula for the sensitivity of population growth rate to changes in life history parameters. *Theoretical Population Biology* **14**: 215–230.

Caswell, H. 1997. Methods of matrix population analysis. In S. Tuljapurkar and H. Caswell, editors, *Structures: Population Models in Marine, Terrestrial, and Freshwater Systems*. Chapman and Hall, New York.

Caswell, H. 2000. Prospective and retrospective perturbation analysis: Their roles in conservation biology. *Ecology* **81**: 619–627.

Caswell, H. 2001. *Matrix Population Models*. Sinauer Associates, Sunderland, MA.

Caswell, H. 2012. Matrix models and sensitivity analysis of populations classified by age and stage: A vec-permutation matrix approach. *Theoretical Ecology* **5**: 403–417.

Caswell, H., and R. Salguero-Gómez. 2013. Age, stage and senescence in plants. *Journal of Ecology* **101**: 585–595.

Catts, E. P., and M. L. Goff. 1992. Forensic entomology in criminal investigations. *Annual Review of Entomology* **37**: 253–272.

Caughley, G. 1977. *Analysis of Vertebrate Populations*. John Wiley & Sons, Chichester, UK.

Caughley, G. 1981. Overpopulation. Pages 7–19 *in* P. A. Jewell, S. Holt, and D. Hart, editors, *Problems in Management of Locally Abundant Wild Mammals*. Academic Press, New York.

Caughley, G. 1983. Dynamics of large mammals and their relevance to culling. Pages 115–126 *in* R. N. Owen-Smith, editor, *Management of Large Mammals in African Conservation Areas*. Haum Educational Publishers, Pretoria.

Chamberlain, A. T. 2006. *Demography in Archaeology*. Cambridge University Press, Cambridge, UK.

Charlesworth, B. 1994. *Evolution in Age-Structured Populations*. Cambridge University Press, Cambridge, UK.

Cheung, S. L. K., J.-M. Robine, E. J.-C. Tu, and G. Caselli. 2005. Three dimensions of the survival curve: Horizontalization, verticalization, and longevity extension. *Demography* 42: 243–258.

Chiang, C. L. 1984. *The Life Table and Its Applications*. Robert E. Krieger Publishing, Malabar, FL.

Chisumpa, V. H., and C. O. Odimegwu. 2018. Decomposition of age- and cause-specific adult mortality contributions to the gender gap in life expectancy from census and survey data in Zambia. *SSM—Population Health* 5: 218–226.

Choquet, R., L. Rouan, and R. Pradel. 2009. Program E-Surge: A software application for fitting multievent models. Pages 845–865 *in* D. L. Thomson, E. G. Cooch, and M. J. Conroy, editors, *Modeling Demographic Processes in Marked Populations*. Springer US, Boston.

Christensen, K., T. E. Johnson, and J. W. Vaupel. 2006. The quest for genetic determinants of human longevity: Challenges and insights. *Nature Reviews Genetics* 7: 436–447.

Clark, G. 2014. *The Son Also Rises*. Princeton University Press, Princeton, NJ.

Cleveland, W. S. 1984. Graphical methods for data presentation: Full scale breaks, dot charts, and multibased logging. *American Statistician* 38: 270–280.

Coale, A. J. 1957. How the age distribution of a human population is determined. *Cold Spring Harbor Symposium on Quantitative Biology* 22: 83–89.

Coale, A. J. 1972. *The Growth and Structure of Human Populations*. Princeton University Press, Princeton, NJ.

Coale, A. J., and T. J. Trussell. 1974. Model fertility schedules: Variations in the age of childbearing in human populations. *Population Index* 40: 185–258.

Cochran, M. E., and S. Ellner. 1992. Simple methods for calculating age-based life history parameters for stage-structured populations. *Ecological Monographs* 62: 345–364.

Cohen, J. E. 1979. Ergodic theorems in demography. *Bulletin of the American Mathematical Society* 1: 275–295.

Cohen, J. E. 1984. Demography and morbidity: A survey of some interactions. Pages 199–222 *in* N. Keyfitz, editor, *Population and Biology*. Ordina Editions, Liege, Belgium.

Cohen, J. E. 1995. *How Many People Can the Earth Support?* W. W. Norton, New York.

Colchero, F., and J. S. Clark. 2012. Bayesian inference on age-specific survival for censored and truncated data. *Journal of Animal Ecology* 81: 139–149.

Colchero, F., O. R. Jones, and R. Maren. 2012. BaSTA: An R package for Bayesian estimation of age-specific survival from incomplete mark-recapture/recovery data with covariates. *Methods in Ecology and Evolution* 3: 466–470.

Colchero, F., R. Rau, O. R. Jones, J. A. Barthold, D. A. Conde, A. Lenart, L. Nemeth et al. 2016. The emergence of longevous populations. *Proceedings of the National Academy of Sciences* 113: E7681–E7690.

Cole, L. C. 1954. The population consequences of life history phenomena. *Quarterly Review of Biology* 29: 103–137.

Collett, D. 2015. *Modelling Survival Data in Medical Research*. 3rd ed. CRC Press, Boca Raton, FL.

ColorBrewer. 2018. http://www.personal.psu.edu/cab38/ColorBrewer/ColorBrewer_intro.html.

Conde, D. A., F. Colchero, M. Gusset, P. Pearce-Kelly, O. Byers, N. Flesness, R. K. Browne, and O. R. Jones. 2013. Zoos through the lens of the IUCN Red List: A global metapopulation approach to support conservation breeding programs. *PLoS ONE* 8: e80311.

Conde, D. A., N. Flesness, F. Colchero, O. R. Jones, and A. Scheuerlein. 2011. An emerging role of zoos to conserve biodiversity. *Science* 331: 1390.

Conde, D. A., J. Staerk, F. Colchero, R. da Silva, J. Schöley, H. M. Baden, L. Jouvet, J. E. Fa, H. Syed, E. Jongejans, S. Meiri, J.-M. Gaillard, S. Chamberlain, J. Wilcken, O. R. Jones,

J. P. Dahlgren, U. K. Steiner, L. M. Bland, I. Gomez-Mestre, J.-D. Lebreton, J. González Vargas, N. Flesness, V. Canudas-Romo, R. Salguero-Gómez, O. Byers, T. Bjørneboe Berg, A. Scheuerlein, S. Devillard, D. S. Schigel, O. A. Ryder, H. P. Possingham, A. Baudisch, and J. W. Vaupel. 2019. Data gaps and opportunities for comparative and conservation biology. *Proceedings of the National Academy of Sciences* **116**: 9658–64.

Connelly, J. W., J. H. Gammonley, and T. W. Keegan. 2012. Harvest management. Pages 202–231 *in* N. J. Silvy, editor, *The Wildlife Techniques Manual: Research*. Johns Hopkins University Press, Baltimore.

Conway, W. G. 1986. The practical difficulties and financial implications of endangered species breeding programmes. *International Zoo Yearbook* **24**: 210–219.

Cook, D., E.-K. Lee, and M. Majumder. 2016. Data visualization and statistical graphics in big data analysis. *Annual Review of Statistics and Its Application* **3**: 133–159.

Cook, P. E., L. E. Hugo, I. Iturbe-Ormaetxe, C. R. Williams, S. F. Chenoweth, S. A. Ritchie, P. A. Ryan, et al. 2006. The use of transcriptional profiles to predict adult mosquito age under field conditions. *Proceedings of the National Academy of Sciences* **103**: 18060–18065.

Cook, P. E., C. J. McMeniman, and S. L. O'Neil. 2008. Modifying insect population age structure to control vector-borne disease. *Advances in Experimental Medicine and Biology* **627**: 126–140.

Cook, P. E., and S. P. Sinkins. 2010. Transcriptional profiling of *Anopheles gambiae* mosquitoes for adult age estimation. *Insect Molecular Biology* **19**: 745–751.

Cooley, D. M., et al. 2003. Exceptional longevity in pet dogs is accompanied by cancer resistance and delayed onset of major diseases. *Journal of Gerontology: Biological Sciences* **58A**: 1078–1084.

Cordes, E. E., D. C. Bergquist, M. L. Redding, and C. R. Fisher. 2007. Patterns of growth in cold-seep vestimenferans including *Seepiophila jonesi*: A second species of long-lived tubeworm. *Marine Ecology* **28**: 160–168.

Cormack, R. M. 1964. Estimates of survival from the sighting of marked animals. *Biometrika* **51**: 429–438.

Costa, D. L. 1998. *The Evolution of Retirement. An American Economic History, 1880–1990*. University of Chicago Press, Chicago.

Costa, J. T. 2017. *Darwin's Backyard: How Small Experiments Led to a Big Theory*. W. W. Norton, New York.

Coulson, T. 2012. Integral projection models, their construction and use in posing hypotheses in ecology. *Oikos* **121**: 1337–1350.

Coulson, T., G. M. Mace, E. Hudson, and H. Possingham. 2001. The use and abuse of population viability analysis. *Trends in Ecology and Evolution* **16**: 219–221.

Coumans, A. M., M. Cruyff, P. G. M. VanderHeijden, J. Wold, and H. Schmeets. 2017. Estimating homelessness in the Netherlands using a capture-recapture approach. *Social Indicator Research* **130**: 189–212.

Cousins, J. A., J. P. Sadler, and J. Evans. 2008. Exploring the role of private wildlife ranching as a conservation tool in South Africa: Stakeholder perspectives. *Ecology and Society* **13**: 43. https://www.ecologyandsociety.org/vol13/iss2/art43/.

Crafts, N. F. R. 1978. Average age at first marriage for women in the mid-nineteenth-century England and Wales: A cross-section study. *Population Studies* **32** :21–25.

Crick, F. 1986. The challenge of biotechnology. *Humanist* **46**: 8–9, 32.

Crimmins, E. M. 2015. Lifespan and healthspan: Past, present, and promise. *Gerontologist* **55**: 901–911.

Crimmins, E. M., and H. Beltran-Sanchez. 2011. Mortality and morbidity trends: Is there compression of morbidity? *Journal of Gerontology: Social Sciences* **66B**: 75–86.

Crimmins, E. M., M. D. Hayward, and Y. Saito. 1994. Changing mortality and morbidity rates and the health status and life expectancy of the older population. *Demography* **31**: 159–175.

Crimmins, E. M., M. D. Hayward, and Y. Saito. 1996. Differentials in active life expectancy in the older population of the United States. *Journal of Gerontology: Social Sciences* **51B**: S111–S120.

Croft, D. P., L. J. N. Brent, D. W. Franks, and M. A. Cant. 2015. The evolution of prolonged life after reproduction. *Trends in Ecology & Evolution* **30**: 407–416.

Crone, E. E., et al. 2013. Ability of matrix models to explain the past and predict the future of plant populations. *Conservation Biology* **27**: 968–978.

Crone, E. E., E. S. Menges, M. M. Ellis, T. Bell, P. Bierzychudek, J. Ehrlén, T. N. Kaye, et al. 2011. How do plant ecologists use matrix population models? *Ecology Letters* **14**: 1–8.

Crouse, D. T., L. B. Crowder, and H. Caswell. 1987. A stage-based population model for loggerhead sea turtles and implications for conservation. *Ecology* **68**: 1412–1423.

Curtsinger, J. W. 2015. The retired fly: Detecting life history transition in individual *Drosophila melanogaster* females. *Journals of Gerontology: Series A* **70**: 1455–1460.

Curtsinger, J. W. 2016. Retired flies, hidden plateaus, and the evolution of senescence in *Drosophila melanogaster*. *Evolution* **70**: 1297–1306.

Curtsinger, J. W., H. H. Fukui, D. R. Townsend, and J. W. Vaupel. 1992. Demography of genotypes: Failure of the limited life-span paradigm in *Drosophila melanogaster*. *Science* **258**: 461–463.

Damschen, E. I., D. V. Baker, G. Bohrer, R. Nathan, J. L. Orrock, J. R. Turner, L. A. Brudvig, et al. 2014. How fragmentation and corridors affect wind dynamics and seed dispersal in open habitats. *Proceedings of the National Academy of Sciences* **111**: 3484–3489.

Darwin, Charles. 1859. *On the Origin of Species by Means of Natural Selection*. J. Murray, London.

Dau, B. K., K. V. K. Gilardi, F. M. Gulland, A. Higgins, J. B. Holcomb, J. S. Leger, and M. H. Ziccardi. 2009. Fishing gear-related injury in California marine wildlife. *Journal of Wildlife Diseases* **45**: 355–362.

Davis, M. A. 2011. Invasion biology. Pages 364–369 *in* D. Simberloff and M. Rejmanek, editors, *Encyclopedia of Biological Invasions*. University of California Press, Berkeley.

Dawe, E. G., J. M. Hoenig, and X. Xu. 1993. Change-in-ratio and index-removal methods for population assessment and their application to snow crab (*Chionoecetes opilio*). *Canadian Journal of Fisheries and Aquatic Science* **50**: 1467–1476.

Dawid, H., G. Feichtinger, J. R. Goldstein, and V. M. Veliov. 2009. Keeping a learned society young. *Demographic Research* **20**: 541–558.

Dawkins, R. 2004. *The Ancestor's Tale*. Houghton Mifflin, Boston.

Deevey, E. S. J. 1947. Life tables for natural populations of animals. *Quarterly Review of Biology* **22**: 283–314.

deKroon, H. A., A. Plaisier, J. v. Groenendael, and H. Caswell. 1986. Elasticity: The relative contribution of demographic parameters to population growth rate. *Ecology* **67**: 1427–1431.

DeLury, D. B. 1954. The assumptions underlying estimates of mobile populations. In O. Kempthorne, editor, *Statistics and Mathematics in Biology*. Iowa State College Press, Ames.

Demetrius, L. 1978. Adaptive value, entropy and survivorshop. *Nature* **275**: 213–214.

Desena, M. L., J. D. Edman, J. M. Clark, S. B. Symington, and T. W. Scott. 1999. *Aedes aegypti* (Diptera: Culicidae) age determination by cuticular hydrocarbon analysis of female legs. *Journal of Medical Entomology* **36**: 824–830.

deVos, V., R. G. Bengis, and H. J. Coetzee. 1983. Population control of large mammals in Kruger National Park. Pages 213–231 *in* R. N. Owen-Smith, editor, *Management of Large Mammals in African Conservation Areas*. Haum Educational Publishers, Pretoria.

Dharmalingam, A. 2004. Reproductivity. Pages 407–428 *in* J. S. Siegel and D. A. Swanson, editors, *The Methods and Materials of Demography*. Elsevier Academic Press, Amsterdam.

Djibouti, T. 2013. The greatest walk. https://www.nationalgeographic.org/projects/out-of-eden -walk/media/2013-03-the-greatest-walk/.

Doblhammer, G. 2004. *The Late Life Legacy of Very Early Life*. Springer, Berlin.

Dobzhansky, T. 1973. Nothing in biology makes sense except in the light of evolution. *American Biology Teacher* **35**: 125–129.

Dong, X., B. Milholland, and J. Vijg. 2016. Evidence for a limit to human lifespan. *Nature* **538**: 257–259.

Dorland, S. G. H. 2018. The touch of a child: An analysis of fingernail impressions on Late Woodland pottery to identify childhood material interactions. *Journal of Archaeological Science: Reports* **21**: 298–304.

Duarte, N. 2008. *Slideology: The Art and Science of Creating Great Presentations*. O'Reilly, Beijing.

Duarte, N. 2010. *Resonate: Present Visual Stories That Transform Audiences*. John Wiley & Sons, New York.

Dublin, H. T., and J. O. Ogutu. 2015. Population regulation of African buffalo in the Mara Serengeti ecosystem. *Wildlife Research* **42**: 382–393.

Dublin, L. I., and A. J. Lotka. 1925. On the true rate of natural increase. *Journal of the American Statistical Association* **20**: 305–339.

Dudasik, S. W. 1980. Victimization in natural disaster. *Disasters* **4**: 329–338.

duToit, R., editor. 2006. *Guidelines for Implementing SADC Rhino Conservation Strategies*. SADC Regional Programme for Rhino Conservation, Harare, Zimbabwe.

Dyer, R. J. 2015. Population graphs and landscape genetics. *Annual Review of Ecology, Evolution, and Systematics* **46**: 327–342.

Easterling, M. R., S. P. Ellner, and P. M. Dixon. 2000. Size-specific sensitivity: Applying a new structured population model. *Ecology* **81**: 694–708.

Eberhart, L. L. 1969. Population estimates from recapture frequencies. *Journal of Wildlife Management* **33**: 28–39.

Edmonson, A. J., I. J. Lean, I. D. Weaver, T. Farver, and G. Webster. 1989. A body condition scoring chart for Holstein dairy cows. *Journal of Dairy Science* **72**: 68–78.

Efford, M. 2004. Density estimation in live-trapping studies. *Oikos* **106**: 598–610.

Egidi, V., and L. Frova. 2006. Chapter 47: Relationship between morbidity and mortality by cause. Pages 81–92 *in* G. Caselli, J. Vallin, and G. Wunsch, editors, *Demography: Analysis and Synthesis*. Elsevier, Amsterdam.

Ekamper, P., F. van Poppel, A. D. Stein, and L. H. Lumey. 2014. Independent and additive association of prenatal famine exposure and intermediary life conditions with adult mortality between age 18–63 years. *Social Science & Medicine* **119**: 232–239.

Elandt-Johnson, R. C. 1980. *Survival Models and Data Analysis*. John Wiley and Sons, New York.

Ellison, P. T., and M. T. O'Rourke. 2000. Population growth and fertility regulation. Pages 553–586 *in* S. Stinson, B. Bogin, R. Huss-Ashmore, and D. O. Rourke, editors, *Human Biology: An Evolutionary and Biocultural Perspective*. Wiley-Liss, New York.

Ellner, S. P., and M. Rees. 2006. Integral projection models for species with complex demography. *American Naturalist* **167**: 410–428.

Engelman, M., C. L. Seplaki, and R. Varadhan. 2017. A quiescent phase in human mortality? Exploring the ages of least vulnerability. *Demography* **54**: 1097–1118.

Enright, N. J., M. Franco, and J. Silvertown. 1995. Comparing plant life-histories using elasticity analyses: The importance of life-span and the number of life-cycle stages. *Oecologia* **104**: 79–84.

Ergon, T., Ø. Borgan, C. R. Nater, and Y. Vindenes. 2018. The utility of mortality hazard rates in population analyses. *Methods in Ecology and Evolution*. doi: 10.1111/2041-210X.13059.

Estes, R. D. 1991. *The Behavior Guide to African Mammals, Including Hoofed Mammals, Carnivores, Primates*. University of California Press, Berkeley.

Euler, L. 1760. A general investigation into the mortality and multiplication of the human species. Translated to English from French by N. and B. Keyfitz. *Theoretical Population Biology* **1** (1970): 307–314.

Exter, T. G. 1986. How to think about age. *American Demographics* **8**: 50–51.

Ezard, T. H. G., J. M. Bullock, H. J. Dalgleish, A. Millon, F. Pelletier, A. Ozgul, and D. N. Koons. 2010. Matrix models for a changeable world: The importance of transient dynamics in population management. *Journal of Applied Ecology* **47**: 515–523.

Ezenwa, V. O., A. E. Jolles, and M. P. O'Brien. 2009. A reliable body condition scoring technique for estimating condition in African buffalo. *African Journal of Ecology* **47**: 476–481.

Fa, J. E., S. M. Funk, and D. O'Connell. 2011. *Zoo Conservation Biology*. Cambridge University Press, Cambridge, UK.

Fa, J. E., J. Gusset, N. Flesness, and D. A. Conde. 2014. Zoos have yet to unveil their full conservation potential. *Animal Conservation* **17**: 97–100.

Fazio, S. 2014. The impacts of African elephant (*Loxodonta africana*) on biodiversity within protected areas of Africa and a review of management options. Grand Valley State University, Grand Rapids, MI. Honors project. http://scholarworks.gvsu.edu/honorsprojects /271.

Feeney, G. 1983. Population dynamics based on birth intervals and parity progression. *Population Studies* **37**: 75–89.

Feeney, G. 2003. Lexis diagram. Pages 586–588 *in* P. Demeny and G. McNicoll, editors, *Encyclopedia of Population*. Macmillan Reference, New York.

Feeney, G. 2006. Increments to life and mortality tempo. *Demographic Research* **14**: 27–46.

Feeney, G. 2013. What is data? *Demography-Statistics-Information Technology Letter* **3** (October 31). http://demographer.com/dsitl/03-what-is-data/.

Feeney, G. 2014. Dot chart or bar graph? The demography-statistics information technology letter. Letter No. 8. http://demographer.com/dsitl/08-cleveland-dot-plots/.

Feinstein, A. R., D. M. Sosin, and C. K. Wells. 1985. The Will Rogers phenomenon: Stage migration and new diagnostic techniques as a source of misleading statistics for survival in cancer. *New England Journal of Medicine* **312**: 1604–1608.

Feller, W. 1950. *An Introduction to Probability Theory and Its Applications*. Wiley, New York.

Ferreira, S. M., C. Greaver, G. A. Knight, M. H. Knight, I. P. J. Smit, and D. Pienaar. 2015. Disruption of rhino demography by poachers may lead to population decline in Kruger National Park, South Africa. *PLoS ONE* **10**: e0127783. https://doi.org/10.1371/journal .pone.0127783.

Few, S. 2013. *Show Me the Numbers: Designing Tables and Graphs to Enlighten*. Analytical Press, Burlingame, CA.

Fienberg, S. E. 1972. The multiple recapture census for closed populations and incomplete 2k contingency tables. *Biometrica* **59**: 591–599.

Finch, C. E. 1990. *Longevity, Senescence, and the Genome*. University of Chicago Press, Chicago.

Finch, C. E. 2012. Evolution of the human lifespan, past, present, and future: Phases in the evolution of human life expectancy in relation to the inflammatory load. *Proceedings of the American Philosophical Society* **156**: 9–44.

Finney, D. J. 1964. *Probit Analysis*. Cambridge University Press, Cambridge, UK.

Fisher, R. A. 1930. *The Genetical Theory of Natural Selection*. Dover Publications, New York.

Fisher, R. A. 1958. *The Genetical Theory of Natural Selection*. 2nd ed. Dover Publications, New York.

Foster, E. A., D. W. Franks, S. Mazzi, S. K. Darden, K. C. Balcomb, J. K. B. Ford, and D. P. Croft. 2012. Adaptive prolonged postreproductive life span in killer whales. *Science* **337**: 1313.

Frank, S. A. 2007. *Dynamics of Cancer*. Princeton University Press, Princeton, NJ.

Frazer, B. D. 1972. Population dynamics and recognition of biotypes in the pea aphid (Homoptera: Aphididae). *Canadian Entomologist* **104**: 1717–1722.

Freedman, V. A., J. D. Kasper, J. C. Cornman, et al. 2011. Validation of new measures of disability and functioning in the National Health and Aging Trends study. *Journal of Gerontology: Medical Sciences* **66A**: 1013–1021.

Freedman, V. A., B. C. Spillman, P. M. Andreski, et al. 2013. Trends in late-life activity limitations in the United States: An update from five national surveys. *Demography* 50: 661–671.

Frey, B. S., D. A. Savage, and B. Torgler. 2009. Surviving the *Titanic* disaster: Economic, natural and social determinants. CESifo Working Paper No. 2551, Category 2: Public Choice. CESifo Group, Munich.

Fries, J. F. 1980. Aging, natural death, and the compression of morbidity. *New England Journal of Medicine* 303: 130–135.

Frisbie, W. P. 2005. Infant mortality. Pages 251–282 *in* D. Poston and M. Micklin, editors, *Handbook of Population*. Springer, New York.

Fulton, W. C. 2012. The population growth and control of African elephants in Kruger National Park, South Africa: Modeling, managing, and ethics concerning a threatened species. Master's thesis, Regis University, Denver. https://epublications.regis.edu/theses/560.

Gage, T. B. 2001. Age-specific fecundity of mammalian populations: A test of three mathematical models. *Zoo Biology* 20: 487–499.

Garrett-Jones, C. 1964. Prognosis for interruption of malaria transmission through assessment of the mosquito's vectorial capacity. *Nature* 204: 1173–1175.

Gavrilov, L. A., and N. S. Gavrilova. 2006. Reliability theory of aging and longevity. Pages 3–42 *in* E. J. Masoro and S. N. Austad, editors, *Handbook of the Biology of Aging*. Elsevier/Academic Press, San Diego.

Gavrilov, L. A., and N. S. Gavrilova. 2012. Mortality measurement at advanced ages: A study of the Social Security Administration death master file. *North American Actuarial Journal* 15: 432–447.

Gavrilov, L. A., and N. S. Gavrilova. 2015. New developments in the biodemography of aging and longevity. *Gerontology* 61: 364–371.

Gavrilov, L. A., and N. S. Gavrilova. 2019. Late-life mortality is underestimated because of data errors. *PLOS Biology* 17: e3000148.

Gay, H. 2012. Before and after *Silent Spring*: From chemical pesticides to biological control and integrated pest management—Britain, 1945–1980. *AMBIX* 59: 88–108.

Gee, K. L., J. H. Holman, M. K. Causey, A. N. Rossi, and J. B. Armstrong. 2002. Aging white-tailed deer by tooth replacement and wear: A critical evaluation of a time-honored technique. *Wildlife Society Bulletin* (1973–2006) 30: 387–393.

Gehlenborg, N., and B. Wong. 2012a. Heat maps. *Nature Methods* 9: 213.

Gehlenborg, N., and B. Wong. 2012b. Into the third dimension. *Nature Methods* 9: 851.

Gehlenborg, N., and B. Wong. 2012c. Mapping quantitative data to color. *Nature Methods* 9: 769.

Gehlenborg, N., and B. Wong. 2012d. Power of the plane. *Nature Methods* 9: 935.

Geist, V. 1966. Validity of horn segment counts in aging bighorn sheep. *Journal of Wildlife Management* 30: 634–646.

Gennard, D. 2012. *Forensic Entomology*. 2nd ed. Wiley-Blackwell, West Sussex, UK.

Gerade, B. B., S. H. Lee, T. W. Scott, J. D. Edman, L. C. Harrington, S. Kitthawee, J. W. Jones, and J. M. Clark. 2004. Field validation of *Aedes aegypti* (Diptera: Culicidea) age estimation by analysis of cuticular hydrocarbons. *Journal of Medical Entomology* 41: 231–238.

Gershon, N., and W. Page. 2001. What storytelling can do for information visualization. *Communications of the Association of Computer Machinery* 44: 31–37.

Getz, W. M. 1984. Population dynamics: A per capita resource approach. *Journal of Theoretical Biology* 108: 623–643.

Getz, W. M., and R. G. Haight. 1989. *Population Harvesting: Demographic Models of Fish, Forest, and Animal Resources*. Princeton University Press, Princeton, NJ.

Gill, A. M., and J. Foley. 1996. Predicting educational attainment for a minor child: Some further evidence. *Journal of Forensic Economics* 9: 101–112.

Gill, S. P. 1986. The paradox of prediction. *Daedalus* 115: 17–48.

Goldenberg, D. 2015. Why the oldest person in the world keeps dying. *New York Times*, May 25, 2015.

Goldman, N., and G. Lord. 1986. A new look at entropy and the life table. *Demography* 23: 275–282.

Goldstein, J. R. 2009. Life lived equals life left in stationary populations. *Demographic Research* 20: 3–6.

Goldstein, J. R., and K. W. Wachter. 2006. Relationships between period and cohort life expectancy: Gaps and lags. *Population Studies* 60: 257–269.

Gompertz, B. 1825. On the nature of the function expressive of the law of human mortality, and on a new mode of determining the value of life contingencies. *Philosophical Transactions of the Royal Society of London* 115: 513–585.

Goodman, D. 1978. Demographic intervention for closely managed populations. Pages 171–195 *in* M. E. Soule and B. A. Wilcox, editors, *Conservation Biology: An Evolutionary-Ecological Perspective*. Sinauer Associates, Sunderland, MA.

Goodman, L. A. 1953. Population growth of the sexes. *Biometrics* 9: 212–225.

Goodman, L. A. 1967. On the age-sex composition of the population that would result from given fertility and mortality conditions. *Demography* 4: 423–441.

Goodman, L. A. 1971. On the sensitivity of the intrinsic growth rate to changes in the age-specific birth and death rates. *Theoretical Population Biology* 2: 339–354.

Gourbin, C., and G. Wunsch. 2006. Chapter 40: Health, illness, and death. Pages 5–12 *in* G. Caselli, J. Vallin, and G. Wunsch, editors, *Demography: Analysis and Synthesis*. Elsevier, Amsterdam.

Grant, A. 1997. Selection pressures on vital rates in density dependent populations. *Proceedings of the Royal Society B* 264: 303–306.

Grant, A., and T. G. Benton. 2000. Elasticity analysis for density-dependent populations in stochastic environments. *Ecology* 81: 680–693.

Grayson, D. K. 1990. Donner party deaths: A demographic assessment. *Journal of Anthropological Research* 46: 223–242.

Griffith, A. B., R. Salguero-Gómez, C. Merow, and S. McMahon. 2016. Demography beyond the population. *Journal of Ecology* 104: 271–280.

Grotewiel, M. S., I. Martin, P. Bhandari, and E. Cook-Wiens. 2005. Functional senescence in *Drosophila melanogaster*. *Ageing Research Reviews* 4: 372–397.

Guillot, M. 2003. The cross-sectional average length of life (CAL): A cross-sectional mortality measure that reflects the experience of cohorts. *Population Studies* 57: 41–54.

Guillot, M. 2005. Life tables. Pages 594–602 *in* D. Poston and M. Micklin, editors, *Handbook of Population*. Springer, New York.

Gurven, M., and R. Walker. 2006. Energetic demand of multiple dependents and the evolution of slow human growth. *Proceedings of the Royal Society B: Biological Sciences* 273: 835–841.

Hafez, E. S. E., and M. W. Schein. 1962. The behavior of cattle. Pages 256–296 *in* E. S. E. Hafez, editor, *The Behavior of Domestic Animals*. Bailliere, Tindall & Cox, London.

Hamerman, D. 2010. Can biogerontologists and geriatricians unite to apply aging science to health care in the decade ahead? *Journal of Gerontology: Biological Sciences* 65A: 1193–1197.

Hammer, M., and R. Foley. 1996. Longevity, life history and allometry: How long did hominids live? *Human Evolution* 11: 61–66.

Hanski, I. 1998. Metapopulation dynamics. *Nature* 396: 41–49.

Hanski, I., and M. E. Gilpin. 1997. *Metapopulation Biology: Ecology, Genetics, and Evolution*. Academic Press, San Diego.

Haramis, G. M., J. D. Nichols, K. H. Pollock, and J. E. Hines. 1986. The relationship between body mass and survival of wintering canvasbacks. *Auk* 103: 506–514.

Hargrove, J. W., R. Ouifki, and J. E. Ameh. 2011. A general model for mortality in adult tsetse, *Medical and Veterinary Entomology* 25: 385–94.

Harper, J. L., and J. White. 1974. The demography of plants. *Annual Review of Ecology and Systematics* 5: 419–463.

Harper, S. 2018. Demography. A very short introduction. Oxford University Press, Oxford, UK.

Harvey, P. H., R. D. Martin, and T. H. Clutton-Brock. 1987. Life histories in comparative perspective. Pages 181–196 *in* B. B. Smuts, D. L. Cheney, R. M. Seyfarth, R. W. Wrangham, and T. T. Struhsaker, editors, *Primate Societies*. University of Chicago Press, Chicago.

Hauer, M., J. Baker, and W. Brown. 2013. Indirect estimates of total fertility rate using child woman/ratio: A comparison with the Bogue-Palmore method. *PLoS ONE* 8: e67226.

Hauser, P. M., and O. D. Duncan, editors. 1959. *The Study of Population*. University of Chicago Press, Chicago.

Haussmann, M. F., and C. M. Vleck. 2002. Telomere length provides a new technique for aging animals. *Oecologia* 130: 325–328.

Hawkes, K. 2003. Grandmothers and the evolution of human longevity. *American Journal of Human Biology* 15: 380–400.

Hawkes, K. 2004. The grandmother effect. *Nature* 428: 128–129.

Hawkes, K., J. O'Connell, and N. Blurton-Jones. 2001. Hunting and nuclear families. *Current Anthropology* 42: 681–709.

Hayes, B. 2012. Methuselah's choice. Bit-player: An amateur's outlook on computation and mathematics. Accessed October 21, 2018 http://bit-player.org/2012/methuselahs-choice.

Hayward, M., E. M. Crimmins, and Y. Saito. 1998. Cause of death and active life expectancy in the older population of the United States. *Journal of Aging and Health* 10: 192–213.

Hayward, M. D., and B. K. Gorman. 2004. The long arm of childhood: The influence of early-life social conditions on men's mortality. *Demography* 41: 87–107.

Hayward, M. D., and D. F. Warner. 2005. The demography of population health. Pages 809–825 *in* D. Poston and M. Micklin, editors, *Handbook of Population*. Springer, New York.

Healy, K., and J. Moody. 2014. Data visualization in sociology. *Annual Review of Sociology* 40: 105–128.

Healy, J. 2019. *Data Visualization: A Practical Introduction*. Princeton University Press, Princeton, NJ.

Heimpel, G. E., and N. J. Mills. 2017. *Biological Control: Ecology and Applications*. Cambridge University Press, Cambridge, UK.

Herzman, C. 1999. The biological embedding of early experience and its effects on health in adulthood. *Annals of the New York Academy of Sciences* 896: 85–95.

Hertz, R., and M. K. Nelson. 2019. *Random Families. Genetic Strangers, Sperm Donor Siblings, and the Creation of New Kin*. Oxford University Press, Oxford, UK.

Hjelmborg, J. v., I. Iachine, A. Skytthe, J. W. Vaupel, M. McGue, M. Koskenvuo, J. Kaprio, et al. 2006. Genetic influence on human lifespan and longevity. *Human Genetics* 119: 312–321.

Ho, J. Y., E. Frankenberg, C. Sumantri, and D. Thomas. 2017. Adult mortality five years after a natural disaster. *Population and Development Review* 43: 467–490.

Hobcraft, J., J. Menken, and S. Preston. 1982. Age, period, and cohort effects in demography: A review. *Population Index* 48: 4–43.

Hoffman, R. 2014. The tensions of storytelling. *American Scientist* 102: 250–253.

Hoffmann, M., C. Hilton-Taylor, A. Angulo, M. Böhm, et al. 2010. The impact of conservation on the status of the world's vertebrates. *Science* 330: 1503.

Höhn, C. 1987. The family life cycle: Needed extensions of the concept. Pages 65–80 *in* J. Bongaarts, T. K. Burch, and K. W. Wachter, editors, *Family Demography: Methods and Their Application*. Clarendon Press, Oxford, UK.

Hook, E. B., and R. R. Regal. 1995. Capture-recapture methods in epidemiology: Methods and limitations. *Epidemiologic Reviews* 17: 243–264.

Horiuchi, S. 2003. Age patterns of mortality. Pages 649–654 *in* P. Demeny and G. McNicoll, editors, *Encyclopedia of Population*. Gale Group, New York.

Horiuchi, S., and A. J. Coale. 1990. Age patterns of mortality for older women: An analysis using the age-specific rate of mortality change with age. *Mathematical Population Studies* 2: 245–267.

Horvitz, C. C. 2011. Demography. Pages 147–150 *in* D. Simberloff and M. Rejmanek, editors, *Encyclopedia of Biological Invasions*. University of California Press, Berkeley.

Horvitz, C. C. 2016. Life history theory: Basics. Pages 384–389 *in* R. M. Kliman, editor, *Encyclopedia of Evolutionary Biology*. Academic Press, Oxford, UK.

Horvitz, C. C., and D. W. Schemske. 1986. Seed dispersal of a neotropical Myrmecochore: Variation in removal rates and dispersal distance. *Biotropica* 18: 319–323.

Horvitz, C., D. Schemske, and H. Caswell. 1997. The relative "importance" of life-history stages to population growth: Prospective and retrospective analyses. Pages 247–271 *in* S. Tuljapurkar and H. Caswell, editors, *Structures: Population Models in Marine, Terrestrial, and Freshwater Systems*. Chapman and Hall, New York.

Hosmer, D. W. J., S. Lemeshow, and S. May. 2008. *Applied Survival Analysis*. Wiley, Hoboken, NJ.

Howell, N. 1979. *The Demography of the Dobe !Kung*. Academic Press, New York.

Huffaker, C. B., and C. E. Kennett. 1966. Biological control of *Parlatoria oleae* (Colvee) through the compensatory action of two introduced parasites. *Hilgardia* 37: 283–335.

Hughes, B. G., and S. Hekimi. 2017. Many possible maximum lifespan trajectories. *Nature* 546: E8–E9.

HumanMortalityDatabase. 2018. Human Mortality Database. University of California, Berkeley (USA), and Max Planck Institute for Demographic Research (Germany). Accessed January 15, 2018. www.mortality.org or www.humanmortality.de.

Hunnicutt, B. J., and M. Krzywinski. 2016a. Neural circuit diagrams. *Nature Methods* 13: 189.

Hunnicutt, B. J., and M. Krzywinski. 2016b. Pathways. *Nature Methods* 13: 5.

Ivanov, S., and V. Kandiah. 2003. Fertility, age-patterns. In P. Demeny and G. McNicoll, editors, *Encyclopedia of Population*. Gale Group, New York.

Jacquard, A. 1984. Concepts of genetics and concepts of demography: Specificities and analogies. Pages 29–40 *in* N. Keyfitz, editor, *Population and Biology*. Ordina Editions, Liege, Belgium.

Jdanov, D. A., V. M. Shkolnikov, A. A. van Raalte, and E. M. Andreev. 2017. Decomposing current mortality differences into initial differences and differences in trends: The contour decomposition method. *Demography* 54: 1579–1602.

Jegou, B., and M. Skinner. 2018. Volume I: Male reproduction. *Encyclopedia of Reproduction*. 2nd ed. Academic Press, New York.

Jennions, M. D., and D. W. Macdonald. 1994. Cooperative breeding in mammals. *Trends in Ecology and Evolution* 9: 89–93.

Jeronimus, B. F., N. Stavrakakis, R. Veenstra, and A. J. Oldehinkel. 2015. Relative Age Effects in Dutch Adolescents: Concurrent and Prospective Analyses. *PLoS ONE* 10: e0128856.

Jolles, A. E. 2007. Population biology of African buffalo (*Syncerus caffer*) at Hluhluwe-iMfolozi Park, South Africa. *African Journal of Ecology* 45: 398–406.

Jolles, A. E., V. O. Ezenwa, R. S. Etienne, W. C. Turner, and H. Olff. 2008. Interactions between macroparasites and microparasites drive infection patterns in free-ranging African buffalo. *Ecology* 89: 2239–2250.

Jolly, G. M. 1965. Explicit estimates from capture-recapture data with both death and immigration-stochastic model. *Biometrika* 52: 225–247.

Jones, S., R. Martin, and D. Pilbeam. 1992. *The Cambridge Encyclopedia of Human Evolution*. Cambridge University Press, Cambridge, UK.

Jones, W. R., A. Scheuerlein, et al. 2013. Diversity of ageing across the tree of life. *Nature* 505: 169–173.

Jonzén, N., P. Lundberg, and A. Gårdmark. 2001. Harvesting spatially distributed populations. *Wildlife Biology* 7: 197–203.

Jordan, C. W. 1967. *Life Contingencies*. Society of Actuaries, Chicago.

Judge, D. S., and J. R. Carey. 2000. Post-reproductive life predicted by primate patterns. *Journal of Gerontology: Biological Sciences* **55A**: B201–B209.

Kallmann, F. J., and J. D. Rainer. 1959. Physical anthropology and demography. Pages 759–790 *in* P. M. Hauser and O. D. Duncan, editors, *The Study of Population*. University of Chicago Press, Chicago.

Kamp, K. 2001. Where have all the children gone? The archaeology of childhood. *Journal of Archaeological Method and Theory* **8**: 1–34.

Kannisto, V. 1994. *Development of Oldest-Old Mortality, 1950–1990: Evidence from 28 Developed Countries*. Odense University Press, Odense, Denmark.

Kannisto, V. 1996. *The Advancing Frontier of Survival*. Odense University Press, Odense, Denmark.

Kaplan, E. L., and P. Meier. 1958. Nonparametric estimation from incomplete observations. *Journal of the American Statistical Association* **53**: 457–481.

Kaplan, H. S. 2002a. Evolutionary demography. Pages 329–336 *in* M. Pagel, editor, *Encyclopedia of Evolution*. Oxford University Press, Oxford, UK.

Kaplan, H. S. 2002b. Life history theory: Human life history. Pages 627–631 *in* M. Pagel, editor, *Encyclopedia of Evolution*. Oxford University Press, Oxford, UK.

Katz, S., L. G. Branch, M. H. Branson, J. A. Papsidero, J. C. Beck, and D. S. Greer. 1983. Active life expectancy. *New England Journal of Medicine* **309**: 1218–1224.

Kawachi, I., and S. V. Subramanian. 2005. Health demography. Pages 787–808 *in* D. Poston and M. Micklin, editors, *Handbook of Population*. Springer, New York.

Kawata, K. 2012. Exorcising of a cage: A review of American zoo exhibits. Part III. *Der Zoologische Garten* **81**: 132–146.

Kelker, G. H. 1940. Estimating deer populations by a different hunting loss in the sexes. *Proceedings of the Utah Academy of Science, Arts and Letters* **17**: 6–69.

Key, C. A., L. C. Aiello, and T. Molleson. 1994. Cranial suture closure and its implications for age estimation. *International Journal of Oseoarchaeology* **4**: 193–207.

Keyfitz, N. 1971. On the momentum of population growth. *Demography* **8**: 71–80.

Keyfitz, N. 1977. *Introduction to the Mathematics of Population with Revisions*. Addison-Wesley, Reading, PA.

Keyfitz, N. 1984a. Introduction: Biology and demography. Pages 1–7 *in* N. Keyfitz, editor, *Population and Biology*. Ordina Editions, Liege, Belgium.

Keyfitz, N., editor. 1984b. *Population and Biology*. Ordina Editions, Liege, Belgium.

Keyfitz, N. 1985. *Applied Mathematical Demography*. 2nd ed. Springer-Verlag, New York.

Keyfitz, N., and J. A. Beekman, editors. 1984. *Demography through Problems*. Springer-Verlag, New York.

Keyfitz, N., and H. Caswell. 2010. *Applied Mathematical Demography*. Springer, New York.

Keyfitz, N., D. Nagnur, and D. Sharma. 1967. On the interpretation of age distributions. *American Statistical Association Journal* **62**: 862–874.

Kientz, J. L., M. E. Barnes, and D. Durben. 2017. Concentration of stocked trout catch and harvest by small number of recreational angles. *Journal of Fisheries Sciences* **11**: 69–76.

Kim, Y. J. 1986. Examination of the generalized age distribution. *Demography* **23**: 451–461.

Kim, Y. J., and J. L. Aron. 1989. On the equality of average age and average expectation of remaining life in a stationary population. *SIAM Review* **31**: 110–113.

Kim, Y. J., R. Schoen, and P. S. Sarma. 1991. Momentum and the growth-free segment of a population. *Demography* **28**: 159–173.

King, R. 2012. A review of Bayesian state-space modelling of capture-recapture-recovery data. Interface Focus. http://rsfs.royalsocietypublishing.org/content/royfocus/early/2012/01/24/rsfs.2011.0078.full.pdf.

Kintner, H. J. 2004. The life table. Pages 301–340 *in* J. S. Siegel and D. A. Swanson, editors, *The Methods and Materials of Demography*. Elsevier/Academic Press, Amsterdam.

Kirkland, J. L., and C. Peterson. 2009. Healthspan, translation, and new outcomes for animal studies of aging. *Journals of Gerontology Series A: Biological Sciences and Medical Sciences* **64A**: 209–212.

Kleinbaum, D. G., and M. Klein. 2012. *Survival Analysis: A Self-Learning Text*. Springer, New York.

Klevezal, G. A. 1996. *Recording Structures of Mammals: Determination of Age and Reconstruction of Life History*. A. S. Balkema, Rotterdam.

Knaflic, C. N. 2015. *Story Telling with Data: A Data Visualization Guide for Business Professionals*. John Wiley & Sons, New York.

Knobil, E., and J. D. Neil, editors. 1998. *Encyclopedia of Reproduction*. Academic Press, San Diego.

Koons, D. N., M. Gamelon, J.-M. Gaillard, L. M. Aubry, R. F. Rockwell, F. Klein, R. Choquet, and O. Gimenez. 2014. Methods for studying cause-specific senescence in the wild. *Methods in Ecology and Evolution* **5**: 924–933.

Kosara, R., and J. Mackinlay. 2013. Storytelling: The next step for visualization. *Computer* **46**: 44–50.

Kouloussis, N. A., N. T. Papadopoulos, B. I. Katsoyannos, H.-G. Müller, J.-L. Wang, Y.-R. Su, F. Molleman, and J. R. Carey. 2011. Seasonal trends in *Ceratitis capitata* reproductive potential derived from live-caught adult females in Greece. *Entomologia Experimentalis et Applicata* **140**: 181–188.

Kouloussis, N. A., N. T. Papadopoulos, H.-G. Müller, J.-L. Wang, M. Mao, B. I. Katsoyannos, P.-F. Duyck, and J. R. Carey. 2009. Life table assay for assessing relative age bias in Medfly capture methods. *Entomologia Experimentalis et Applicata* **132**: 172–181.

Krafsur, E. S., R. D. Moon, and Y. Kim. 1995. Age structure and reproductive composition of summer *Musca autumnalis* (Diptera: Muscidae) populations estimated by pterin concentrations. *Journal of Medical Entomology* **32**: 685–696.

Kramer, K. L. 2010. Cooperative breeding and its significance to the demographic success of humans. *Annual Review of Anthropology* **39**: 417–436.

Kramer, K. L. 2011. The evolution of human parental care and recruitment of juvenile help. *Trends in Ecology and Evolution* **28**: 533–540.

Kramer, K. L. 2014. Why what juveniles do matters in the evolution of cooperative breeding. *Human Nature* **25**: 49–65.

Krebs, C. J. 1999. *Ecological Methodology*. 2nd ed. Benjamin Cummings, Menlo Park, CA.

Kreitlow, K. L. T. 2010. Insect succession in a natural environment. Pages 251–269 *in* J. H. Byrd and J. L. Castner, editors, *Forensic Entomology: The Utility of Arthropods in Legal Investigations*. CRC Press, Boca Raton, FL.

Krzywinski, M. 2013a. Axes, ticks and grids. *Nature Methods* **10**: 275.

Krzywinski, M. 2013b. Elements of visual style. *Nature Methods* **10**: 371.

Krzywinski, M. 2013c. Labels to callouts. *Nature Methods* **10**: 275.

Krzywinski, M., and A. Cairo. 2013. Storytelling. *Nature Methods* **10**: 687.

Krzywinski, M., and E. Savig. 2013. Multidimensional data. *Nature Methods* **10**: 595.

Krzywinski, M., and B. Wong. 2013. Plotting symbols. *Nature Methods* **10**: 451.

Lamb, V. L., and J. S. Siegel. 2004. Health demography. Pages 341–370 *in* J. S. Siegel and D. A. Swanson, editors, *The Methods and Materials of Demography*. Elsevier/Academic Press, Amsterdam.

Land, K. C., and A. Rogers, editors. 1982. *Multidimensional Mathematical Demography*. Academic Press, New York.

Land, K. C., Y. Yang, and Z. Yi. 2005. Mathematical demography. Pages 659–717 *in* D. Poston and M. Micklin, editors, *Handbook of Population*. Springer, New York.

LeBourg, E., B. Thon, J. Legare, B. Desjardins, and H. Charbonneau. 1993. Reproductive life of French-Canadians in the 17–18th centuries: A search for a trade-off between early fecundity and longevity. *Experimental Gerontology* **28**: 217–232.

Lebreton, J.-D., K. P. Burnham, J. Clobert, and D. R. Anderson. 1992. Modeling survival and testing biological hypotheses using marked animals: A unified approach with case studies. *Ecological Monographs* **62**: 67–118.

Lebreton, J.-D., and J.-M. Gaillard. 2016. Wildlife demography: Population processes, analytical tools and management applications. Pages 29–54 *in* R. Mateo, B. Arroyo, and J. T. Garcia, editors, *Current Trends in Wildlife Research*. Springer, Cham, Switzerland..

Lebreton, J. D., J. D. Nichols, R. J. Barker, R. Pradel, and J. A. Spendelow. 2009. Chapter 3: Modeling individual animal histories with multistate capture-recapture models. *Advances in Ecological Research* **41**: 87–173.

Ledent, J., and Y. Zeng. 2010. Multistate demography. Pages 137–163 *in* E. Z. Yi, *Encyclopedia of Life Support Systems*. United Nations, Singapore.

Lee, E. T. 1992. *Statistical Methods for Survival Data*. 2nd ed. Wiley-Interscience, New York.

Lefkovitch, L. P. 1965. The study of population growth in organisms grouped by stages. *Biometrics* **21**: 1–18.

Lehane, M. J. 1985. Determining the age of an insect. *Parasitology Today* **1**: 81–85.

Lenart, A., and J. W. Vaupel. 2017. Questionable evidence for a limit to human lifespan. *Nature* **546**: E13–E14.

Lenner, P. 1990. The excess mortality rate: A useful concept in cancer epidemiology. *Acta Oncologica* **29**: 573–576.

Lennox, J. G. 1991. Darwinian thought experiments: A function for just-so stories. Pages 223–246 *in* T. Horowitz and G. J. Massey, editors, *Thought Experiments in Science and Philosophy*. Rowman and Littlefield, Savage, MD.

Leridon, H. 1984. Selective effects of sterility and fertility. Pages 83–98 *in* N. Keyfitz, editor, *Population and Biology*. Ordina Editions, Liege, Belgium.

Leridon, H. 2004. The demography of a learned society. *Population-E* **59**: 81–114.

Leslie, P. H. 1945. On the use of matrices in certain population mathematics. *Biometrika* **33**: 183–217.

Leslie, P. H., and R. M. Ransom. 1940. The mortality, fertility and rate of natural increase of the vole (*Microtus agrestis*) as observed in the laboratory. *Journal of Animal Ecology* **9**: 27–52.

Lester, L. A., H. W. Avery, A. S. Harrison, and E. A. Standora. 2013. Recreational boats and turtles: Behavioral mismatches result in high rates of injury. *PLoS ONE*. doi: 10.1371/journal.pone.0082370.

Levins, R. 1969. Some demographic and genetic consequences of environmental heterogeneity for biological control. *Bulletin of Entomological Society of America* **15**: 237–240.

Levins, R. 1970. Extinction. Pages 77–107 *in* M. Gerstenhaber, editor, *Some Mathematical Problems in Biology*. American Mathematical Society, Providence, RI.

Lewontin, R. C. 1965. Selection for colonizing ability. Pages 77–94 *in* H. G. Baker and G. L. Stebbins, editors, *The Genetics of Colonizing Species*. Academic Press, New York.

Lewontin, R. C. 1984. Laws of biology and laws in social science. Pages 19–28 *in* N. Keyfitz, editor, *Population and Biology*. Ordina Editions, Liege, Belgium.

Lex, A., and N. Gehlenborg. 2014. Sets and intersections. *Nature Methods* **11**: 779.

Lidal, E. M., M. Natali, D. Patel, H. Hauser, and I. Viola. 2013. Geological storytelling. *Computers & Graphics* **37**: 445–459.

Limburg, K. E., T. A. Hayden, W. E. Pine III, M. D. Yard, R. Kozdon, and J. W. Valley. 2013. Of travertine and time: Otolith chemistry and microstructure detect provenance and demography of endangered humpback chub in Grand Canyon, USA. *PLoS ONE* **8**: e84235.

Lincoln, F. C. 1930. Calculating waterfowl abundance on the basis of banding returns. *USDA Circular* **118**: 1–4.

Lindsey, P. A., G. A. Balme, V. R. Booth, and N. Midlane. 2012. The significance of African lions for the financial viability of trophy hunting and the maintenance of wild land. *PLoS ONE* **7**: e29332.

Lindsey, P. A., G. A. Balme, P. Funston, P. Henschel, L. Hunter, H. Madzikanda, N. Midlane, and V. Nyirenda. 2013. The trophy hunting of African lions: Scale, current management practices and factors undermining sustainability. *PLoS ONE* 8: ve73808.

Linklater, W. L., K. Adcock, P. DePreez, R. R. Swaisgood, P. R. Law, M. H. Knight, and G. I. H. Kerley. 2011. Guidelines for herbivore translocation simplified: Black rhinoceros case study. *Journal of Applied Ecology* 48: 493–502.

Liu, X. 2012. *Survival Analysis.* Wiley, West Sussex, UK.

Livi-Bacci, M. 1984. Introduction: Autoregulating mechanisms in human populations. Pages 109–116 *in* N. Keyfitz, editor, *Population and Biology.* Ordina Editions, Liege, Belgium.

Lopez, A. 1961. *Problems in Stable Population Theory.* Office of Population Research, Princeton, NJ.

Lotka, A. J. 1907. The progeny of a population element. *Science* 26: 21–22.

Lotka, A. J. 1922. The stability of the normal age distribution. *Proceedings of the National Academy of Sciences* 8: 339–345.

Lotka, A. J. 1924. *Elements of Physical Biology.* Williams & Wilkins, Baltimore.

Lotka, A. J. 1928. The progeny of a population element. *American Journal of Hygiene* 8: 875–901.

Lotka, A. J. 1934. Part I. Principes. *Theorie analytique des Associations Biologiques.* Hermann et Cie, Paris.

Lundquist, J. H., D. L. Anderton, and B. Yaukey. 2015. *Demography: The Study of Human Population.* Waveland Press, Long Grove, IL.

Lynn, J., and D. M. Adamson. 2003. Living well at the end of life. White Paper. RAND Health, Santa Monica, CA.

Lyons, E. D., M. A. Schroeder, and L. A. Robb. 2012. Criteria for determining sex and age of birds and mammals. Pages 207–229 *in* N. J. Silvy, editor, *The Wildlife Techniques Manual: Research.* Johns Hopkins University Press, Baltimore.

Ma, K.-L., I. Liao, J. Frazier, H. Hauser, and H.-N. Kostis. 2012. Scientific storytelling using visualization. *IEEE Computer Graphics and Applications* 32: 12–19.

MacArthur, R. H., and E. O. Wilson. 1967. *The Theory of Island Biogeography.* Princeton University Press, Princeton, NJ.

MacDonald, G. 1952. The analysis of the sporozoite rate. *Tropical Disease Bulletin* 49: 569–586.

MacDonald, G. 1957. *The Epidemiology and Control of Malaria.* Oxford University Press, Oxford, UK.

Maeda, J. 2006. *The Laws of Simplicity: Design, Technology, Business, Life.* MIT Press, Cambridge, MA.

Makeham, W. 1860. On the law of mortality and the construction of annuity tables. *Assurance Magazine and Journal of the Institute of Actuaries* 8: 301–310.

Makeham, W. M. 1867. On the law of mortality. *Journal of the Institute of Actuaries* 13: 325–367.

Malthus, T. R. 1798. *Population: The First Essay.* Reprinted 1959 by University of Michigan Press, Ann Arbor.

Manton, K. G., and E. Stallard. 1984. *Recent Trends in Mortality Analysis.* Academic Press, Orlando, FL.

Manton, K. G., and E. Stallard. 1991. Cross-sectional estimates of active life expectancy for the U.S. elderly and oldest-old populations. *Journal of Gerontology* 46: S170–S182.

Markowska, A. L., and S. J. Breckler. 1999. Behavioral biomarkers of aging: Illustration of a multivariate approach for detecting age-related behavioral changes. *Journal of Gerontology: Biological Sciences* 12: B549–B566.

Marzolin, G. 1988. Polyginie du Cincle plongeur (*Cinclus cinclus*) dans les cotes de Lorrain. *L'oiseau et al revue Francaise d'ornithologie* 58: 277–286.

Mathis, F. H. 1991. A generalized birthday problem. *SIAM Review* **33**: 265–270.

Mazza, R. 2009. *Introduction to Information Visualization.* Springer-Verlag, London.

Mbizah, M. M., G. Steenkamp, and R. J. Groom. 2016. Evaluation of the applicability of different age determination methods for estimating age of the endangered African wild dog (*Lycaon Pictus*). *PLoS ONE* **11**: e0164676.

McCurdy, S. A. 1994. Epidemiology of disaster: The Donner party (1846–1847). *Western Journal of Medicine* **160**: 338–342.

McDermid, V. 2014. *Forensics: What Bugs, Burns, Prints, DNA and More Tell Us about Crime.* Grove Press, New York.

McDonald, T. L., S. C. Amstrup, E. V. Regehr, and B. J. J. Manly. 2005. Examples. Pages 196–265 *in* S. C. Amstrup, T. L. McDonald, and B. J. J. Manly, editors, *Handbook of Capture-Recapture Analysis.* Princeton University Press, Princeton, NJ.

McGraw, J. B., and H. Caswell. 1996. Estimation of individual fitness from life-history data. *American Naturalist* **147**: 47–64.

McGue, M., J. W. Vaupel, N. Holm, and B. Harvald. 1993. Longevity is moderately heritable in a sample of Danish twins born 1870–1880. *Journal of Gerontology* **48**: B237–B244.

McInerny, G., and M. Krzywinski. 2015. Unentangling complex plots. *Nature Methods* **12**: 591.

Medawar, P. B. 1981. *The Uniqueness of the Individual.* 2nd ed. Dover Publications, New York.

Meine, C., M. Soulé, and R. F. Noss. 2006. A Mission-driven discipline: The growth of conservation biology. *Conservation Biology* **20**: 631–651.

Merow, C., J. P. Dahlgren, J. C. E. Metcalf, et al. 2014. Advancing population ecology with integral projection models: A practical guide. *Methods in Ecology and Evolution* **5**: 99–110.

Mesle, F. 2006. Chapter 42: Medical causes of death. Pages 29–44 *in* G. Caselli, J. Vallin, and G. Wunsch, editors, *Demography: Analysis and Synthesis.* Elsevier, Amsterdam.

Mesle, F., J. Vallin, J.-M. Robine, G. Desplanques, and A. Cournil. 2010. Is it possible to measure life expectancy at 110 in France? Pages 231–246 *in* H. Maier, J. Gampe, B. Jeune, J.-M. Robine, and J. W. Vaupel, editors, *Supercentenarians.* Springer, Heidelberg, Germany.

Metcalf, C. J. E., and S. Pavard. 2007. Why evolutionary biologists should be demographers. *Trends in Ecology and Evolution* **22**: 205–212.

Mikaberidze, A. 2007. *The Battle of Borodino: Napoleon versus Kutuzov.* Pen & Sword, London.

Miller, R. J. 2014. Big data curation. 20th International Conference on Management of Data (COMAD), Hyderabad, India.

Mills, L. S. 2013. *Conservation of Wildlife Populations: Demography, Genetics and Management.* 2nd ed. Wiley-Blackwell, Oxford, UK.

Missov, T. I., A. Lenart, L. Nemeth, V. Canudas-Romo, and J. W. Vaupel. 2015. The Gompertz force of mortality in terms of the modal age of death. *Demographic Research* **32**: 1031–1048.

Mloszewski, J. J. 1983. *The Behavior and Ecology of the African Buffalo.* Cambridge University Press, Cambridge, UK.

Molla, M. T., D. K. Wagener, and J. H. Madans. 2001. Summary measures of population health: Methods for caluculating healthy life expectancy. Department of Health and Human Services, Washington, DC.

Molleman, F., B. J. Zwaan, P. M. Brakefield, and J. R. Carey. 2007. Extraordinary long life spans in fruit-feeding butterflies can provide window on evolution of life span and aging. *Experimental Gerontology* **42**: 472–482.

Moore, H. E., J. L. Pechal, M. E. Benbow, and F. P. Drijfhout. 2017. The potential use of cuticular hydrocarbons and multivariate analysis to age empty puparial cases of *Calliphora vicina* and *Lucilia sericata*. *Scientific Reports* **7**: 1933.

Morehouse, A. T., and M. S. Boyce. 2016. Grizzly bears without borders: Spatially explicit capture-recapture in southwestern Alberta. *Journal of Wildlife Management* 80: 1152–1166.

Morgan, S. P., and K. J. Hagewen. 2005. Fertility. Pages 229–249 *in* D. Poston and M. Micklin, editors, *Handbook of Population*. Springer, New York.

Morris, R. F. 1959. Single-factor analysis in population dynamics. *Ecology* 40: 580–588.

Morris, W. F., and D. F. Doak. 2002. Quantitative conservation biology: Theory and practice of population viability analysis. Sinauer Associates, Sunderland, MA.

Müller, H.-G., J. R. Carey, D. Wu, and J. W. Vaupel. 2001. Reproductive potential determines longevity of female Mediterranean fruit flies. *Proceedings of the Royal Society, London B* 268: 445–450.

Müller, H.-G., J.-L. Wang, W. B. Capra, P. Liedo, and J. R. Carey. 1997. Early mortality surge in protein-deprived females causes reversal of sex differential of life expectancy in Mediterranean fruit flies. *Proceedings of the National Academy of Sciences* 94: 2762–2765.

Müller, H.-G., J.-L. Wang, J. R. Carey, E. P. Caswell-Chen, C. Chen, N. Papadopoulos, and F. Yao. 2004. Demographic window to aging in the wild: Constructing life tables and estimating survival functions from marked individuals of unknown age. *Aging Cell* 3: 125–131.

Müller, H.-G., J.-L. Wang, W. Yu, A. Delaigle, and J. R. Carey. 2007. Survival in the wild via residual demography. *Theoretical Population Biology* 72: 513–522.

Munroe, R. 2015. *Thing Explainer: Complicated Stuff in Simple Words*. Houghton Mifflin/Harcourt, New York.

Murdock, S. H., and D. R. Ellis. 1991. *Applied Demography: An Introduction to Basic Concepts, Methods, and Data*. Westview Press, Boulder, CO.

Murphy, T. E., L. Han, H. G. Allore, P. N. Peduzzi, T. M. Gill, and H. Lin. 2011. Treatment of death in the analysis of longitudinal studies of gerontological outcomes. *Journal of Gerontology: Medical Sciences* 66A: 109–114.

Murray, C. J. L., J. Salomon, and C. Mathers. 1999. A critical examination of summary measures of population health. In *WHO Global Programme on Evidence for Health Policy*. World Health Organization, Geneva. https://www.scielosp.org/scielo.php?pid=S0042-9686200000080008&script=sci_arttext&tlng=es.

Namboodiri, K., and C. M. Suchindran. 1987. *Life Table Techniques and Their Applications*. Academic Press, Orlando, FL.

Nenecke, M. 2001. A brief history of forensic entomology. *Forensic Science International* 120: 2–14.

Neves, R. J., and S. N. Moyer. 1988. Evaluation of techniques for age determination of freshwater mussels (Unionidae). *American Malacological Bulletin* 6: 179–188.

Nicholls, H. 2017. Darwin's domestic discoveries (review of *Darwin's Backyard*). *Nature* 548: 389–390.

Norbert, R. 2002. Basic life insurance mathematics. http://www.math.ku.dk/~mogens/lifebook.pdf.

Novoseltsev, V. N., J. R. Carey, J. A. Novoseltseva, N. T. Papapopoulos, S. Blay, and A. I. Yashin. 2004. Systemic mechanisms of individual reproductive life history in female Medflies. *Mechanisms of Ageing and Development* 125: 77–87.

Novoseltsev, V. N., A. I. Michalski, J. A. Novoseltseva, A. I. Yashin, J. R. Carey, and A. M. Ellis. 2012. An age-structured extension to the vectorial capacity model. *PLoS ONE* 7: e39479. doi: 39410.31371/journal.pone.0039479.

Nowak, R. M. 1991. *Walker's Mammals of the World*. Johns Hopkins University Press, Baltimore.

Oksuzyan, A., K. Juel, J. W. Vaupel, and K. Christensen. 2008. Men: Good health and high mortality: Sex differences in health and aging. *Aging Clinical and Experimental Research* 20: 91–102.

Oksuzyan, A., I. Petersen, H. Stovring, P. Bingley, J. W. Vaupel, and K. Christensen. 2009. The male-female health-survival paradox: A survey and register study of the impact of sex-specific selection and information bias. *Annals of Epidemiology* **19**: 504–511.

Olshansky, S. J., B. A. Carnes, and C. Cassel. 1990. In search of Methuselah: Estimating the upper limits to human longevity. *Science* **250**: 634–639.

Olshansky, S. J., B. A. Carnes, and A. Desesquelles. 2001. Prospects for human longevity. *Science* **291**: 1491–1492.

Owen-Smith, N., G. I. H. Kerley, B. Page, R. Slotow, and R. J. vanAarde. 2006. A scientific perspective on the management of elephants in the Kruger National Park and elsewhere. *South African Journal of Science* **102**: 389–394.

Papadopoulos, N., J. R. Carey, C. Ioannou, H. Ji, H.-G. Müller, J.-L. Wang, S. Luckhart, and E. Lewis. 2016. Seasonality of post-capture longevity in a medically-important mosquito (*Culex pipiens*). *Frontiers in Ecology and Evolution* **4**: 63. doi: 10.3389/fevo.20016.00063.

Papadopoulos, N. T., J. R. Carey, B. I. Katsoyannos, N. A. Kouloussis, H.-G. Müller, and X. Liu. 2002. Supine behaviour predicts time-to-death in male Mediterranean fruit flies. *Proceedings of the Royal Society of London: Biological Sciences* **269**: 1633–1637.

Papadopoulos, N., R. Plant, and J. R. Carey. 2013. From trickle to flood: The large-scale, cryptic invasion of California by tropical fruit flies. *Proceedings of the Royal Society of London.* http://dx.soi.org/10.1098/rspb.2013.1466.

Parmenter, R. R., T. L. Yates, D. R. Anderson, K. P. Burnham, et al. 2003. Small-mammal density estimation: A field comparison of grid-based vs. web-based density estimators. *Ecological Monographs* **73**: 1–26.

Partridge, L., and K. Fowler. 1992. Direct and correlated responses to selection on age at reproduction in *Drosophila melanogaster*. *Evolution* **46**: 76–91.

Patronek, G. J., D. J. Waters, and L. T. Glickman. 1997. Comparative longevity of pet dogs and humans: Implications for gerontology research. *Journal of Gerontology* **52A**: B171–B178.

Pearl, R. 1924. *Studies in Human Biology.* Williams & Wilkins, Baltimore.

Pearl, R. 1925. *The Biology of Population Growth.* Alfred A. Knopf, New York.

Pearl, R. 1928. *The Rate of Living.* Knopf, New York.

Pearl, R., and L. J. Reed. 1920. On the rate of growth of the population of the United States since 1790 and its mathematical representation. *Proceedings of the National Academy of Sciences* **6**: 275–288.

Peeples, R., and C. T. Harris. 2015. What is a life worth in North Carolina? A look at wrongful-death awards. *Review of Litigation* **37**: 497–518.

Perks, W. 1932. On some experiments in the graduation of mortality statistics. *Journal of the Institute of Actuaries* **63**:12–27.

Perz, S. G. 2004. Population change. Pages 253–264 *in* J. S. Siegel and D. A. Swanson, editors, *The Methods and Materials of Demography.* Elsevier/Academic Press, Amsterdam.

Petersen, C. G. T. 1896. The yearly immigration of young plaice into the Limfjord from the German Sea. *Report of the Danish Biological Station* **6**: 1–48.

Peterson, M. J., and P. J. Ferro. 2012. Wildife health and disease: Surveillance, investigation, and management. Pages 181–206 *in* N. J. Silvy, editor, *The Wildlife Techniques Manual: Research.* Johns Hopkins University Press, Baltimore.

Peterson, R. K. D., R. S. Davis, L. G. Higley, and O. A. Fernandes. 2009. Mortality risk in insects. *Environmental Entomology* **38**: 1–10.

Phillips, J. 2012. Storytelling in earth sciences: The eight basic plots. *Earth-Science Reviews* **115**: 153–162.

Pianka, E. R. 1978. *Evolutionary Ecology.* 2nd ed. Harper & Row, New York.

Pielou, A. C. 1979. *Mathematical Ecology.* John Wiley & Sons, New York.

Pierce, B. L., R. R. Lopez, and N. J. Silvy. 2012. Estimating animal abundance. Pages 284–310 *in* N. J. Silvy, editor, *The Wildlife Techniques Manual: Research.* Johns Hopkins University Press, Baltimore.

Pol, L. G., and R. K. Thomas. 1992. *The Demography of Health and Health Care*. Kluwer Academic/Plenum Publishers, Dordrecht, Netherlands.

Polanowski, A. M., J. Robbins, D. Chandler, and S. N. Jarman. 2014. Epigenetic estimation of age in humpback whales. *Molecular Ecology Resources* 14: 976–987.

Pollack, K. H., J. D. Nichols, C. Brownie, and J. E. Hines. 1990. Statistical inference for capture-recapture experiments. *Wildlife Monographs* 107: 1–97.

Pollak, R. A. 1986. A reformulation of the two-sex problem. *Demography* 23: 247–259.

Pollak, R. A. 1987. The two-sex problem with persistent unions: A generalization of the birth matrix-mating rule model. *Theoretical Population Biology* 32:176–187.

Pollard, J. H. 1982. The expectation of life and its relationship to mortality. *Journal of the Institute of Actuaries* 109: 225–240.

Pontius, J. S., J. E. Noyer, and M. L. Deaton. 1989. Estimation of stage transition time: Application to entomological studies. *Annals Entomological Society of America* 82: 135–148.

Posamentier, A. S., and I. Lehmann. 2007. *The Fabulous Fibonacci Numbers*. Prometheus Books, New York.

Poston, D. L., and L. F. Bouvier. 2010. *Population and Society: An Introduction to Demography*. Cambridge University Press, Cambridge, UK.

Poston, D. L., and W. P. Frisbie. 2005. Ecological demography. Pages 601–624 *in* D. Poston and M. Micklin, editors, *Handbook of Population*. Springer, New York.

Poston, D., and M. Micklin, editors. 2005. *Handbook of Population*. Springer, New York.

Pressat, R., and C. Wilson. 1987. *Dictionary of Demography*. Blackwell, New York.

Preston, S., and M. Guillot. 1997. Population dynamics in an age of declining fertility synthesis. *Genus* 53: 15–31.

Preston, S. H., P. Heuveline, and M. Guillot. 2001. *Demography: Measuring and Modeling Population Processes*. Blackwell Publishers, Malden, MA.

Preston, S. H., N. Keyfitz, and R. C. Schoen. 1972. *Causes of Death: Life Tables for National Populations*. Seminar Press, New York.

Promislow, D., M. Tatar, S. Pletcher, and J. R. Carey. 1999. Below-threshold mortality: Implications for studies in evolution, ecology and demography. *Journal of Evolutionary Biology* 12: 314–328.

Rao, A. R. R. S., and J. R. Carey. 2015. Generalization of Carey's equality and a theorem on stationary population. *Journal of Mathematical Biology* 71: 583–594.

Rao, A. R. R. S., and J. R. Carey. 2019. On three properties of stationary populations and knotting with non-stationary populations, *Journal of Mathematical Biosciences*, doi.org/10.1007/s11538-019-00652-7.

Rau, R., C. Bohk-Ewald, M. M. Muszynska, and J. W. Vaupel. 2018. Visualizing mortality dynamics in the Lexis diagram. Springer Series on Demographic Methods and Population Analysis. Springer Open, Cham, Switzerland. doi: 10.1007/978-3-319-64820-0.

Rau, R., M. Ebeling, T. Missoy, and J. Cohrn. 2018. How often does the oldest person alive die? A demographic application of queueing theory. Population Association of America annual meeting, April 26, 2018, Denver.

Rees, M., D. Z. Childs, and S. P. Ellner. 2014. Building integral projection models: A user's guide. *Journal of Animal Ecology* 83: 528–545.

Revkin, A. C. 2012. *New York Times Dot Earth Blog*. Posted January 31, 2012.

Ricketts, T. H., E. Dinerstein, T. Boucher, T. M. Brooks, S. H. M. Butchart, M. Hoffmann, J. F. Lamoreux, et al. 2005. Pinpointing and preventing imminent extinctions. *Proceedings of the National Academy of Sciences* 102: 18497.

Ricklefs, R. E., and A. Scheuerlein. 2002. Biological implications of the Weibull and Gompertz models of aging. *Journal of Gerontology: Biological Science* 57A: B69–B76.

Riffe, T., J. Scholey, and F. Villavicencio. 2017. A unified framework of demographic time. *Genus* 73(x): 710.1186/s41118-41017-40024-41114.

Rivas, A. E., M. C. Allender, M. Mitchell, and J. K. Whittington. 2014. Morbidity and mortality in reptiles presented to a wildlife care facility in Central Illinois. *Human-Wildlife Interactions* 8: 78–87.

Rivers, J. P. W. 1982. Women and children last: An essay on sex discrimination in disasters. *Disasters* 6: 256–267.

Roach, D. A., and J. R. Carey. 2014. Population biology of aging in the wild. *Annual Review of Ecology, Evolution and Systematics* 45: 421–443.

Roam, D. 2009. *The Back of the Napkin*. Penguin Books, London.

Roam, D. 2014. *Show and Tell: How Everybody Can Make Extraordinary Presentations*. Penguin Books, London.

Robine, J.-M. 2001. Redefining the stages of the epidemiological transition by a study of the dispersion of life spans: The case of France. *Population: An English Selection* 13: 173–194.

Robinson, J. G., and K. H. Redford. 1991. Sustainable harvest of neotropical forest animals. Pages 415–429 *in* J. G. Robinson and K. H. Redford, editors, *Neotropical Wildlife Use and Conservation*. University of Chicago Press, Chicago.

Roff, D. A. 1992. *The Evolution of Life Histories*. Chapman & Hall, New York.

Roff, D. A. 2002. *Life History Evolution*. Sinauer Associates, Inc., Sunderland, MA.

Rogers, A. 1984. *Introduction to Multiregional Mathematical Demography*. John Wiley & Sons, New York.

Rogers, A. 1995. *Multiregional Demography: Principles, Methods and Extensions*. John Wiley, New York.

Rogers, A., and L. J. Castro. 1981. Model migration schedules. International Institute for Applied Systems Analysis (IIASA) Research Report, Laxenburg, Austria.

Rogers, R. G., R. A. Hummer, and P. M. Krueger. 2005. Adult mortality. Pages 283–309 *in* D. Poston and M. Micklin, editors, *Handbook of Population*. Kluwer Academic/Plenum Publishers, New York.

Rota, J.-C. 1996. Ten lessons I wish I had been taught. http://alumni.media.mit.edu/~cahn/life/gian-carlo-rota-10-lessons.html.

Rowland, D. 2003. *Demographic Methods and Concepts*. Oxford University Press, Oxford, UK.

Royama, T. 1996. A fundamental problem in key factor analysis. *Ecology* 77: 87–93.

Royle, J. A., R. B. Chandler, R. Sollmann, and B. Gardner. 2014. *Spatial Capture-Recapture*. Elsevier, Amsterdam.

Rozing, M. P., T. B. L. Kirkwood, and R. G. J. Westendorp. 2017. Is there evidence for a limit to human lifespan? *Nature* 546: E11–E12.

Ruggles, S. 2012. The future of historical family demography. *Annual Review of Sociology* 38: 423–441.

Ryder, N. B. 1973. Two cheers for ZPG. *Daedalus* 102: 45–62.

Ryder, N. B. 1975. Notes on stationary populations. *Population Index* 41: 3–28.

Salinari, G. 2018. Rethinking mortality deceleration. *Biodemography and Social Biology* 64:127–138.

San Diego Zoo Global Library. 2016. African and Asian Lions (*Panthera leo*) Fact Sheet. Accessed October 26, 2017. http://ielc.libguides.com/sdzg/factsheets/lions.

Sartor, F. 2006. Chapter 50: The environmental factors of mortality. Pages 129–142 *in* G. Caselli, J. Vallin, and G. Wunsch, editors, *Demography: Analysis and Synthesis*. Academic Press, Amsterdam.

Schenk, A. N., and M. J. Souza. 2014. Major anthropogenic causes for and outcomes of wild animal presentation to a wildlife clinic in East Tennessee, USA, 2000–2011. *PLoS ONE* 9: e93517.

Sear, R. 2015. Evolutionary demography: A Darwinian renaissance in demography. Pages 406–412 *in* J. D. Wright, editor, *International Encyclopedia of the Social & Behavioral Sciences*. Elsevier, Oxford, UK.

Seber, G. A. F. 1965. A note on the multiple-recapture census. *Biometrika* 52: 249–259.

Seber, G. A. F. 1970. Estimating time-specific survival and reporting rates for adult birds from band returns. *Biometrika* **57**: 313–318.

Sermet, C., and E. Camboi. 2006. Chapter 41: Measuring the state of health. Pages 13–28 *in* G. Caselli, J. Vallin, and G. Wunsch, editors, *Demography: Analysis and Synthesis*. Elsevier, Amsterdam.

Sharpe, F. R., and A. J. Lotka. 1911. A problem in age-distribution. *Philosophical Magazine* **21**: 435–438.

Shoresh, N., and B. Wong. 2012. Data exploration. *Nature Methods* **9**: 5.

Shoumatoff, A. 1985. *The Mountain of Names: A History of the Human Family*. Vintage Books, New York.

Sibly, R. M. 2002. Life history theory: An overview. Pages 623–627 *in* M. Pagel, editor, *Encyclopedia of Evolution*. Oxford University Press, Oxford, UK.

Siegel, J. S., and D. A. Swanson, editors. 2004. *The Methods and Materials of Demography*. 2nd ed. Elsevier/Academic Press, Amsterdam.

Siler, W. 1979. A competing-risk model for animal mortality. *Ecology* **60**: 750–757.

Silvy, N. J., editor. 2012. *The Wildlife Techniques Manual: Research*. Johns Hopkins University Press, Baltimore.

Silvy, N. J., R. R. Lopez, and M. J. Peterson. 2012. Techniques for marking wildlife. Pages 230–257 *in* N. J. Silvy, editor, *The Wildlife Techniques Manual: Research*. Johns Hopkins University Press, Baltimore.

Simberloff, D., and M. Rejmanek. 2011. *Encyclopedia of Biological Invasions*. University of California Press, Berkeley.

Sinclair, A. R. E. 1977. *The African Buffalo: A Study of Resource Limitation of Populations*. University of Chicago Press, Chicago.

Skalski, J., K. Ryding, and J. Millspaugh. 2005. *Wildlife Demography*. Elsevier, Burlington, MA.

Skiadas, C. H., and C. Skadas, editors. 2018. *Demography and Health Issues*. Springer, Dordrecht, Netherlands.

Slade, B., M. L. Parrott, A. Paproth, M. J. L. Magrath, G. R. Gillespie, and T. S. Jessop. 2014. Assortative mating among animals of captive and wild origin following experimental conservation releases. *Biology Letters* **10**(11): 20140656. http://dx.doi.org/20140610.20141098/rsbl.20142014.20140656.

Smith, D., and N. Keyfitz. 1977. *Mathematical Demography*. Springer-Verlag, Berlin.

Smith, J. M. 1982. Storming the fortress. *New York Review of Books*, May 1982, 5.

Snell, T. W. 1978. Fecundity, developmental time, and population growth rate. *Oecologia* **32**: 119–125.

Sormani, M. P., M. Tintore, M. Rovaris, A. Rovira, X. Vidal, P. Bruzzi, M. Filippi, and X. Montalban. 2008. Will Rogers phenomenon in multiple sclerosis. *Annals of Neurobiology* **64**: 428–433.

Species360. 2018. Zoological information management system (ZIMS). https://www.species360.org/.

Speer, J. H. 2010. *Fundamentals of Tree-Ring Research*. University of Arizona Press, Tucson.

Spencer, T., and J. Flaws. 2018. Volume II: Female reproduction. *Encyclopedia of Reproduction*. 2nd ed. Academic Press, New York.

Spizman, L. M. 2016. Estimating educational attainment and earning capacity of a minor child. In F. D. Tinary, editor, *Forensic Economics*. Palgrave MacMillan, New York.

Spuhler, J. N. 1959. Physical anthropology and demography. Pages 728–758 *in* P. M. Hauser and O. D. Duncan, editors, *The Study of Population*. University of Chicago Press, Chicago.

Starrs, D., B. C. Ebner, and C. J. Fulton. 2016. All in the ears: Unlocking the early life history biology and spatial ecology of fishes. *Biological Reviews* **91**: 86–105.

Stearns, S. C. 2002. *The Evolution of Life Histories*. Oxford University Press, Oxford, UK.

Stevenson, R. D., and W. A. Woods Jr. 2006. Condition indices for conservation: New uses for evolving tools. *Integrative and Comparative Biology* **46**: 1169–1190.

Streit, M., and N. Gehlenborg. 2015. Temporal data. *Nature Methods* **12**: 97.

Styer, L. M., J. R. Carey, J.-L. Wang, and T. W. Scott. 2007. Mosquitoes do senesce: Departure from the paradigm of constant mortality. *American Journal of Tropical Medicine and Hygiene* **76**: 111–117.

Sullivan, D. F. 1971. A single index of mortality and morbidity. *HSMHA Health Reports* **86**: 347–354.

Swanson, D. A., T. K. Burch, and L. M. Tedrow. 1996. What is applied demography? *Population Research and Policy Review* **15**: 403–418.

Tamborini, C. R., C. Kim, and A. Sakamoto. 2015. Education and lifetime earnings in the United States. *Demography* **52**: 1383–1407.

Tatar, M. 2009. Can we develop genetically tractable models to assess healthspan (rather than life span) in animal models? *Journals of Gerontology Series A: Biological Sciences and Medical Sciences* **64A**: 161–163.

Tatar, M., and J. R. Carey. 1994. Sex mortality differentials in the bean beetle: Reframing the question. *American Naturalist* **144**: 165–175.

Thacker, E. T., R. L. Hamm, J. Hagen, C. A. Davis, and F. Guthery. 2016. Evaluation of the Surrogator® system to increase pheasant and quail abundance. *Wildlife Society Bulletin* **40**: 310–315.

Tomberlin, J. K., R. Mohr, M. E. Benbow, A. M. Tarone, and S. VanLaerhoven. 2011. A roadmap for bridging basic and applied research in forensic entomology. *Annual Review of Entomology* **56**: 401–421.

Tufte, E. R. 2001. *The Visual Display of Quantitative Information.* Graphics Press, Cheshire, CT.

Tuljapurkar, S. 1984. Demography in stochastic environments. I. Exact distributions of age structure. *Journal of Mathematical Biology* **19**: 335–350.

Tuljapurkar, S. 1989. An uncertain life: Demography in random environments. *Theoretical Population Biology* **35**: 227–294.

Tuljapurkar, S., editor. 1990. *Lecture Notes in Biomathematics: Population Dynamics in Variable Environments.* Springer-Verlag, New York.

Tuljapurkar, S. 2003. Renewal theory and the stable population model. Pages 839–843 *in* P. Demeny and G. McNicoll, editors. *Encyclopedia of Population.* Gale Group, New York.

Tuljapurkar, S., and C. C. Horvitz. 2006. From stage to age in variable environments: Life expectancy and survivorship. *Ecology* **87**: 1497–1509.

Tuljapurkar, S. D., and S. H. Orzack. 1980. Population dynamics in variable environments I. Long-run growth rates and extinction. *Theoretical Population Biology* **18**: 314–342.

Tuljapurkar, S., U. K. Steiner, and S. H. Orzack. 2009. Dynamic heterogeneity in life histories. *Ecology Letters* **12**: 93–106.

Tyndale-Biscoe, M. 1984. Age-grading methods in adult insects: A review. *Bulletin of Entomological Research* **74**: 341–377.

Vallin, J. 2006a. Chapter 2: Population: Replacement and change. Pages 9–14 *in* G. Caselli, J. Vallin, and G. Wunsch, editors, *Demography: Analysis and Synthesis.* Academic Press, Amsterdam.

Vallin, J. 2006b. Chapter 53: Mortality, sex and gender. Pages 177–194 *in* G. Caselli, J. Vallin, and G. Wunsch, editors, *Demography: Analysis and Synthesis.* Academic Press, Amsterdam.

Vallin, J., and G. Berlinguer. 2006. Chapter 48: From endogenous mortality to the maximum human life span. Pages 95–116 *in* G. Caselli, J. Vallin, and G. Wunsch, editors, *Demography: Analysis and Synthesis.* Academic Press, Amsterdam.

Vallin, J., and G. Caselli. 2006a. Chapter 11: Cohort life table. Pages 103–129 *in* G. Caselli, J. Vallin, and G. Wunsch, editors, *Demography: Analysis and Synthesis.* Academic Press, Amsterdam.

Vallin, J., and G. Caselli. 2006b. Chapter 14: The hypothetical cohort as a tool for demographic analysis. Pages 163–195 *in* G. Caselli, J. Vallin, and G. Wunsch, editors, *Demography: Analysis and Synthesis.* Academic Press, Amsterdam.

Vallin, J., and G. Caselli. 2006c. Chapter 19: Population replacement. Pages 239–248 *in* G. Caselli, J. Vallin, and G. Wunsch, editors, *Demography: Analysis and Synthesis*. Academic Press, Amsterdam.

Vallin, J., S. D'Souza, and A. Palloni, editors. 1990. *Measurement and Analysis of Mortality: New Approaches*. Clarendon Press, Oxford, UK.

van der Heijden, P. G. M., E. Zwane, and D. Hessen. 2009. Structurally missing data problems in multiple list capture-recapture data. *AstA Advances in Statistical Analysis* 93: 5–21.

van Laerhoven, S. L. 2010. Ecological theory and its application in forensic entomology. Pages 493–517 *in* J. H. Byrd and J. L. Castner, editors, *Forensic Entomology: The Utility of Arthropods in Legal Investigations*. CRC Press, Boca Raton, FL.

van Tienderen, P. H. 1995. Life cycle trade-offs in matrix population models. *Ecology* 76: 2482–2489.

van Wyk, P. 2017. Cape buffalo: At the mercy of lions, drought and disease. *MalaMala Game Reserve Blog*. https://blog.malamala.com/index.php/2016/2001/cape-buffalo/.

Varley, G. C., and G. R. Gradwell. 1960. Key factors in population studies. *Journal of Animal Ecology* 29: 399–401.

Vaupel, J. W. 1986. How change in age-specific mortality affects life expectancy. *Population Studies* 40: 147–157.

Vaupel, J. W. 2009. Life lived and left: Carey's equality. *Demographic Research* 20: 7–10.

Vaupel, J. W. 2010. Biodemography of human ageing. *Nature* 464: 536–542.

Vaupel, J. W., and J. R. Carey. 1993. Compositional interpretations of Medfly mortality. *Science* 260: 1666–1667.

Vaupel, J. W., J. R. Carey, K. Christensen, T. E. Johnson, et al. 1998. Biodemographic trajectories of longevity. *Science* 280: 855–860.

Vaupel, J. W., K. G. Manton, and E. Stallard. 1979. The impact of heterogeneity in individual frailty on the dynamics of mortality. *Demography* 16: 439–454.

Vaupel, J. W., and F. Villavicencio. 2018. Life lived and left: Estimating age-specific mortality in stable populations with unknown ages. *Demographic Research* (forthcoming).

Vaupel, J. W., Z. Wang, K. Andreev, and A. I. Yashin. 1997. *Population Data at a Glance: Shaded Contour Maps of Demographic Surfaces over Age and Time*. University Press of Southern Denmark, Odense.

Vaupel, J. W., and A. I. Yashin. 1985. Heterogeneity's ruses: Some surprising effects of selection on population dynamics. *American Statistician* 39: 176–185.

Wachter, K. 2003. Stochastic population theory. Pages 921–924 *in* P. Demeny and G. McNicoll, editors, *Encyclopedia of Population*. Gale Group, New York.

Wachter, K. W. 2014. *Essential Demographic Methods*. Harvard University Press, Cambridge, MA.

Wachter, K. W., and R. A. Bulatao, editors. 2003. *Offspring: Human Fertility Behavior in Biodemographic Perspective*. National Academies Press, Washington, DC.

Wachter, K., and C. Finch, editors. 1997. *Between Zeus and the Salmon: The Biodemography of Longevity*. National Academies Press, Washington, DC.

Wallis, M. 2017. *The Best Land under Heaven: The Donner Party in the Age of Manifest Destiny*. Liveright Publishing, New York.

Wang, J.-L., H.-G. Müller, and W. B. Capra. 1998. Analysis of oldest-old mortality: LIfe tables revisited. *Annals of Statistics* 26: 126–163.

Watkins, S., J. Menken, and J. Bongaarts. 1987. Demographic foundations of family change *American Sociological Review* 52: 346–358.

Watson, T. 2018. Prehistoric children toiled at tough tasks. *Nature* 561: 445–446.

Weibull, W. 1951. A statistical distribution on wide applicability. *Journal of Applied Mechanics* 18: 293–297.

Weinbaum, K. Z., J. S. Brashares, C. D. Golden, and W. M. Getz. 2013. Searching for sustainability: Are assessments of wildlife harvests behind the times? *Ecological Letters* 16: 99–111.

Weissgerber, T. L., N. M. Milic, S. J. Winham, and V. D. Garovic. 2015. Beyond Bar and Line Graphs: Time for a New Data Presentation Paradigm. *PLoS Biol* **13**: e1002128.

Wells, J. D., and L. R. Lamotte. 2010. Estimating postmortem interval. Pages 367–388 *in* J. H. Byrd and J. L. Castner, editors, *Forensic Entomology: The Utility of Arthropods in Legal Investigations*. CRC Press, Boca Raton, FL.

White, G. C., and K. P. Burnham. 1999. Program MARK: Survival estimation from populations of marked animals. *Bird Study* **46**: 120–138.

White, M. J., and D. P. Lindstrom. 2005. Internal migration. Pages 311–346 *in* D. Poston and M. Micklin, editors, *Handbook of Population*. Springer, New York.

Whitham, J. C., and N. Wielebnowski. 2013. New directions for zoo animal welfare science. *Applied Animal Behaviour Science* **147**: 247–260.

Whitman, K., A. M. Starfield, H. S. Quadling, and C. Packer. 2004. Sustainable trophy hunting of African lions. *Nature* **428**: 175–178.

WHO. 2001. International classification of functioning, disability and health. World Health Organization, Geneva.

WHO. 2014. WHO methods for life expectancy and healthy life expectancy. World Health Organization, Geneva.

Widdowson, E. M. 1976. The response of the sexes to nutritional stress. *Nutritional Society Proceedings* **35**: 175–180.

Widdowson, E. M., and R. A. McAnce. 1963. The effect of finite periods of undernutrition at different ages on the composition and subsequent development of the rat. *Proceedings of the Royal Society of London* **158**: 329–341.

Willekens, F. 2003. Multistate demography. In P. Demeny and G. McNicoll, editors, *Encyclopedia of Population*. Gale Group, New York.

Willekens, F. 2005. Biographic forecasting: Bridging the micro-macro gap in population forecasting. *New Zealand Population Review* **31**: 77–124.

Willekens, F. 2014. *Multistate Analysis of Life Histories with R*. Springer, Cham, Switzerland.

Wilson, C., editor. 1985. *The Dictionary of Demography*. Basil Blackwell, Paris.

Wilson, D. L. 1994. The analysis of survival (mortality) data: Fitting Gompertz, Weibull, and logistic functions. *Mechanisms of Ageing and Development* **74**: 15–33.

Wilson, E. O. 1971. *The Insect Societies*. Belknap Press, Cambridge, MA.

Wilson, E. O. 1975. *Sociobiology: The New Synthesis*. Belknap Press, Cambridge, MA.

Wilson, E. O. 1984. New approaches to the analysis of social systems. Pages 41–52 *in* N. Keyfitz, editor, *Population and Biology*. Ordina Editions, Liege, Belgium.

Wilson, E. O. 1998. *Consilience: The Unity of Knowledge*. Alfred A. Knopf, New York.

Wilson, E. O. 2012. *The Social Conquest of Earth*. Liveright Publishing Corporation, New York.

Wingard, D. L. 1984. The sex differential in morbidity, mortality, and lifestyle. *Annual Review of Public Health* **5**: 433–458.

Woese, C. R. 2004. A new biology for a new century. *Microbiology and Molecular Biology Reviews* **68**: 173–186.

Wolfe, R. 2014. Data visualisation: A practical guide to producing effective visualisations for research communication. London School of Hygiene & Tropical Medicine, London.

Wong, B. 2010a. Color coding. *Nature Methods* **7**: 573.

Wong, B. 2010b. Design of data figures. *Nature Methods* **7**: 665.

Wong, B. 2010c. Gestalt principles (Part 1). *Nature Methods* **7**: 863.

Wong, B. 2010d. Gestalt principles (Part 2). *Nature Methods* **7**: 941.

Wong, B. 2010e. Layout. *Nature Methods* **8**: 783.

Wong, B. 2010f. Points of view: Color coding. *Nature Methods* **7**: 573.

Wong, B. 2011a. Arrows. *Nature Methods* **8**: 701.

Wong, B. 2011b. The design process. *Nature Methods* **8**: 987.

Wong, B. 2011c. Negative space. *Nature Methods* **8**: 1.

Wong, B. 2011d. The overview figure. *Nature Methods* **8**: 365.

Wong, B. 2011e. Points of review (Part 1). *Nature Methods* 8: 101.

Wong, B. 2011f. Points of review (Part 2). *Nature Methods* 8: 189.

Wong, B. 2011g. Salience to relevance. *Nature Methods* 8: 889.

Wong, B. 2011h. Simplify to clarify. *Nature Methods* 8: 611.

Wong, B. 2011i. Typography. *Nature Methods* 8: 277.

Wong, B. 2012. Visualizing biological data. *Nature Methods* 9: 1131.

Wong, B., and R. S. Kjaergaard. 2012. Pencil and paper. *Nature Methods* 9: 1037.

Wong, D. M. 2010. *The Wall Street Journal Guide to Information Graphics*. W. W. Norton, New York.

Wood, J. W., editor. 1994. *Dynamics of Human Reproduction: Biology, Biometry, Demography*. Aldine De Gruyter, New York.

Wunsch, G. 2006. Chapter 44: Dependence and independence of causes of death. Pages 57–60 *in* G. Caselli, J. Vallin, and G. Wunsch, editors, *Demography: Analysis and Synthesis*. Academic Press, Amsterdam.

Wunsch, G., J. Vallin, and G. Caselli. 2006. Chapter 3: Population increase. Pages 15–22 *in* G. Caselli, J. Vallin, and G. Wunsch, editors, *Demography: Analysis and Synthesis*. Academic Press, Amsterdam.

Yashin, A. I., I. A. Iachine, and A. S. Begun. 2000. Mortality modeling: A review. *Mathematical Population Studies* 8: 305–332.

Zajitschek, F., C. E. Brassil, R. Bonduriansky, and R. C. Brooks. 2009. Sex effects on life span and senescence in the wild when dates of birth and death are unknown. *Ecology* 90: 1698–1707.

Zamoyski, A. 2005. *1812: Napoleon's Fatal March on Moscow*. Harper Perennial, London.

Zhao, Z.-H., C. Hui, R. E. Plant, M. Su, T. E. Carpenter, N. T. Papadopoulos, Z. Li, and J. R. Carey. 2019a. Life table invasion models: Spatial progression and species-specific partitioning. *Ecology* e02682.

Zhao, Z.-H., C. Hui, R. E. Plant, N. T. Papadopoulos, T. E. Carpenter, Z. Li, and J. R. Carey. 2019b. The failure of success: Continuous eradication-recurrence cycles of a globally invasive pest, *Ecological Applications* (in press).

Zug, G. R. 1993. *Herpetology: An Introductory Biology of Amphibians and Reptiles*. Academic Press, San Diego.

Zuo, W., S. Jiang, Z. Guo, M. Feldman, and S. Tuljapurkar. 2018. An advancing front of old age human survival. *Proceedings of the National Academy of Sciences*. https://doi.org/10.1073/pnas.1812337115.

INDEX